IET ENERGY ENGINEERING 130

Wind and Solar Based Energy Systems for Communities

Other volumes in this series:

Volume 1	**Power Circuit Breaker Theory and Design** C.H. Flurscheim (Editor)
Volume 4	**Industrial Microwave Heating** A.C. Metaxas and R.J. Meredith
Volume 7	**Insulators for High Voltages** J.S.T. Looms
Volume 8	**Variable Frequency AC Motor Drive Systems** D. Finney
Volume 10	**SF_6 Switchgear** H.M. Ryan and G.R. Jones
Volume 11	**Conduction and Induction Heating** E.J. Davies
Volume 13	**Statistical Techniques for High Voltage Engineering** W. Hauschild and W. Mosch
Volume 14	**Uninterruptible Power Supplies** J. Platts and J.D. St Aubyn (Editors)
Volume 15	**Digital Protection for Power Systems** A.T. Johns and S.K. Salman
Volume 16	**Electricity Economics and Planning** T.W. Berrie
Volume 18	**Vacuum Switchgear** A. Greenwood
Volume 19	**Electrical Safety: A guide to causes and prevention of hazards** J. Maxwell Adams
Volume 21	**Electricity Distribution Network Design, 2nd Edition** E. Lakervi and E.J. Holmes
Volume 22	**Artificial Intelligence Techniques in Power Systems** K. Warwick, A.O. Ekwue and R. Aggarwal (Editors)
Volume 24	**Power System Commissioning and Maintenance Practice** K. Harker
Volume 25	**Engineers' Handbook of Industrial Microwave Heating** R.J. Meredith
Volume 26	**Small Electric Motors** H. Moczala *et al.*
Volume 27	**AC–DC Power System Analysis** J. Arrillaga and B.C. Smith
Volume 29	**High Voltage Direct Current Transmission, 2nd Edition** J. Arrillaga
Volume 30	**Flexible AC Transmission Systems (FACTS)** Y.-H. Song (Editor)
Volume 31	**Embedded Generation** N. Jenkins *et al.*
Volume 32	**High Voltage Engineering and Testing, 2nd Edition** H.M. Ryan (Editor)
Volume 33	**Overvoltage Protection of Low-Voltage Systems, Revised Edition** P. Hasse
Volume 36	**Voltage Quality in Electrical Power Systems** J. Schlabbach *et al.*
Volume 37	**Electrical Steels for Rotating Machines** P. Beckley
Volume 38	**The Electric Car: Development and future of battery, hybrid and fuel-cell cars** M. Westbrook
Volume 39	**Power Systems Electromagnetic Transients Simulation** J. Arrillaga and N. Watson
Volume 40	**Advances in High Voltage Engineering** M. Haddad and D. Warne
Volume 41	**Electrical Operation of Electrostatic Precipitators** K. Parker
Volume 43	**Thermal Power Plant Simulation and Control** D. Flynn
Volume 44	**Economic Evaluation of Projects in the Electricity Supply Industry** H. Khatib
Volume 45	**Propulsion Systems for Hybrid Vehicles** J. Miller
Volume 46	**Distribution Switchgear** S. Stewart
Volume 47	**Protection of Electricity Distribution Networks, 2nd Edition** J. Gers and E. Holmes
Volume 48	**Wood Pole Overhead Lines** B. Wareing
Volume 49	**Electric Fuses, 3rd Edition** A. Wright and G. Newbery
Volume 50	**Wind Power Integration: Connection and system operational aspects** B. Fox *et al.*
Volume 51	**Short Circuit Currents** J. Schlabbach
Volume 52	**Nuclear Power** J. Wood
Volume 53	**Condition Assessment of High Voltage Insulation in Power System Equipment** R.E. James and Q. Su
Volume 55	**Local Energy: Distributed generation of heat and power** J. Wood
Volume 56	**Condition Monitoring of Rotating Electrical Machines** P. Tavner, L. Ran, J. Penman and H. Sedding
Volume 57	**The Control Techniques Drives and Controls Handbook, 2nd Edition** B. Drury
Volume 58	**Lightning Protection** V. Cooray (Editor)
Volume 59	**Ultracapacitor Applications** J.M. Miller
Volume 62	**Lightning Electromagnetics** V. Cooray
Volume 63	**Energy Storage for Power Systems, 2nd Edition** A. Ter-Gazarian
Volume 65	**Protection of Electricity Distribution Networks, 3rd Edition** J. Gers
Volume 66	**High Voltage Engineering Testing, 3rd Edition** H. Ryan (Editor)
Volume 67	**Multicore Simulation of Power System Transients** F.M. Uriate
Volume 68	**Distribution System Analysis and Automation** J. Gers
Volume 69	**The Lightening Flash, 2nd Edition** V. Cooray (Editor)
Volume 70	**Economic Evaluation of Projects in the Electricity Supply Industry, 3rd Edition** H. Khatib
Volume 72	**Control Circuits in Power Electronics: Practical issues in design and implementation** M. Castilla (Editor)
Volume 73	**Wide Area Monitoring, Protection and Control Systems: The enabler for smarter grids** A. Vaccaro and A. Zobaa (Editors)
Volume 74	**Power Electronic Converters and Systems: Frontiers and applications** A.M. Trzynadlowski (Editor)
Volume 75	**Power Distribution Automation** B. Das (Editor)
Volume 76	**Power System Stability: Modelling, analysis and control** B. Om P. Malik
Volume 78	**Numerical Analysis of Power System Transients and Dynamics** A. Ametani (Editor)
Volume 79	**Vehicle-to-Grid: Linking electric vehicles to the smart grid** J. Lu and J. Hossain (Editors)
Volume 81	**Cyber–Physical–Social Systems and Constructs in Electric Power Engineering** Siddharth Suryanarayanan, Robin Roche and Timothy M. Hansen (Editors)
Volume 82	**Periodic Control of Power Electronic Converters** F. Blaabjerg, K. Zhou, D. Wang and Y. Yang
Volume 86	**Advances in Power System Modelling, Control and Stability Analysis** F. Milano (Editor)
Volume 87	**Cogeneration: Technologies, Optimisation and Implementation** C.A. Frangopoulos (Editor)
Volume 88	**Smarter Energy: From smart metering to the smart grid** H. Sun, N. Hatziargyriou, H.V. Poor, L. Carpanini and M.A. Sánchez Fornié (Editors)
Volume 89	**Hydrogen Production, Separation and Purification for Energy** A. Basile, F. Dalena, J. Tong and T.N. Veziroğlu (Editors)
Volume 90	**Clean Energy Microgrids** S. Obara and J. Morel (Editors)
Volume 91	**Fuzzy Logic Control in Energy Systems with Design Applications in Matlab/Simulink®** İ.H. Altaş
Volume 92	**Power Quality in Future Electrical Power Systems** A.F. Zobaa and S.H.E.A. Aleem (Editors)
Volume 93	**Cogeneration and District Energy Systems: Modelling, analysis and optimization** M.A. Rosen and S. Koohi-Fayegh
Volume 94	**Introduction to the Smart Grid: Concepts, technologies and evolution** Salman K. Salman
Volume 95	**Communication, Control and Security Challenges for the Smart Grid** S.M. Muyeen and S. Rahman (Editors)
Volume 97	**Synchronized Phasor Measurements for Smart Grids** M.J.B. Reddy and D.K. Mohanta (Editors)
Volume 98	**Large Scale Grid Integration of Renewable Energy Sources** Antonio Moreno-Munoz (Editor)
Volume 100	**Modeling and Dynamic Behaviour of Hydropower Plants** N. Kishor and J. Fraile-Ardanuy (Editors)
Volume 101	**Methane and Hydrogen for Energy Storage** R. Carriveau and David S-K. Ting
Volume 108	**Fault Diagnosis of Induction Motors** Jawad Faiz, Vahid Ghorbanian and Gojko Joksimović
Volume 110	**High Voltage Power Network Construction** K. Harker
Volume 124	**Power Market Transformation** B. Murray
Volume 131	**Metaheuristic Optimization in Power Engineering** J. Radosavljević
Volume 905	**Power System Protection, 4 volumes**

Wind and Solar Based Energy Systems for Communities

Edited by
Rupp Carriveau and David S-K. Ting

The Institution of Engineering and Technology

Published by The Institution of Engineering and Technology, London, United Kingdom

The Institution of Engineering and Technology is registered as a Charity in England & Wales (no. 211014) and Scotland (no. SC038698).

First published 2018

The Institution of Engineering and Technology
Michael Faraday House
Six Hills Way, Stevenage
Herts, SG1 2AY, United Kingdom

www.theiet.org

British Library Cataloguing in Publication Data
A catalogue record for this product is available from the British Library

ISBN 978-1-78561-544-3 (hardback)
ISBN 978-1-78561-545-0 (PDF)

Typeset in India by MPS Limited

Contents

Chapter 1

Introduction

Rupp Carriveau[1] and David S-K. Ting[1]

According to dictionary.com [1], a community is a social group of any size whose members dwell in a specific locality under one government, and they share the same cultural and historical heritage. Accordingly, both big cities and small towns are communities. Their future leans on, among other things, the proper management of sustainable energy and natural resources that they feed on [2]. For the more progressive localities, there is already in place some solid share of renewable energy; wind, in particular. Other than the promise of including a greater extent of solar, these wind-fed communities must look at ways to further their current and future wind.

To improve performance and maintenance of the existing, particularly wind, infrastructure alone would render significant stability in the community, renewable energy systems. The resulting benefits include both a cleaner environment and healthier economy. In Chapter 2, Morshedizadeh *et al.* describe practical, data-driven methods for predicting commercial wind turbine performance. A comprehensive set of supervisory control and data acquisition data was utilized to delineate the entire process, including the proper treatment of missing data and outliers, training of the intelligent algorithm, pertinent feature selection, and modelling of the design networks. The superiority of Bruce Lee's Jeet Kone Do is largely attributed to the combination and adaptation of the most potent moves from differing styles [3,4]. Likewise, it is illustrated that fusion of appropriately-weighted, different networks can enhance the decision-making process, for best prediction of wind turbine performance. Furthermore, wind farms, large or small, can be optimized to best benefit the communities. Vasel-Be-Hagh outlines essential elements of wind farm layout optimization in Chapter 3. Several wake loss models and different search methods, ranging from primitive to sophisticated ones, are analysed, compared, and contrasted. The base location, number, rotor diameter, hub height, rotational direction, and yaw angle of wind turbines, along with the shape of the wind farm are fully considered in the optimization process. Wake models, from the computationally intensive large eddy simulation to purely empirical models are explored. For optimization, Generic Algorithm, Greedy Algorithm, Particle Swarm

[1]Turbulence and Energy Laboratory, Department of Engineering, University of Windsor, Canada

Optimization, Ant Colony Search Algorithm, Simulated Annealing, and Definite Point Selection Algorithm are presented. The chapter concludes with 13 distinct, interesting cases for the readers to sweat over. The experiential exercise aims at bringing readers up from 'interested' to 'involved', or, 'amateurs' to 'new practitioners', concerning wind farm optimization.

Admit it or not, money, to a large extent, does make the world go around. Miller and Carriveau attempt to underplay the outstanding challenge with renewable energy implementation, the very high capital cost, in Chapter 4. So, why don't we leave renewable energy developments to private enterprise? Among other benefits, community energy provides power for the community, potential income for the community, promotes social cohesion, provides local jobs, and certain control regarding the protection of our environment. While wind and solar successes in some smaller communities, both in overseas and in North America, can alleviate significant uncertainty, the transferability of these to other communities lingers on the specific available incentives and legalities. Their newness to the industry also induces extra costs for learning, which is absorbed into the raising transaction costs. To alleviate the multitude of challenges faced by newcomers, small renewable energy developers financing options are not only provided, but the readers are guided through the process with specific successful examples. These cases also have the potential to be extended into commercial, renewable energy project finance.

How can we exploit solar? The most direct way is to harness the heat, i.e. solar thermal. And, this solar thermal energy is most commonly used for domestic heating. In Chapter 5, Bhalla *et al.* present the potential application of various solar thermal energy technologies at community-level. While the chapter is written based largely on research and experiences gathered in India, it has an international context, that is, the conveyed expertise can be put to practice all over planet earth, where there is sufficient sunshine. They describe the main attributes and key differences between non-concentrating and concentrating solar thermal collectors. The newer, volumetric-based solar energy absorption approach via the use of nanoparticles is presented and calculated numerically for a ten-household community. The various parameters, especially those associated with the nanoparticles, influencing the performance of the solar collector are clearly disseminated.

Many places where the sun is scorching, there is a shortage of drinkable water. When salt water is available, it can be desalinated. However, commercial water desalination processes are extremely energy intensive and are limited to a few big and wealthy cities, as conveyed by Ameen *et al.* in Chapter 6. For smaller communities, especially those in remote places where there is a lack of grid energy supply, the abundantly available solar energy can be harnessed for water desalination, to quench the thirst of our financially less able fellow human beings. The chapter starts with a general categorization of various types of existing water desalination systems, and focuses in on those suited for small communities and remote areas. A single slope solar still in Pasni, District Gwadar, Balochistan of Pakistan is given as a case study. The involved design, fabrication, and cost are presented to illustrate the practical operation of the technology.

The gracious energy lavished via the sun can provide benefit beyond the heat or thermal aspect. In Chapter 7, Singh *et al.* provide an overview of a community level solar photovoltaic (PV) installation in the form of a solar garden, or solar farm, or solar power plant, to partially or fully meet the electricity needs. Three ownership models, along with their benefits and barriers, are presented. The realization of these models are illustrated with the help of actual examples, that is, completed, conceptualized, and under construction projects in Canada, the United States of America, Europe, South America, and East Africa. It is not surprising to see that no single model works in all situations. The important point is that countries, provinces, states, and cities must develop policies that promote investment in community renewable energy projects, notably, solar PV projects.

We definitely do not want to see any part of solar PV systems flying around during gusty days. As such, it is critical that the wind loads on solar PV installations are properly assessed, and to ensure that the installations can withstand the beatings of the wind. Kazmirowicz *et al.* examine the wind loads on rooftop solar PV installations in an urban setting, in Chapter 8. The renewed concern on the vitality of these installations is, in part, due to the continuous cost reduction, drastic rise in rooftop units in crowded urban areas, and the global rise in damaging wind conditions such as typhoons and hurricanes. The induced loads by these and other extreme winds are of particular interest. Surprising or not, there is a lack of field measurements of wind loads on solar PV structures. The existing Australian standard is presented, prior to introducing Urban Wind Environment, and this is logically followed by Australian mounting system design practices. Both wind tunnel test methods and computational fluid dynamic simulations are useful in securing the structural health of current and future PV installations.

Maroufmashat *et al.* suggest using a network of energy hubs for simultaneous reduction of greenhouse gases and improvement of energy efficiency, in Chapter 9. These smart energy systems or hubs integrate multiple resources and technologies; including energy conversion technologies and transmission utilities, operators, and service providers. In this way, energy generation and usage can be directly linked with economic growth, climate change, and energy security. The inclusion of energy storage systems in the network ensure its versatility, considering the very diverse forms of renewable energy integrated into the power grid. An urban area consisting of four buildings in Ontario, Canada is used to illustrate the workings of such a network, where each building is an energy hub. For the future, they forecast the inclusion of hydrogen, carried through the existing natural gas line. Their model is unique in the sense that the influence of a dynamic emission factor on the optimal design of the network has been considered.

With renewable energy, one cannot undervalue its better half, energy storage; it is worth noting that energy storage is also useful for mitigating the instantaneous load–demand variation. As conveyed by Zhu *et al.* in Chapter 10, batteries deserve a comprehensive, up-to-date coverage. Community energy storage, operating on distributed energy storage systems at end-user homes and businesses, can help in alleviating the pressure of local transformers. Conventional and advanced battery technologies are covered, followed by electric vehicle penetration, and the economics of battery storage. For proper battery selection, the energy consumption pattern is a priori. The selection process safety considerations are discussed. It is somewhat awakening to learn that safety policies and standards are a key challenge in keeping the industry on the right track. Like solar PV installation regulations, according to these Australian authors, Australia, once again, stands as a pioneer in regulatory intervention to drive adoption of energy storage.

There will be no future if conservation is undermined. And conservation goes hand-in-hand with demand management. Ma and Venkatesh propose a conservation model which targets conservation at the most price-sensitive consumers but only during peak energy-consumption periods, in Chapter 11. Other than peak-usage disincentives, they also propose incentives for consumption during off-peak periods. This disincentive–incentive conservation approach works by flattening the demand curve, delaying base expansions, and lowering unit costs. The target is on the localized community energy systems, like the battery storage imparted by Zhu *et al.* in the previous chapter, to absorb momentary fluctuations in supply and demand. Their philosophy differs from the traditional one, which aims at realizing conservation by lowering the variable costs through behavioural change and efficiency. Presumably, Ma and Venkatesh understand our great struggle when it comes to changing habits; like the idiom, it is easier to change mountains and rivers then to alter one's character. Wisely, they tackle the challenge head-on, by reducing fixed costs, which requires no behavioural change of the consumers.

The book ends with Smart Multi-Energy Microgrids by Capuder and Dragičević. In agreement with Ma and Venkatesh, Capuder and Dragičević confirm the need of a flexible system capable of responding to variables of the renewable sources and consumer usage. Microgrids are entities within a community which integrate flexible load resources with on-site generation and storage. They can be as small as residential homes, called smart homes, or larger commercial buildings and industrial facilities. Conventionally, only electricity systems are considered under microgrids. A limited amount of flexibility can be introduced, but only with expansive energy storage systems. What is presented in the chapter, on the other hand, is the concept of flexible multi-energy microgrids which efficiently utilize the existing sources and infrastructures. The enhanced flexibility is enabled by exchange of energy between different energy vectors; which encompass electricity, heat, gas, hydrogen, cooling, and other forms of energy. The futuristic promise lies in flexible operation of distributed multi-energy systems.

References

[1] www.dictionary.com accessed on October 20, 2017.

[2] Carriveau, R., and Ting, D.S-K. Monograph of Energy and Natural Resources. International Journal of Environmental Studies, in press.

[3] https://en.wikipedia.org/wiki/Bruce_Lee accessed on October 23, 2017.

[4] Lee, L., Bruce Lee. The Man Only I Knew. New York, NY: Warner Books Inc.; 1975.

Chapter 2

Data-driven methods for prediction of small-to-medium wind turbines performance

Majid Morshedizadeh[1], *Rupp Carriveau*[1] *and David S-K. Ting*[1]

Abstract

The growth in the wind energy is rapidly increasing. Accurate modelling of wind turbines performance as targeted by ongoing research studies can escalate wind energy production capabilities, reliability, and expand its potential to replace fossil fuels. In addition, optimisation of turbines will considerably expand the profit margins and garner the attraction of investors.

2.1 Introduction

One of the most beneficial applications of wind energy is in the form of producing electrical energy by small-to-medium wind turbines in rural communities. Wind turbines must be installed in those areas which are away from any wind-blocking obstacles that normally exist in urban areas, as such areas with open land that provide this condition are excellent choices. According to American Wind Energy Association, in excess of environmental benefits, development of this industry helps individuals to generate power and decrease their energy bills. Performance optimisation of these turbines is expected to increase the number of installed machines and such growth will also eventually decrease the costs of turbines for manufacturers.

Two major approaches to analysing and optimisation of any system including the wind turbines are model-based and data-driven methods. A model-based or a knowledge-based analysis is derived from the analytical understanding of the system, whereas the data obtained from the history of the machine performance is the basis of data-driven models. These methods can also be sub-categorised into other groups that each includes its advantages and drawbacks as summarised in Table 2.1.

Various factors play parts to determine the most suitable and efficient method and there is no unique answer. An undeniably important factor is the existing

[1]Turbulence and Energy Laboratory, Department of Engineering, University of Windsor, Canada

Table 2.1 A comparison of various approaches to analyse wind turbines

Type	Categories	Advantages	Drawbacks
Model based	Experimental [1,2]	Reliable results	Dynamic scaling, expensive
	Numerical [3,4]	Not expensive	Validation required
Data driven	Parametric [5,6]	Simple	High error
	Non-parametric [7]	Low error	Susceptible to noisy data

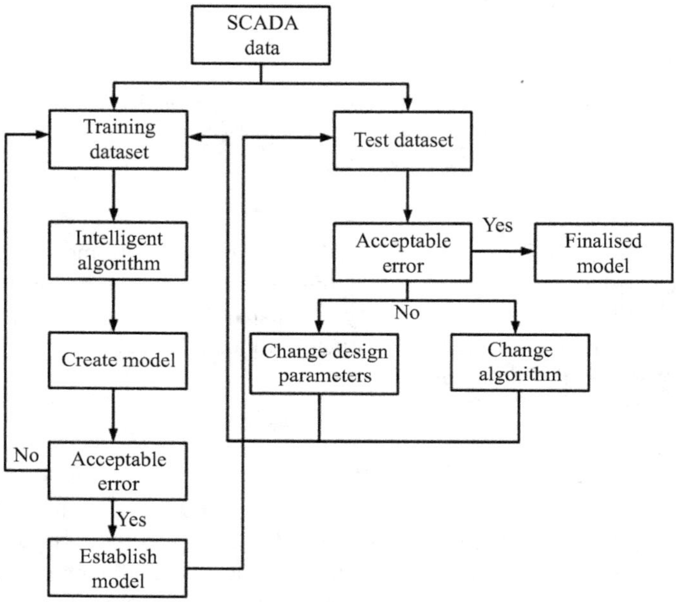

Figure 2.1 Wind turbines data-driven modelling process

complicated correlation between key components of a wind turbine. Such characteristic that creates major uncertainties highlights the difficulties associated with the model-based approach. On the other hand, wind farms often have access to supervisory control and data acquisition (SCADA) data that contain valuable information about performance history of the turbines. The procedure to finalise a data-driven modelling process using an intelligent algorithm is presented in Figure 2.1. As can be observed in this figure, SCADA data should first be divided into two subsets of training and test datasets. The size of training dataset should be such that this portion fully represents the entire data particularly in terms of boundaries and trend. This set is then applied to create a model based on the selected intelligent algorithm. Continuous calculation of test error after each adjustment of modifiable parameters in the utilised network allows achieving the acceptable test error and establishment of the model. The next step is to evaluate the performance of the model by testing it using a new set of data (testing data). If the result of the test (test error) is not acceptable, either the design parameter or the algorithm itself should be changed. Otherwise, the modelling is finalised.

In this chapter, a comprehensive procedure to successfully establish data-driven models for a variety of purposes is presented followed by an industrial case study targeted to design a power-production monitoring system in which some of the proposed methods are practically applied.

2.2 SCADA data treatment

All data-driven modelling begin with a procedure called data pre-processing in which the raw or primary data collected by the system are prepared to be applied by the desired machine-learning technique. The necessity of the process is due to likely incident of sensor malfunction that causes recording noisy and incorrect values. Such records harm the network training accuracy and may even terminate the process. Generally, the steps required to achieve a perfect and smooth pre-processed data vary depending on the type and size of the acquired dataset.

The first step to pre-process the SCADA data of a wind turbine is to ensure that all recorded values of all the features are bounded between the anticipated minimum and maximum range. For instance, a recorded value of 200 °C for oil temperature is obviously false and must be taken care of.

The next step is the detection of outliers. Outliers are defined as the data points that do not follow the same trend as the great majority. To achieve this goal, extreme care must be taken to avoid categorising abnormal behaviour or underperformance of turbines as being an outlier. More specifically, too many unusual records of irregularity at a relatively short period of time are an indication of abnormal performance, not outlier data points.

Out-of-range and outlier values can generally be treated in two manners. They can either be removed or considered as missing values. In this chapter, it is explained how to replace the missing values with some acceptable substitutes, and therefore, such data points will be regarded as missing.

In any data-driven modelling, the treatment method of missing values should be clarified. The most simple and yet common method in conducted studies is to remove the missing values. This approach might be suitable provided that there is not a substantial number of missing values in the dataset. Due to customary sensor malfunction observed in SCADA system, this is not always the case and removing all the data points in a row just because there is one missing leads to useful information loss that is collected by other sensors that can be up to 150 signals [8,9]. In addition, having access to consecutive data points makes the predictive models considerably more reliable. This in enlightened by the fact that the wind pattern as the major influential factor on wind turbine performance is subjected to drastic changes in time. Thus, consecutive data make it possible to capture such alterations.

The process of replacing missing values with some reasonable data is called data imputation [10]. Generally, application, type, and the size of datasets, and missing value pattern determine the most suitable method of imputation. These methods can be categorised into two groups of statistical learning theory and machine-learning tools [11].

Machine-learning tools that are generally more advanced than their statistical counterparts are also often more computationally extensive. Considering the large datasets collected by SCADA system of a wind turbine, it is preferred to employ simpler tools, as possible. However, there should be a balance between the outcome of the applied algorithm for the imputation and the time and memory it requires for the process.

Following sections explain a number of more suitable approaches to replace the missing values of SCADA data. Depending on the function and pattern, each may be considered the most applicable.

2.2.1 Mean or median value

The simplest method of imputation is to substitute the missing values in records of a parameter with the mean or median value of that feature. This works best for the features with very low standard deviations and also smaller datasets. For the parameters in the SCADA data, however, this is not normally the best option although it does not require heavy computations.

2.2.2 K-Nearest neighbour

This is also a simple and yet very effective method for data imputation. In this technique, the K most similar rows of data with the one containing the missing value are found and then the imputed record will be the mean value of those parameters in the recognised neighbouring rows. The similarity measurement can be performed using a variety of distance functions.

Despite its simplicity, one major disadvantage of this method is the requirement to search through the entire dataset to find the most suitable neighbours which can at times be very computational for large datasets. Moreover, the optimum number of neighbours cannot be found unless various numbers are examined and the performance of each is studied. This too requires heavy computations.

2.2.3 Expectation–maximisation

The other powerful algorithm for imputation is expectation–maximisation algorithm, which has been applied in several studies [12,13]. In this method, regression parameters in incomplete features are computed for complete datasets (where their missing records are deleted) with available values from the estimation of mean and covariance matrix. After that, a conditional expectation value replaces the missing ones and then covariance matrix will be re-estimated to minimise the imputation error. After several iterations when there is none or a negligible update in imputed records, the process is finalised and the completed dataset is generated. Although this method usually leads to an acceptable level of imputation errors, the iterative pattern of this method consumes non-trivial quantities of time and memory for large datasets.

2.2.4 Decision tree

Decision trees split the datasets into smaller groups and place them in the leaf nodes. It is performed from top to the bottom starting from the root node. In the beginning, all the training instances are at the root. Then, based on a selected feature, the dataset is partitioned into lower nodes. Similarly, the dataset is further divided using another feature. This process continues till it is reached to the targeted feature (the one including the missing points). The selection of features can be based on a variety of statistical measures. One of the most popular ones is information gain, which is itself based on the concept of Gibbs entropy as expressed in the following equations:

$$E(T) = \sum_{i=1}^{c} -p_i log_2 p_i \tag{2.1}$$

$$E(T,X) = \sum_{c \in X} p(c)E(c) \tag{2.2}$$

where E is the entropy, c is the number of possible outcomes, and p_i is the possibility of each outcome. T and X represent the features in the dataset, where T is the target feature. Figure 2.2 illustrates a schematic of a decision tree when the data is divided into different groups. In this schematic, besides the root node, there are two other node layers (N_1 and N_2) that act as rules to direct to the correct leaf where data points are placed as sub-datasets of d_1, d_2, $\ldots$, d_j. After splitting the datasets

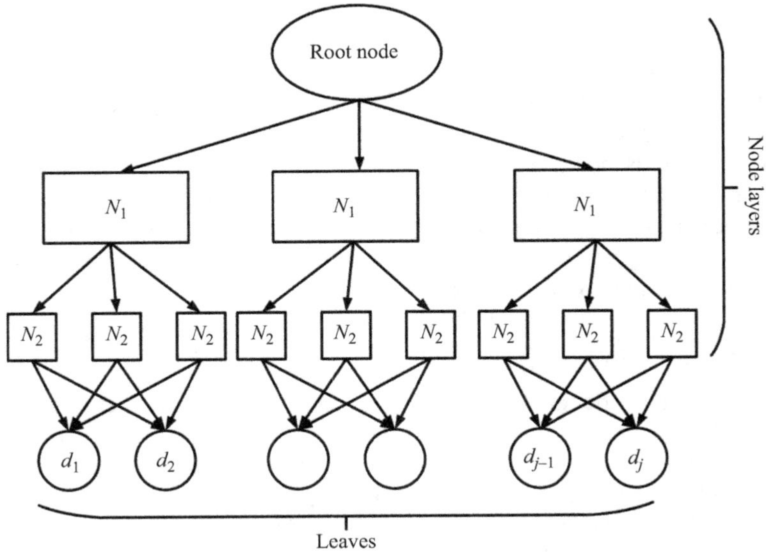

Figure 2.2 Schematic of a decision tree

into smaller groups each containing the most similar data points, a simple algorithm can be applied to missing the missing records. In Section 2.4, missing values in each sub-dataset is imputed using mean value of each feature.

The final step of data pre-processing is normalisation of the range of the parameters. This step is essential when more than one input is used to train the model. Applying inputs with different data ranges undermines lower range inputs. This normalisation is normally done according to the following formula:

$$V = \frac{X - X_{min}}{X_{max} - X_{min}} \tag{2.3}$$

where V is the normalised variable and X is the variable.

2.3 Feature selection

There are a substantial number of monitored parameters in the wind turbine SCADA systems. Taking them all into consideration for performance, evaluation is not practical and may not even be helpful. Selection of appropriate features initially depends on the desired modelling type and investigated target. In knowledge-based modelling, it is very common to choose parameters solely based on the physical knowledge of systems [14]. In this approach, parameters are chosen on the basis of conceptual importance and influence. For instance, if power generation is being monitored, the following equation can be utilised:

$$P = \frac{1}{2}\rho A C_p V^3 \tag{2.4}$$

where ρ is the air density, A is the rotor sweep area, V is the wind speed, and C_p is the power coefficient. This coefficient is the ratio of real electric power produced by a wind turbine divided by the total wind power flowing into the turbine blades at a specific wind speed.

However, it is now clear that other factors should also be taken into consideration to create more realistic models particularly because (2.4) is not accurate at wind speeds larger than rated wind speed according to the power curve of a wind turbine. In this chapter, two powerful methods are explained that either of them can be employed based upon the modelling criteria.

2.3.1 Correlation coefficients

To distinguish all influential factors on the desired output parameter, correlation analysis can be carried out. In this method, parameters that may initially seem irrelevant but can potentially play a positive role in the establishment of reliable models will be recognised. It should also be noted that machine-learning algorithms will not perform well when the number of input parameters is relatively high. Therefore, to find this optimum number, statistical correlation method is a suitable choice in addition to the physical knowledge of the system. The correlation coefficient between parameters (r) is calculated by means of the following equation:

$$r = \frac{\sum_{i=1}^{n}(x_i - \bar{x})(y_i - \bar{y})}{\sqrt{\sum_{i=1}^{n}(x_i - \bar{x})^2}\sqrt{\sum_{i=1}^{n}(y_i - \bar{y})^2}} \tag{2.5}$$

where x_i and y_i are the two variables that are investigated for correlation that can be any of the potential input parameters and the output, n is the number of data points and $\bar{x}$ and $\bar{y}$ represent the mean values. This coefficient, which has a value from -1 to 1, shows the strength of the relationship between the two parameters. The absolute value of 1 indicates perfect correlation while zero means no correlation. Negative values show reverse relationship [14]. The greater value of r for each signal indicates more influence of that signal on the output and consequently, that signal is a more suitable input candidate.

2.3.2 Principal component analysis

Principal component analysis (PCA) is a valuable statistical method applied in a variety of research areas to find a hidden pattern in high dimensional data. It helps express and highlight similarities and differences in the dataset and can be utilised for condition monitoring, pattern recognition, anomaly detection, etc. for various industrial processes. In such applications, PCA, formulated as a multivariate statistical process control task, extracts a few independent components from highly correlated process data and use them to monitor the operation of the process more efficiently [15–17]. The following sections explain the mathematical basis for off-line and recurrent PCA.

2.3.2.1 Off-line PCA

To create an off-line PCA model, the mean value of the data points should be zero. To do so, the mean value of the recorded data points should be subtracted from their original values. Afterwards, the covariance matrix is calculated and then its eigenvalues and eigenvectors are determined. The next step is to choose the components and form the features. To achieve this goal, eigenvectors are arranged based on the eigenvalues from the highest to the lowest. This is to ultimately put the components into an order of importance. Then, the items of lesser significance can be removed. Finally, the new dataset is generated and that finishes the PCA transformation. Once the significant components are kept and the feature vector is created, the transpose of that vector is calculated and multiplied by the original dataset on the left of the transposed.

Despite this method's simplicity and popularity, it is only applicable to stationary models due to its off-line structure. To address this issue, the recursive PCA as explained in the following section can be employed to track the dynamic behaviour of a system.

2.3.2.2 Recurrent PCA

When PCA models are established from the data according to the aforementioned procedure, they are time-invariant, unlike most actual industrial processes that tend to change over time. The main characteristics of the industrial processes that are time-varying include the changes in the mean, variance, and the correlation structure between variables that can affect the number of significant principal components (PCs). In the case of applying a time-invariant PCA model for the condition monitoring tasks in processes that might characteristically change over time, false alarms will occur that obviously question the reliability of the model. Slow time-varying behaviours are normally observed in industrial processes as a result of equipment ageing, sensors or process drifting, and preventive maintenance and cleaning. To address this issue, adaptive or recursive PCA models have been developed [18,19].

The intent here is to develop a method to update the PCA whenever a new block of data becomes accessible. Considering X_1^0 as the initial block of data, the mean of each column is given in the vector

$$b_1 = \frac{1}{n_1}\left(X_0^1\right)^T 1_{n_1} \tag{2.6}$$

where $1_{n_1} = [1, 1, \ldots, 1]^T \in R^{n_1}$

The data is scaled to zero mean and unit variance is given by

$$\Sigma_1 = diag(\sigma_{1.1}, \ldots, \sigma_{1.m}) \tag{2.7}$$

$$X_1 = \left(X_1^0 - 1_{n_1} b_1^T\right)\Sigma_1^{-1} \tag{2.8}$$

whose ith element is the standard deviation of the ith sensor; $i = 1, \ldots, m$. The correlation matrix is

$$R_1 = \frac{1}{n_1 - 1} X_1^T X_1 \tag{2.9}$$

The new block of data is expected to augment the data matrix and calculate the correlation matrix recursively. Assume that b_k, X_k and R_k have been calculated when the kth block of data is collected. The task for recursive calculation is to calculate b_{k+1}, X_{k+1} and R_{k+1} when the next block of data $X_{n_{k+1}}^n \in R^{n_{k+1} \times m}$ is available. Denoting

$$X_{k+1}^0 = \begin{bmatrix} X_k^0 \\ X_{n_{k+1}}^0 \end{bmatrix} \tag{2.10}$$

for all the $k+1$ block of data, the mean vector b_{k+1} is related to b_k by the following relation:

$$\left(\sum_{i=1}^{k+1} n_i\right) b_{k+1} = \left(\sum_{i=1}^{k} n_i\right) b_k + \left(X_{n_{k+1}}^0\right)^T 1_{n_{k+1}} \tag{2.11}$$

Denoting $N_k = \sum_{i=1}^{k} n_i$, the above equation yields to the following recursive calculation:

$$b_{k+1} = \frac{N_k}{N_{k+1}} b_k + \frac{1}{N_{k+1}} \left(X^0_{n_{k+1}}\right)^T 1_{n_{k+1}} \tag{2.12}$$

The recursive calculation of X_{k+1} is given by

$$\begin{aligned} X_{k+1} &= \left[X^0_{k+1} - 1_{k+1} b^T_{k+1}\right] \Sigma^{-1}_{k+1} = \left[\begin{bmatrix} X^0_k \\ X^0_{n_{k+1}} \end{bmatrix} - 1_{k+1} b^T_{k+1}\right] \Sigma^{-1}_{k+1} \\ &= \begin{bmatrix} X^0_k - 1_k \Delta b^T_{k+1} - 1_k b^T_k \\ X^0_{n_{k+1}} - 1_{k+1} b^T_{k+1} \end{bmatrix} \Sigma^{-1}_{k+1} = \begin{bmatrix} X_k \Sigma_k \Sigma^{-1}_{k+1} - 1_k \Delta b^T_{k+1} \Sigma^{-1}_{k+1} \\ X_{n_{k+1}} \end{bmatrix} \end{aligned} \tag{2.13}$$

where

$$X_k = \left(X^0_k - 1_k b^T_k\right) \Sigma^{-1}_k \tag{2.14}$$

$$X_{n_{k+1}} = \left(X_{n_{k+1}} 1_{n_{k+1}} b^T_{k+1}\right) \Sigma^{-1}_{k+1} \tag{2.15}$$

$$\Sigma_j = diag(\sigma_{j.1}, \ldots, \sigma_{j.m}), \quad j = k, k+1 \tag{2.16}$$

$$\Delta b_{k+1} = b_{k+1} - b_k \tag{2.17}$$

The recursive computation of the standard deviation has the following relationship [19]:

$$(N_{k+1} - 1)\sigma^2_{k+1.i} = (N_k - 1)\sigma^2_{k.i} + N_k \Delta b^2_{k+1}(i) + \|X^0_{n_{k+1}}(:, i) - 1_{n_{k+1}} b_{k+1}(i)\|^2 \tag{2.18}$$

where $X^0_{n_{k+1}}(:, i)$ is the ith column of the associated matrix. $b_{k+1}(i)$ and $\Delta b_{k+1}(i)$ are the ith elements of the associated vectors and $\|A\|$ is the norm of the matrix A. Similarly, the recursive calculation of the correlation matrix has the following form:

$$\begin{aligned} R_{k+1} &= \frac{1}{N_{k+1} - 1} X^T_{k+1} X_{k+1} - \frac{N_k - 1}{N_{k+1} - 1} \Sigma^{-1}_{k+1} \Sigma_k R_k \Sigma_k \Sigma^{-1}_{k+1} \\ &\quad - \frac{N_k}{N_{k+1} - 1} \Sigma^{-1}_{k+1} \Delta b_{k+1} \Delta b^T_{k+1} \Sigma^{-1}_{k+1} + \frac{1}{N_{k+1} - 1} X^T_{n_{k+1}} X_{n_{k+1}} \end{aligned} \tag{2.19}$$

Therefore, (2.12), (2.13), and (2.19) create the recursive manner for the PCA model.

2.4 Modelling design networks

This section introduces the design methodology to establish a monitoring system. It is aimed to create an intelligent monitoring tool to fully track any desired parameter of wind turbines. There are a substantial number of machine-learning tools that can be employed to achieve this goal. Based on the popularity in conducted studies, achieved reliable results and the potential applicability in various types of studies, two powerful and efficient networks of multi-layer perceptron (MLP) neural network (NN) and adaptive neuro-fuzzy inference system (ANFIS) are selected to be utilised here. The two states of static and dynamic models using such networks are then explained. In addition, it is shown how these networks can be fused together to benefit from complementary inference due to each individual MLP and ANFIS networks. To do so, each network is individually trained over the entire training dataset and a final decision is then made based upon an aggregation space [20].

To develop this technique for the purpose of accurately monitoring the machine performance, ordered weighted averaging (OWA) is introduced as the fusion method to integrate inferences of MLP and ANFIS techniques. Finally, a clarification of estimation and prediction concepts is presented followed by a performance evaluation method for the established networks.

2.4.1 Multi-layer perceptron

MLP networks are widely applied for a variety of purposes in data-driven modelling. They are proven to create accurate networks with fast computational abilities. They are feed forward NNs that create a mapping between input and output spaces. They have been suitably utilised for pattern recognition, condition monitoring, fault diagnosis, function approximation, and many other purposes [21–23]. The MLP network consists of multiple layers of nodes in the forward direction and all nodes in each layer are completely linked to the nodes of the following layer. There are three layers of input, output, and hidden layers in its design. The nodes in the hidden layer are neurones with a non-linear activation function that can take various shapes such as hyperbolic tangent, sigmoid, and so on. The output layers, however, mostly apply nodes with linear functions, whereas input layer acts as a buffer.

During the training of the network, weights of each input $(W_1, W_2, \ldots, W_n)$ and also the bias term (b) are determined and then aggregated to be applied by the activation function. In addition, MLP network employs the supervised learning tool of error back propagation algorithm for training. A typical structure of an MLP with n number of inputs, $X_1, X_2, \ldots, X_n$, is illustrated in Figure 2.3. To train the networks, the gradient descent with momentum method is an acceptable choice. In this method, in addition to error calculation, the general trend of the network error will also be determined. This creates a reduction in the risk of local minima and leads to an enhanced generalisation [24]. The other influential factor in the structure of this network is the number of neurones in the hidden layer. To find the optimum number of neurones, it is recommended to run at least ten times during which only the number of neurones is changed to seek the configuration with the best generalisation [25,26]. This helps to avoid overfitting.

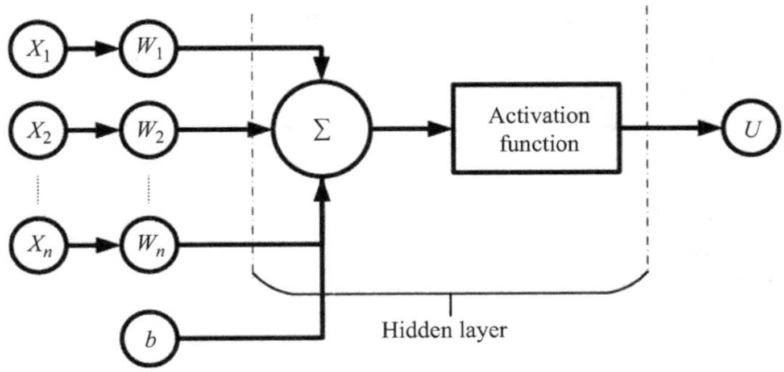

Figure 2.3 A typical MLP network

2.4.2 Adaptive neuro-fuzzy inference system

ANFIS networks were initially introduced by Takagi and Hayeshi [27] and then further developed by Jang [28]. They have been extensively operated in many research works and were first employed by Schlechtingen *et al.* [7] for the purpose of wind turbine monitoring. As the name implies, ANFIS networks are a combination of the quality-based fuzzy approach and NNs adaptive characteristics to attain enhanced network performance.

Fuzzy logic (FL) creates an inference mechanism under cognitive uncertainty, whereas NNs benefit from computational advantages such as adaption, fault tolerance, and generalisation. Therefore, ANFIS method is established as NNs and FL are engaged for dealing with cognitive uncertainty in a more qualitative manner. In practice, ANFIS is a fuzzy model that enables adapting and training in a flexible system framework. As a result, ANFIS does not require expert knowledge to establish the modelling system.

The mathematical transformation of the expert knowledge as required in FL modelling is a time-consuming and demanding task to perform. On the other hand, NNs can efficiently handle non-linear prediction and estimation with high computational capabilities. Therefore, this integration of methods can create a computationally efficient environment while providing an inference mechanism.

The creation of an inference mechanism in FL is done by utilisation of membership functions. Mamdani [29] and Sugeno [30] have proposed two common types of inference systems. In the Sugeno type, the consequent part is a non-fuzzy equation, while a fuzzy linguistic value was proposed in the Mamdani type. The Sugeno type is a more suited choice for wind turbine analysis mainly because of its computational efficiency, continuity of the generated output surface, and also the fact that a non-fuzzy equation represents the consequent part.

To model fuzzy rules into desired outputs, the Sugeno fuzzy model is implemented as follows:

$$\text{If } x_1 = A_i \text{ and } x_n = B_j \quad \text{then} \quad f_i = p_i x_1 + q_i x_n + r_i$$

where p_i, q_i, and r_i are adaptive design parameters that are regulated during the training phase. A typical five-layer ANFIS structure is shown in Figure 2.4. In this feed-forward structure, parameters in adaptive nodes are continuously modified during the training. Mathematical summarisation of the layers is presented herein [31].

Layer 1. The first layer consists of adaptive nodes. All the neurones in this layer correspond to a linguistic label and the resulting output is equal to the membership function of this layer.

$$OL1_i = \mu A_i(x_1) \tag{2.20}$$

Layer 2. Contrasting to the previous layer, nodes in the Layer 2 are not adaptive (fixed). They approximate the firing strength (w_i) of a rule determined by multiplication of incoming signals.

$$OL2_i = w_i = \mu_{A_i(x_1)}\mu_{B_i(x_n)} \tag{2.21}$$

Layer 3. The nodes in this layer are also fixed. The output of each layer is the ratio of the ith rule firing strength over the summation of firing strengths of all rules, as formulated in (2.22). These nodes normalise the firing strength of the previous layer.

$$OL3_i = \bar{w}_i = \frac{w_i}{\sum_{j=1}^{i} w_i} \tag{2.22}$$

Layer 4. All the nodes in this layer are adaptive. The outcome of each node is equal to multiplication of relative firing strength (calculated in the previous layer) and the consequent parameter as

$$OL4_i = \bar{w}_i f_i = \bar{w}_i(p_i x_1 + q_i x_n + r_i) \tag{2.23}$$

Layer 5. The number of nodes is equal to the number of output parameters, in this layer. This node acts as a summer and computes the summation of all signals from Layer 4.

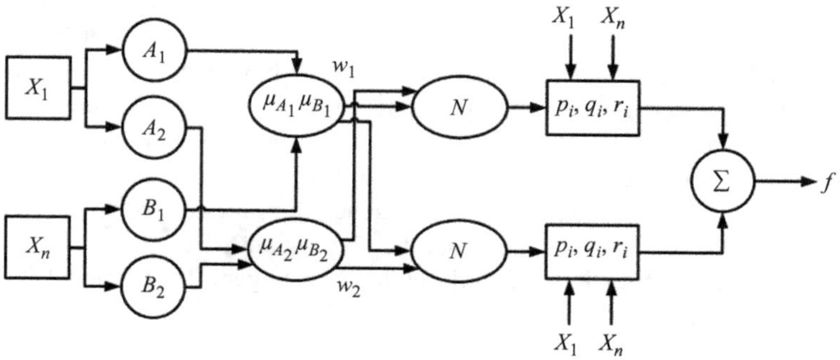

Figure 2.4 A five-layer ANFIS structure

$$OL5_i = \sum_{i=1}^{j} \bar{w}_i f_i = \frac{\sum_i w_i f_i}{\sum_i w_i} \tag{2.24}$$

It should be noted that there is no unique structure for the design of ANFIS networks since a combination of some layers may still produce similar outcomes. According to the structure shown in Figure 2.4, there are two adaptive layers. Parameters in Layer 1, called premise parameters, are adjusted based on the input membership functions. Additionally, in Layer 4, there are three modifiable parameters of p_i, q_i, and r_i which are first-order polynomials that are considered consequent. During the training phase, all these modifiable parameters are tuned in such a way that the network overall error is minimised.

2.4.3 Static and dynamic networks

NNs can be represented in static and dynamic structures. In the static type, the network is simply trained using the selected input parameters as described in the following equation:

$$u(t) = f(x_1(t), x_2(t), \ldots, x_i(t)) \tag{2.25}$$

where $u(t)$ is the network output at the time t and x_i is the ith input of the network.

In the proposed dynamic networks, in addition to the selected parameters as the network inputs ($x_i(t)$), the inputs and the output of the previous iteration time step ($x_i(t-1)$ and $u(t-1)$) are also considered as the network inputs. In other words, the output resulted from each set of inputs plus those inputs are also given to the network for better training. In this case, the output function would be described as follows:

$$u(t) = g(x_1(t), x_1(t-1), x_2(t), x_2(t-1), \ldots, x_i(t), x_i(t-1), u(t-1)) \tag{2.26}$$

2.4.4 Fusion

Fusion schemes are designed to merge different networks together in order to increase overall network accuracy that leads to an enhanced decision-making process. The basic idea is to consider a specific weight for each network's output and then combine the result to benefit from the complementary interference of multiple networks. Figure 2.5 illustrates such design in case of applying two networks. To create such scheme, OWA can be employed. The OWA operator, in general, creates a class of parameterised aggregations operators including minimum, maximum, and the mean. The ideology of this operator has been employed in a variety of applications including data mining, decision-making, approximate reasoning, expert systems, fuzzy systems and control [32–35]. The key advantage of this operator is the ability to encompass various operators bounded between a minimum and a maximum value.

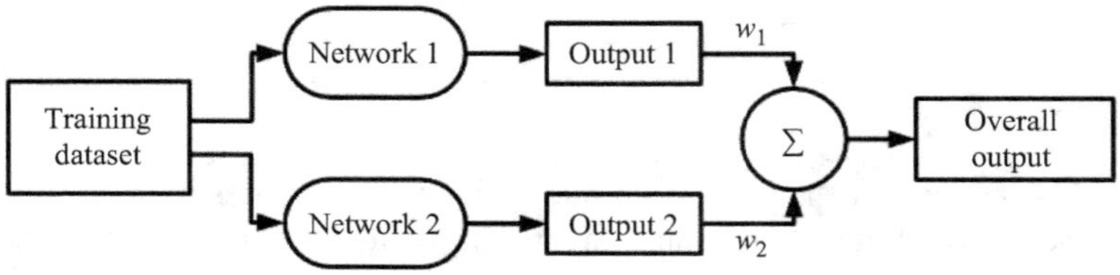

Figure 2.5 Schematic of designing a fusion network

An OWA operator creates a mapping of $F: R^n \rightarrow R$, where n is the space dimension and the associated weight vector $w = (w_1, w_2, \ldots, w_n)^T$ is in such a way that satisfies the conditions expressed in the following equations:

$$w_1 + w_2 + \cdots + w_n = 1, \qquad 0 \leq w_i \leq 1, \quad i = 1, 2, \ldots, n \tag{2.27}$$

$$F(a_1, a_2, \ldots, a_n) = \sum_{i=1}^{n} b_i w_i \tag{2.28}$$

where b_i represents the ith largest of $a_1, a_2, \ldots, a_n$.

The most important item in the application of an OWA operator is the employed method to accurately determine the weights. We consider that there are M number of various data, each of which includes values that are provided by N information sources and also the ideal output which is intended to be estimated. Therefore, each example consists of $N + 1$ values and for the ith example, it can be expressed as $a_1^i, a_2^i, \ldots, a_N^i | b^i$, where a_i denotes the value provided by the jth information source and b_i is the ideal output. For aggregation purposes, the goal is to find the weighting vector w in such a way that the following condition is met:

$$minimise \sum_{j=1}^{M} \left(w\left(a_1^j, \ldots, a_n^j\right) - b^j\right)^2 \tag{2.29}$$

Several methods have been utilised to address (2.29), gradient descent approach being among the most popular, thanks to its high accuracy [36,37].

2.4.5 Estimation and prediction

In any data-driven modelling, networks can be established for two major purposes – either estimation of the output is being targeted, or it is designed to predict the future output value based on the current historical data. Defining the datasets for network training depends on whether it is developed for estimation or prediction. For estimation, at any time during a machine operation, the network is trained to approximate the output value at time t when it is trained by input parameters at the same time t. The mathematical representation of such monitoring is presented as below:

$$U(t) = f(x_1(t), x_2(t), \ldots, x_i(t)) \tag{2.30}$$

where U and x_i represent the output and input parameters, respectively.

In contrast to estimation, when a network is designed to make a prediction of future performance of a machine, input and output signals are not gathered at the same time. In this situation, based on the available input signals at time t, network is designed to predict the output at the time $t+n$. The frequency in which the data have been acquired determines the duration of time for prediction. For instance, if the data is gathered every hour, $n = 1$ means the prediction is being made for 1 h into the future. Similarly, the mathematical expression for prediction is presented as follows:

$$U(t+n) = f(x_1(t), x_2(t), \ldots, x_i(t)) \tag{2.31}$$

2.4.6 Performance evaluation

To determine networks performance mean absolute error (MAE), the following formula can be employed:

$$MAE = \frac{1}{n}\sum_{i=1}^{n}|x_i - y_i| \tag{2.32}$$

where x_i is the predicted value by the established network and y_i is the actual value. MAE indicates the closeness of predicted results with true values and thus creates an appropriate understanding of the model accuracy.

2.5 A case study

In this section, a case study is presented to show how aforementioned techniques and networks can be practically applied to establish a well-designed and accurate monitoring system. In the following sections, the implementation of data pre-processing and establishment of MLP and ANFIS networks in MATLAB® environment are explained.

2.5.1 Data pre-processing

2.5.1.1 Data range

As mentioned before, the first step is to check the data points range. To do so, a reasonable range should be defined for all parameters based on expert knowledge and the recorded values beyond this range are replaced with "Nan". It is imperative to emphasise that these ranges do not represent the normal behaviour of the system and cannot practically be true. For instance, Table 2.2 shows recorded values of wind speed for five different turbines. As can be observed in the table, some recorded values for turbines 3 and 5 are clearly false and need to be removed.

2.5.1.2 Outlier detection

Next step is to remove the data points that, despite falling into a reasonable range, do not follow the same trend as most data points. A common method to detect such values is to visualise the data and based on the results, define new ranges in a

Table 2.2 Recorded values of wind speed in m/s for different turbines

Turbine 1	Turbine 2	Turbine 3	Turbine 4	Turbine 5
5	8	5	9	10
5	7	90000009536743	8	10000038146973
5	8	5	9	12
5	8	5	9	12
5	8	5	9	12

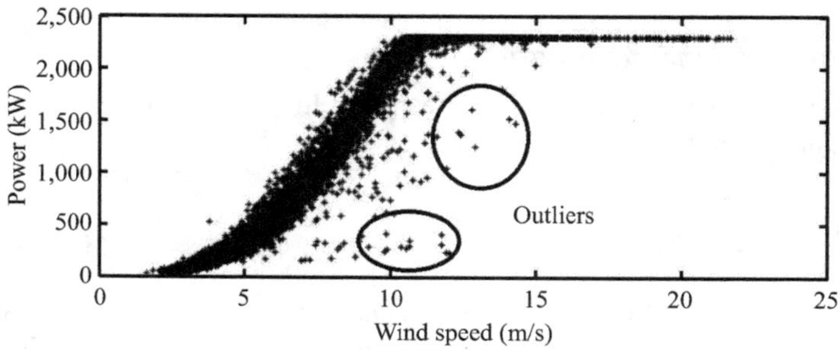

Figure 2.6 Detection of outliers in the power curve

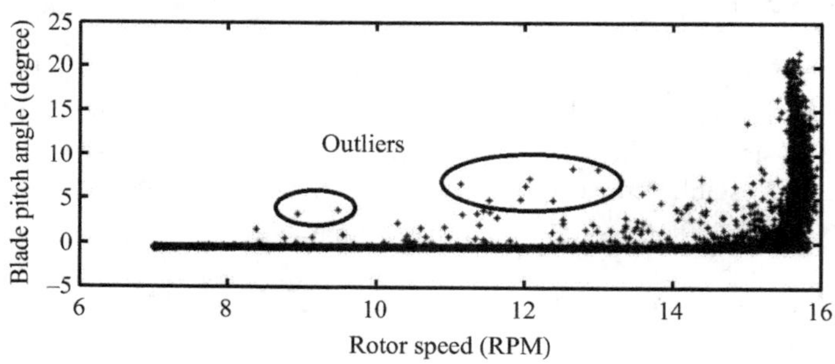

Figure 2.7 Detection of outliers; blade pitch angle vs. rotor speed

conditional manner for such parameters. Figures 2.6 and 2.7 show two examples of such procedure.

2.5.1.3 Missing values

In this stage, the values that are originally missing or have been removed in the previous steps are replaced with some reasonable data. To achieve this goal, a combination of decision trees and mean value has been applied. The procedure is detailed below [12] and summarised in Figure 2.8.

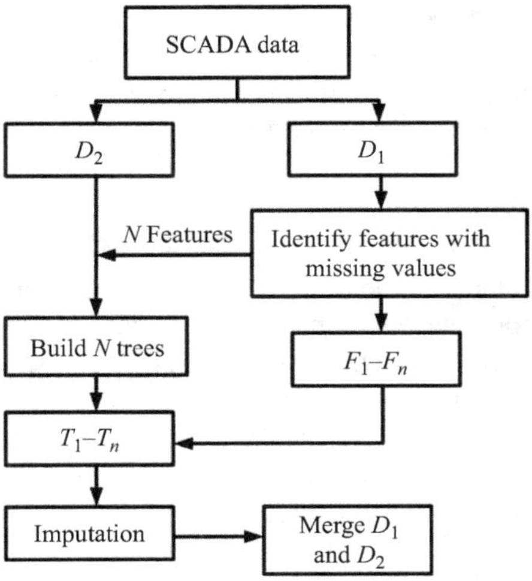

Figure 2.8 Imputation procedure

Step 1: The first step is to divide the dataset into two parts where all instances including a missing value will be placed in one dataset (D_1) and the complete instances are located in the other (D_2).

Step 2: Then, a series of decision trees are built based on the complete dataset (D_2). The target feature of each tree is a feature that contains at least one missing value that was identified in the previous stage. In this step, similar data points are located in the tree leaves or sub-datasets of $d_1, d_2, \ldots, d_j$. To be able to create these sub-datasets, the values are generalised to the n_i number of groups. n_i is equal to the root square of the domain size of feature i.

Step 3: At this stage, the rows of the dataset D_1 are placed in the predicted leaf of the respective tree. For this purpose, the missing feature of each row is first identified and then the tree that has the same feature as its target is used to find the correct leaf. If one instance has more than one missing value, that record will be added to more than one leaf.

Step 4: The imputation is performed here using the mean value of each feature in each leaf. Then, the missing value is replaced by the imputed one in D_1. Finally, D_1 and D_2 are merged together to form the complete imputed dataset.

To calculate the imputation error various methods can be applied. The most common approach is the normalised root mean square (NRMS) according to the following expression:

$$NRMS = \frac{\|X^{estimate} - X^{original}\|}{\|X^{original}\|} \tag{2.33}$$

To validate the proposed method of imputation, a dataset containing 5 per cent random missing values was created and the calculated NRMS was 0.0306. This acceptable error level creates confidence to apply the method when the actual records of the missing values are unknown.

2.5.2 Monitoring networks

In this section, initially, the method of determining the input parameters to monitor output power is explained and then the MLP and ANFIS networks are trained.

To finalise the input space, a combination of PCA and knowledge of the system is applied. Based on common sense and (2.4), wind speed is the most influential parameter on the power production and needs to be included in the input data. The other features in the SCADA data are then applied to PCA modelling to extracts to PCs. Finally, to create a recurrent or dynamic network, the latest available power output is also considered as the final input. As explained in Section 2.3.5, n is considered as the duration of time into the future for which a prediction of power production is intended to be estimated. If t represents the present time, the time of applying each set of inputs to predict the output at the time $t + n$ is presented in Table 2.3. The frequency in which the data points are gathered determines the time intervals into the future that the prediction can be made for. For this study, the obtaining frequency is 1 h which means when $n = 1$, the prediction is for the average power production in the following hour, and $n = 2$ shows the prediction for the next 2 h.

It is also important to note that since the value of n is on the scale of hours, the weather forecast for the wind speed would be precise enough and that is why in Table 2.3, the wind speed is applied at the time $t + n$.

The next step after determining the input and output spaces is to train the MLP and ANFIS networks. To do so, the design parameters according to Table 2.4 are employed.

The test errors of networks are summarised in Table 2.5. The results are based on MAE in terms of kW for 2.3 MW pitch-regulated turbines. The modelling is performed for three different conditions. At first, the networks are designed for power estimation ($n = 0$). Then, two future predictions for the state of $n = 1$ and $n = 2$ have been performed.

Table 2.3 Time of applying each set of inputs to predict the output at the time $t + n$

Inputs	**Time**
Principal component 1	t
Principal component 2	t
Wind speed	$t + n$
Power	$t - 1$

Table 2.4 Networks design criteria

MLP	
Error calculation	Gradient descent
Performance goal	0.004
Learning rate	0.03
Maximum number of epochs	500
Activation function	Sigmoid
Number of neurons in the hidden layer	8
ANFIS	
MF type for inputs	Gaussian
MF type for output	Linear
Number of MFs	Three for each input
Optimisation method	Hybrid
Error tolerance	0.004
Maximum number of epochs	40

Table 2.5 Networks test MAE (kW)

	MLP	ANFIS	Fusion
$n = 0$	38	25	19
$n = 1$	64	51	43
$n = 2$	85	71	56

Although all networks produce acceptable error levels, in all conditions, ANFIS networks outperform the MLP models and the results of merging the networks together are better than each individual network.

2.6 Conclusion

This chapter presented in detail all required steps to finalise a data-driven model for the purpose of condition monitoring of wind turbines. The general advantages of data-mining approach to establishing a reliable model were explained. This modelling was divided into three general steps first of which was the pre-processing of SCADA data. This process made the data applicable for training the networks.

Second, it was explained how correlation coefficient or PC analysis can be applied for the purpose of choosing the most influential parameters as network inputs. That mainly depends on the target feature that determines the monitoring type.

Next, two advanced networks of MLP and ANFIS with proven capabilities in data-driven modelling and various types that they can be applied or combined were described. That finalised the entire modelling process.

Finally, a case study was presented modelling the power curve in which all steps were practically applied to establish well-structured models with acceptable error levels.

Such advanced monitoring system, that can accurately track the performance of the wind turbines, can be a persuasive factor for small communities to further advance the utilisation of wind energy for power production.

References

[1] Yin C, Zhang Z, Wang Z, *et al.* Numerical simulation and experimental validation of ultrasonic de-icing system for wind turbine blade. Applied Acoustics. 2016;114:19–26.

[2] Chen J, Wang Q, Zhang S, *et al.* A new direct design method of wind turbine airfoils and wind tunnel experiment. Applied Mathematical Modelling. 2016;40(3):2002–2014.

[3] Heo YG, Choi NJ, Choi KH, *et al.* CFD study on aerodynamic power output of a 110 kW building augmented wind turbine. Energy and Buildings. 2016;129:162–173.

[4] El-Zahaby AM, Kabeel A, Elsayed S, *et al.* CFD analysis of flow fields for shrouded wind turbine's diffuser model with different flange angles. Alexandria Engineering Journal. 2016.

[5] Lydia M, Selvakumar AI, Kumar SS, *et al.* Advanced algorithms for wind turbine power curve modeling. IEEE Transactions on Sustainable Energy. 2013;4(3):827–835.

[6] Li S, Wunch DC, O'Hair E, and Giesselmann MG. Comparative analysis of regression and artificial neural network models for wind turbine power curve estimation. Journal of Solar Energy Engineering. 2001;123:327–332.

[7] Schlechtingen M, Santos IF, and Achiche S. Using data-mining approaches for wind turbine power curve monitoring: a comparative study. IEEE Transactions on Sustainable Energy. 2013;4(3):671–679.

[8] Frank A. Asuncion. UCI machine learning repository. Irvine, CA: University of California; 2010.

[9] Meng Z, and Shi Z. Extended rough set-based attribute reduction in inconsistent incomplete decision systems. Information Sciences. 2012;204:44–69.

[10] Zhang S. Shell-neighbor method and its application in missing data imputation. Applied Intelligence. 2011;35(1):123–133.

[11] García-Laencina PJ, Sancho-Gómez JL, and Figueiras-Vidal AR. Pattern classification with missing data: a review. Neural Computing and Applications. 2010;19(2):263–282.

[12] Rahman MG, and Islam MZ. Missing value imputation using decision trees and decision forests by splitting and merging records: two novel techniques. Knowledge-Based Systems. 2013;53:51–65.

[13] Yang X, Huang B, Zhao Y, *et al.* Generalized expectation–maximization approach to LPV process identification with randomly missing output data. Chemometrics and Intelligent Laboratory Systems. 2015;148:1–8.

[14] Akande KO, Owolabi TO, and Olatunji SO. Investigating the effect of correlation-based feature selection on the performance of support vector

machines in reservoir characterization. Journal of Natural Gas Science and Engineering. 2015;22:515–522.

[15] Zhou B, Ye H, Zhang H, *et al.* Process monitoring of iron-making process in a blast furnace with PCA-based methods. Control Engineering Practice. 2016;47:1–14.

[16] Cai E, Liang L, Xu G, *et al.* Monitoring of chemical industrial processes using integrated complex network theory with PCA. Chemometrics and Intelligent Laboratory Systems. 2015;140:22–35.

[17] Camacho J, Pérez-Villegas A, García-Teodoro P, *et al.* PCA-based multivariate statistical network monitoring for anomaly detection. Computers & Security. 2016;59:118–137.

[18] Wold S. Exponentially weighted moving principal components analysis and projections to latent structures. Chemometrics and Intelligent Laboratory Systems. 1994;23(1):149–161.

[19] Li W, Yue HH, Valle-Cervantes S, *et al.* Recursive PCA for adaptive process monitoring. Journal of Process Control. 2000;10(5):471–486.

[20] Ng KC, and Abramson B. Consensus diagnosis: a simulation study. IEEE Transactions on Systems, Man, and Cybernetics. 1992;22(5):916–928.

[21] Salahshoor K, Kordestani M, and Khoshro MS. Design of online soft sensors based on combined adaptive PCA and RBF neural networks. In: Computational Intelligence in Control and Automation, 2009. CICA 2009. IEEE Symposium on. IEEE; 2009. p. 89–95.

[22] Shrme AE. Hybrid intelligent technique for automatic communication signals recognition using bees algorithm and MLP neural networks based on the efficient features. Expert Systems with Applications. 2011;38(5):6000–6006.

[23] Waqar T, and Demetgul M. Thermal analysis MLP neural network based fault diagnosis on worm gears. Measurement. 2016;86:56–66.

[24] Schlechtingen M, and Santos IF. Comparative analysis of neural network and regression based condition monitoring approaches for wind turbine fault detection. Mechanical Systems and Signal Processing. 2011;25(5):1849–1875.

[25] Swingler K. Applying neural networks: a practical guide. San Francisco, CA: Morgan Kaufmann; 1996.

[26] Caselitz P, and Giebhardt J. Advanced maintenance and repair for offshore wind farms using fault prediction techniques. In: Proceedings of the World Wind Energy Conference; 2002.

[27] Takagi H, and Hayashi I. NN-driven fuzzy reasoning. International Journal of Approximate Reasoning. 1991;5(3):191–212.

[28] Jang JS. ANFIS: adaptive-network-based fuzzy inference system. IEEE Transactions on Systems, Man, and Cybernetics. 1993;23(3):665–685.

[29] Mamdani EH. Application of fuzzy logic to approximate reasoning using linguistic synthesis. In: Proceedings of the Sixth International Symposium on Multiple-Valued Logic. IEEE Computer Society Press; 1976. p. 196–202.

[30] Sugeno M. Industrial applications of fuzzy control. New York, NY: Elsevier Science Inc.; 1985.

[31] Salahshoor K, Kordestani M, and Khoshro MS. Fault detection and diagnosis of an industrial steam turbine using fusion of SVM (support vector machine) and ANFIS (adaptive neuro-fuzzy inference system) classifiers. Energy. 2010;35(12):5472–5482.

[32] Herrera-Viedma E, Cordón O, Luque M, *et al.* A model of fuzzy linguistic IRS based on multi-granular linguistic information. International Journal of Approximate Reasoning. 2003;34(2–3):221–239.

[33] Kacprzyk J, and Zadrożny S. Computing with words in intelligent database querying: standalone and Internet-based applications. Information Sciences. 2001;134(1):71–109.

[34] Liu X. Some properties of the weighted OWA operator. IEEE Transactions on Systems, Man, and Cybernetics, Part B (Cybernetics). 2006;36(1): 118–127.

[35] Torra V. OWA operators in data modeling and reidentification. IEEE Transactions on Fuzzy Systems. 2004;12(5):652–660.

[36] Filev D, and Yager RR. Learning OWA operator weights from data. In: Fuzzy Systems, 1994. IEEE World Congress on Computational Intelligence, Proceedings of the Third IEEE Conference on. IEEE; 1994. p. 468–473.

[37] Filev D, and Yager RR. On the issue of obtaining OWA operator weights. Fuzzy Sets and Systems. 1998;94(2):157–169.

Chapter 3

Optimization of wind farms for communities

Ahmadreza Vasel-Be-Hagh[1]

Abstract

Energy supplies are moving away from environmentally damaging, finite, and expensive fossil fuels to renewable energy resources through technological innovations. Wind energy is one of the most advanced renewable energy resources due to the extensive research that has been ongoing over the last decades to optimize aerodynamic performance of wind turbines, structural design of wind turbines, control strategies, site selection, and the layout of wind farms. This chapter outlines fundamental elements of wind-farm-layout optimization including optimization parameters, objective functions, wake loss models, and search methods. Optimization parameters include base location, number, rotor diameter, hub height, rotational direction, and yaw angle of wind turbines, as well as shape of wind farm area. In the wake loss models section, all existing wake-loss models including large eddy simulation, nonlinear and linearized Reynolds-averaged Navier–Stokes models, stochastic models, kinematic models, and empirical models are discussed. In addition, different search methods, from simple greedy search algorithms to advanced genetic algorithms (GAs), are briefly reviewed and compared.

3.1 Introduction

The operational performance of a wind turbine sited in a wind farm – of any scale: either a small onshore community-sized wind farm or a commercial-sized offshore wind farm – is negatively affected by the wake of other wind turbines. Hence, under similar wind conditions, the annual energy production (AEP) of a wind turbine sited in a wind farm is always significantly less than that of an identical single isolated wind turbine. To put this into perspective, the relative power production (PP) of a turbine located in the second row of the Nørrekær wind farm, an onshore wind farm in Denmark, is approximately 40%–50% of a single isolated turbine under similar free-stream conditions when wind blows along the column of the turbines. Wind-farm-layout optimization, in its classic definition, is known as

[1]Department of Mechanical Engineering, Tennessee Technological University, USA

optimizing the position of wind turbines in order to minimize the above-described negative wake effect. In more advanced and inclusive analyses, however, several other characteristics of the wind farm, including, number of turbines [1,2], rotor diameter (i.e., turbine type) [3,4], hub height [3,5–10], rotational direction [11], and length of power transmission lines [12–20] are determined simultaneously with the position of wind turbines in order to optimize the AEP of the wind farm, the environmental impacts, and the economic benefits.

The process of optimizing the layout of a wind farm consists of two major steps (Figure 3.1). First, a search algorithm is required to identify all possible layouts over the given wind farm area. Second, a wake-loss model is required to predict the PP of the layout identified in the first step. Depending on the way through which the interaction between these two steps is defined, optimization techniques can be classified as one-way or two-way algorithms.

In optimization algorithms with the one-way structure, the search process is independent of the power prediction step. The search algorithm identifies a layout, and then, the PP of the identified layout is predicted and stored by the power prediction module of the optimization algorithm [Figure 3.1(a)]. Once the search is completed, the layout with the maximum power (or maximum AEP) is selected as

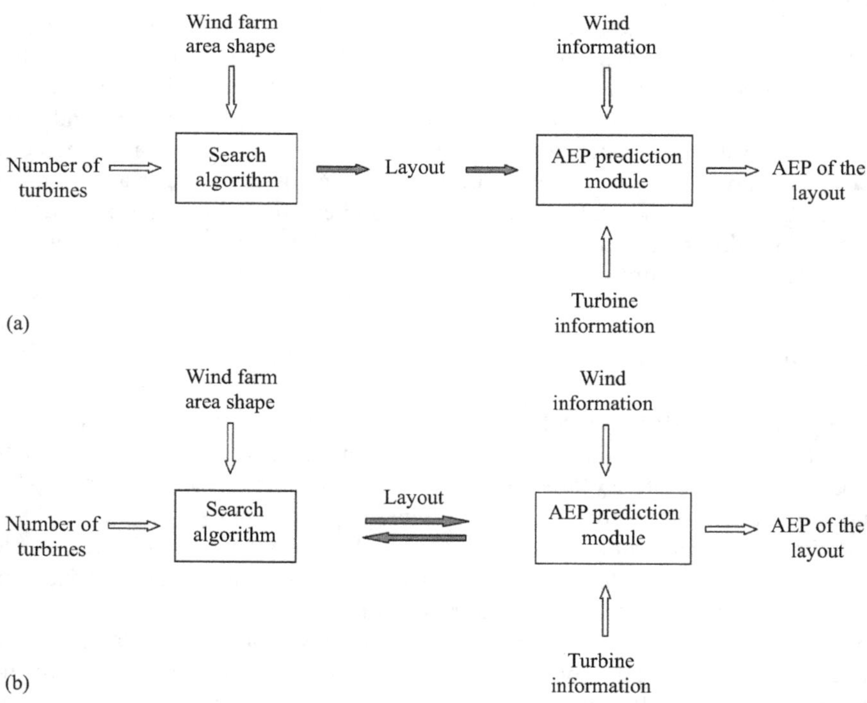

Figure 3.1 Wind-farm-layout optimization algorithm with (a) a one-way and (b) a two-way structures

the optimal layout. One example of a one-way optimization structure is the algorithm developed by Ghaisas and Archer [21]. They described a wind farm layout using four independent design parameters defined as the spacing between consecutive turbines in the *X*-direction (SX), the spacing between consecutive turbines in the *Y*-direction (SY), the staggering of alternate rows in the *Y*-direction (SDY), and the angle between rows and columns (β). Then, hundreds of layouts are identified by assigning ranges of values to those discrete design parameters (i.e., SX, SY, SDY, and β) and PP of each layout is predicted using geometric models (GMs). Finally, the layout with highest PP is chosen as the optimal layout. During this exhaustive search for identifying possible layouts, no information is communicated between the search and the power prediction modules, and all of the layouts are identified upfront.

In optimization algorithms with a two-way structure, however, the search algorithm constantly communicates with the power prediction module and modifies the search process accordingly [Figure 3.1(b)]. Hence, the two-way optimization algorithms are more sophisticated; however, they are smarter and are able to identify more efficient optimal layouts in a shorter period of time. For instance, in the optimization algorithm developed by Vasel-Be-Hagh and Archer [5], first, a turbine-placement grid with N_g grid points is mapped onto the wind farm area, and the optimization algorithm is initialized by placing one turbine at one of the grid points. Then, the second turbine is placed at all $(N_g - 1)$ available locations one by one to determine the base location for which AEP of the two placed turbines is maximized. This procedure continues until n reaches N_T, where N_T is the total number of turbines. This dynamic programing approach identifies the optimal layout by adding one turbine at a time according to the information that is being communicated between the search and the power prediction modules.

In this chapter, first all optimization variables and objective functions that need to be taken into account to develop an efficient design for a community-sized wind farm are discussed (Section 3.2). Then, different kinds of wake-loss models that have been introduced in literature in order to predict PP of a given wind farm are presented (Section 3.3). Finally, available search algorithms that have been developed for wind-farm-layout optimization purposes are presented and discussed (Section 3.4).

3.2 Objective functions and optimization variables

Wind-farm-layout optimization includes identifying not only the optimal positions for the turbines to maximize the power or the AEP of the wind farm but also the optimal hub height, the optimal number of turbines, the optimal rotational direction, and the optimal rotor diameter (i.e., turbine type) to minimize the levelized cost of energy (LCOE), to minimize the adverse environmental impacts such as noise production, and to minimize the fatigue loads acting on the wind turbines. In this section, all of these objective functions and optimization variables are introduced and discussed in details.

3.2.1 Objective functions

A list of the most essential objective functions that are required to be optimized in order to develop an efficient design for a commercial-sized wind farm is presented in Table 3.1. Among all, the AEP of the wind farm is the most critical objective function in a wind-farm-layout optimization analysis. The AEP of a wind farm is calculated as

$$AEP = \sum_{i=1}^{360}\left[fr_i \times \left(\sum_{j=1}^{n_1}\left[p_w(u_j)P_c(u_j)\sum_{k=1}^{n_t}\left[P_{rel(k,i)}\right]\right] + \sum_{j=n_1}^{n_2}\left[p_w(u_j)P_{rated}N_T\right]\right) \times n_h\right] \tag{3.1}$$

where fr_i is the relative frequency of wind in direction i, $p_w(u_j)$ is the probability of having wind at speed of $u_j = (0.5 + j)du$ in direction i and is calculated via Weibull distribution [Eq. (3.2)], du is the wind speed resolution, $P_c(u_j)$ is the power obtained from the power curve of wind turbine at wind speed of u_j, P_{rated} is the rated power of wind turbines, n_1 and n_2 are respectively defined as $n_1 = 1 + u_{rated}/du$ and $n_2 = u_{cut-out}/du$, P_{rel} is the relative power calculated via a wake-loss model, and $n_h = 8{,}760$ denotes number of hours per year. The Weibull distribution, used in (3.1) to represent the wind velocity probability density ranging from the cut-in to the cut-out wind speeds, is defined as

$$P_w(u)du = \left(\frac{k_w}{c_w}\right)\left(\frac{u}{c_w}\right)^{k_w-1} \exp\left[-\left(\frac{u}{c_w}\right)^{k_w}\right]du, \tag{3.2}$$

in which P_w, u, k_w, and c_w are probability density, wind speed, shape factor, and scale factor, respectively.

In many of the wind farm optimization studies, the farm-averaged PP of the wind farm, defined as the total PP of the farm divided by the total number of turbines, is simply used as the objective function of the optimization analysis. It is important to note that a wind farm optimized by maximizing its PP is not necessarily identical to the optimal layout which is obtained based on maximizing the AEP, and as the AEP is what really matters as the total output of a wind farm, it is

Table 3.1 Objective functions considered for wind farm design and development

Objective functions	Objective	Literature
Annual energy production (AEP)	Maximizing	[14,18,22–26]
Power production (PP)	Maximizing	[6,13,27–32]
Levelized cost of energy (LCOE)	Minimizing	[2,24,27,33–38] [12,39–46]
Net present value (NPV)	Maximizing	[47–50]
Noise propagation (NP)	Minimizing	[1,2,51–53]
Loads acting on wind turbines (WTL)	Minimizing	[27,28,54]

recommended to use the AEP as the main objective function of the wind-farm-layout optimization analysis.

Another popular objective of wind-farm-layout optimization algorithms is minimizing the LCOE (\$/kW h). In general, the LCOE is defined as the average total cost to build and operate a wind farm over its lifetime divided by the total energy output of the wind farm over that period of time. Accordingly, the LCOE is calculated as

$$LCOE = \frac{C_{Inv}}{aE_a} + \frac{C_{O\&M}}{E_a} \tag{3.3}$$

where $C_{Inv} = C_{RNA} + C_{SS} + C_{Elect} + C_{Decom}$ is the capital cost in which C_{RNA} is the rotor-nacelle assemblies costs, C_{SS} includes the support and the structure costs, C_{Elect} denotes the electrical interconnection costs, C_{Decom} is the decommissioning cost, $C_{O\&M}$ is the operational and maintenance costs, and a is the annuity factor defined as

$$a = \frac{\left(1 - (1/(1+r))^T\right)}{r} \tag{3.4}$$

where T is the lifetime of the wind farm in years and r is the interest rate in (%) and is defined as the summation of discount rate and inflation rate. In (3.3), E_a is the net effective expected electrical energy of the wind farm and is defined as

$$E_a = \left[\sum_{i=1}^{N_T} \left(E_{WT,i} - E_{WL,i} - E_{CL,i}\right) - E_{LT}\right] \tag{3.5}$$

where $E_{WT,i}$ is the maximum possible energy production of turbine i assuming that turbine i is a front-row turbine, $E_{WL,i}$ is the energy that turbine i loses due to wake effects, $E_{CL,i}$ stands for the energy that turbine i loses through the collection cables, and E_{LT} is the energy loss through the transmission cables.

If the interest rate is low, then it is more efficient to use the net present value (NPV) instead of the LCOE as the objective function for a wind-farm-layout optimization. The NPV is defined to take into account the fact that a given amount of money is more valuable now than it will be in the future as it can be used now to make more money in the future. The NPV is defined as

$$NPV = \frac{P_1}{(1+r)^1} + \frac{P_2}{(1+r)^2} + \cdots + \frac{P_n}{(1+r)^n} \tag{3.6}$$

in which r is the rate of interest and must be given as a decimal (not percent), and P_i stands for the yearly payment of the ith year.

The noise created by a community wind farm, which is located near residential areas, may be annoying to people living nearby; hence, the layout of those wind farms must be developed so that the noise level of the farm is minimized. Noise created by wind farms has two different sources; first, the aerodynamic noise produced by the blades of turbines cutting through the air on their downward motion, and second, noise made by the gearbox system. Employing thinner blades and shifting to direct drive (i.e., gear-less), wind turbine technologies are efficient approaches to reduce noise of wind turbines at the manufacturing stage. At the wind farm design and development stage, however, the total noise production of a wind farm can be reduced by smart placement of wind turbines. In community wind projects, turbines are normally placed in the wind farm based on a trade-off between maximizing the AEP of the wind farm and minimizing its noise propagation. The noise calculations are usually conducted based on the International Standard ISO 9613 which includes a general method of calculation for attenuation of sound during propagation outdoors. Accordingly, sound pressure level of each wind turbine at each receptor location is calculated as

$$L_p = L_w + D_c - A_f \tag{3.7}$$

where $L_w = 100$ dB [1] is sound power level, D_c stands for directivity correction in dB if the source does not emit sound equally in all directions, A_f is the octave-band attenuation defined as

$$A_f = A_{div} + A_{atm} + A_{gr} + A_{bar} + A_{misc} \tag{3.8}$$

in which A_{div} is the attenuation due to geometrical spreading, A_{atm} is the attenuation due to air absorption, A_{gr} is the attenuation due to ground absorption and reflection, A_{bar} is the free field diffraction attenuation of a barrier, and A_{misc} is the attenuation due to miscellaneous effects such as weather variability and dispersion through complex acoustical structures. International Energy Agency has provided the following approximation for (3.7),

$$L_p(d_{ir}) = L_w - 10 \times \log\left(2\Pi d_{ir}^2\right) - \alpha d_{ir} \tag{3.9}$$

where the indices i and r represent turbine and receptor, d is the distance between turbine and receptor, and $\alpha = 0.005 \;\; \mathrm{dB/m}$ is a constant. Individual sound pressure levels calculated via (3.9) are then summed up using the following equation:

$$L_{p,avg} = 10 \times \log\left(\sum_{i=1}^{ns}\left(\sum_{j=1}^{8} 10^{0.1\left(L_p(i,j)+A_f(i,j)\right)}\right)\right), \tag{3.10}$$

in which ns is number of sound sources (i.e., number of turbines), $L_p(i,j)$ is the individual sound pressure level associated with turbine i and octave band j.

Finally, minimizing the fatigue loads acting on the structure of wind turbines can be considered as another objective function for wind-farm-layout optimization analyses as the damage equivalent loads in a wind farm are highly affected by partial wake overlap and can be significantly decreased by smart placement of turbines and smart handling of yaw-misalignment [27,28,54].

3.2.2 Optimization variables

If changing the value of a parameter simultaneously exerts a positive and a negative effect on the objective function of a problem, then that parameter can be considered as an optimization variable and there might be an optimal value for it. The most

Table 3.2 Optimization variables considered for wind farm design and development

Optimization variables	**Studies**
Turbine positions	[12,18,33,55–61]
Number of turbines	[1,13,45,62,63]
Rotor diameter	[3,4]
Hub height	[4–6,8,64]
Electrical cable length	[12–20]
Rotational direction	[11]
Wind farm area	[40,52,65–67]

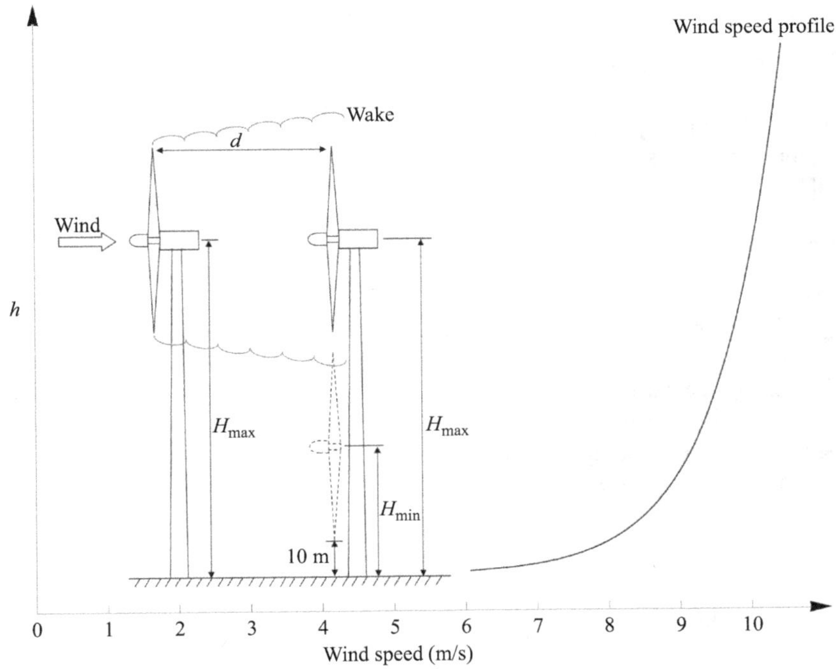

Figure 3.2 Two in-line wind turbines aligned with the wind direction. The upstream turbine is placed at H_{max} while the hub height of the downstream turbine may vary from H_{max} to H_{min} [5]

popular optimization variables for wind-farm-layout optimization analyses are presented in Table 3.2.

For instance, by lowering the hub height of the downstream wind turbine by a specific length equal to $k \times d$, where k is the decay coefficient ($k = 0.04$ and 0.078 for offshore and onshore wind farms, respectively) and d is the axial distance between the turbines, the downstream wind turbine starts to become unexposed to the upstream wake (see Figure 3.2). This *positively* affects the PP of the downstream wind turbine. On the other hand, due to the shear effect, lowering the hub height of the downstream wind turbine causes a reduction in the speed of the wind experienced by this wind turbine, which *negatively* affects its PP. Due to this simultaneous positive and negative effects that are brought about by variation of the hub height of the downwind turbine, the "hub height" can be considered as an optimization variable, and in fact, in many cases, a compromise between the two negative and positive effects can be reached so that the maximum PP of the downstream wind turbine can be achieved at a height lower than the hub height of the upstream wind turbine. Building the downstream wind turbine at this optimal hub height not only increases the PP but also slightly decreases the average height of the wind farm leading to a reduction of the capital and the maintenance costs [5].

3.3 Wake-loss models

A list of the most popular wake loss models used for wind farm design and development purposes is provided in Table 3.3. These models are described in detail in the following sections.

3.3.1 Large eddy simulations

3.3.1.1 Governing equations

Large eddy simulations (LESs) govern dynamics of large eddies by removing those with scales smaller than a filter width from the unsteady Navier–Stokes equations and modeling their effects using a subgrid-scale model. The filter width is defined as $\Delta = \sqrt[3]{\Delta x \Delta y \Delta z}$ where Δx, Δy, and Δz are cell sizes in the x, y, and z directions, respectively. The incompressible formulations of the filtered continuity and momentum equations are as follows:

$$\frac{\partial \bar{u}_i}{\partial x_i} = 0 \tag{3.11}$$

$$\frac{\partial \bar{u}_i}{\partial t} + \frac{\partial \bar{u}_i \bar{u}_j}{\partial x_j} = -\frac{\partial \widehat{p}}{\partial x_i} - \frac{\partial \tau_{ij}^D}{\partial x_j} - \frac{1}{\rho_0}\frac{\partial p_0(x,y)}{\partial x_i} + F_{ext} \tag{3.12}$$

where the bar denotes spatially resolved components; i, j, and k are the indices of the three spatial components x, y, and z; u is the wind speed; t is time; $\widehat{p}$ is the modified pressure defined as $[\bar{p}(x,y,z,t)/\rho_0 - p_0(x,y)/\rho_0 + \rho_0 gz/\rho_0 + (\tau_{kk})/3]$; p and p_0 are the static and mean pressure; ρ_0 is the reference air density; τ_{ij}^D is the traceless part of the wind stress tensor; and F_{ext} stands for the external forces applied to the wind, including those induced by the wind turbines. According to the Boussinesq eddy viscosity assumption, the traceless stress tensor τ_{ij}^D given in (3.12) is defined as

$$\tau_{ij}^D = -2\nu_t \bar{S}_{ij} \tag{3.13}$$

Table 3.3 The most popular wake loss models used in wind energy applications

Wake loss models	
Large eddy simulations (LES)	
	• LES with actuator lines
	• LES with actuator disks
Non-linear Reynolds-averaged Navier–Stokes (RANS) models	
	• $K-\varepsilon$ with actuator lines/disks
	• $K-\omega$ with actuator lines/disks
Stochastic models	
Linearized RANS models	
	• Ainslie
	• Fuga
Kinematic (analytical) models	
	• PARK (Jensen)
	• Bastankhah/Porté Agel (BPA)
	• Xie/Archer
	• Geometric model
	• Frandsen
	• Larsen
Experimental models	• Ishihara

in which the kinematic eddy viscosity ν_t is defined using the subgrid-scale model proposed by Smagorinsky [68] as

$$\nu_t = (c_s\Delta)^2|\bar{S}| \tag{3.14}$$

where $c_s = 0.168$ is the Smagorinsky constant, $\bar{S}_{ij} = \left(\partial\bar{u}_i/\partial x_j + \partial\bar{u}_j/\partial x_i\right)/2$ is the filtered strain rate tensor, and $|\bar{S}| = \sqrt{2\bar{S}_{ij}\bar{S}_{ij}}$ is the norm of the filtered strain rate tensor. The external force F_{ext} term in (3.12) includes the Coriolis force, the buoyancy force, and the force exerted by turbine blades that is calculated using the actuator line model presented in Section 3.3.1.2. Accordingly, the external force F_{ext} can be expressed as

$$F_{ext} = \frac{1}{\rho_0}F_i + g\left(\frac{\bar{\theta}-\theta_0}{\theta_0}\right)\delta_{i3} - \varepsilon_{i3k}f\bar{u}_k \tag{3.15}$$

where F_i is the force generated by the actuator line model, ε_{ijk} is the alternating unit tensor, g stands for the gravitational acceleration, θ is the potential temperature, $\theta_0 = 300$ K is the reference temperature, δ_{ij} is the Kronecker delta, and f is the Coriolis parameter defined as $f = 2\Omega\sin\phi$ in which Ω is the earth rotational speed ($\sim 2.95\times 10^{-5}$ rad/s), and ϕ is the site latitude. The following potential temperature equation needs to be solved coupled with (3.11) and (3.12) to obtain the potential temperature needed to calculate the buoyancy term in (3.15),

$$\frac{\partial \bar{\theta}}{\partial t} + \frac{\partial (\bar{u}_j \bar{\theta})}{\partial x_j} = \frac{\partial q_j}{\partial x_j} \tag{3.16}$$

where q_j represents the temperature flux defined as

$$q_j = -\frac{\nu_t}{Pr_t} \frac{\partial \bar{\theta}}{\partial x_j}, \tag{3.17}$$

and Pr_t is the subgrid turbulent Prandtl number defined as [69]

$$Pr_t = \frac{1}{1 + 2(l/\Delta)} \tag{3.18}$$

in which,

$$l = \begin{cases} \min\left(7.6 \frac{\nu_t}{\Delta} \left(s^{-(1/2)}\right), \Delta\right) & \text{if } s > 0 \\ \Delta & \text{if } s \leq 0 \end{cases} \tag{3.19}$$

and

$$s = \frac{g}{\theta_0} \frac{\partial \bar{\theta}}{\partial z}. \tag{3.20}$$

Usually, $l = \Delta$, and hence, $Pr_t = 1/3$.

3.3.1.2 The actuator line model

The actuator line modeling, proposed by Sørensen and Shen [70], is usually employed along with LESs to model the effect of wind turbines. In this model, the turbine blades are represented by three rotating lines that are discretized into N_{be} blade elements with centers located at (x_n, y_n, z_n). N_{be} is recommended to be at least 40. Using airfoil lookup tables, the aerodynamic forces are calculated for each blade element $f_i^a(x_n, y_n, z_n, t)$. Summation of the aerodynamic forces of blade elements corrected via a regularization kernel yields the body force exerted by the blades onto the flow field,

$$F_i = \sum_{n=1}^{40} \frac{f_i^a(x_n, y_n, z_n, t)}{\pi^{3/2} \varepsilon^3} \exp\left[-\left(\frac{r_n}{\varepsilon}\right)^2\right], \tag{3.21}$$

where $f_i^a(x_n, y_n, z_n, t)$ is the actuator element force, F_i is the force field projected as a body force onto computational fluid dynamics (CFD) grid, r_n is the distance between CFD cell center and the blade element, and ε is used to control the Gaussian width so that it spans from the leading edge to the trailing edge of the blade elements. The value of ε is recommended to be $l_c/4.3$, where l_c indicates the chord length of the blade elements, so at both trailing and leading edges (i.e., $r_n = l_c/2$), the exponential term is reduced to approximately 1% of its maximum [71]. The power calculations are based on the aerodynamic torque that is exerted on the blades. Multiplying the aerodynamic torque by the rotational speed of the rotor yields the power output.

3.3.2 Nonlinear Reynolds-averaged Navier–Stokes (RANS) models

3.3.2.1 Governing equations

The continuity and the momentum equations using the Reynolds-averaged Navier–Stokes (RANS) decomposition are as follows:

$$\frac{\partial U_i}{\partial x_i} = 0 \tag{3.22}$$

$$\rho U_j \frac{\partial U_i}{\partial x_j} = -\frac{\partial P}{\partial x_i} + \frac{\partial}{\partial x_j}\left[\mu\left(\frac{\partial U_i}{\partial x_j} + \frac{\partial U_j}{\partial x_i}\right)\right] + \frac{\partial}{\partial x_j}\left[\mu_t\left(\frac{\partial U_i}{\partial x_j} + \frac{\partial U_j}{\partial x_i}\right)\right] + F_{ext} \tag{3.23}$$

in which μ_t is turbulent viscosity and is defined using a two-equation closure model, such as the $k - \varepsilon$ or the $k - \omega$ models. Turbulent viscosity in the $k - \varepsilon$ model is defined as

$$\mu_t = \rho C_\nu \frac{k^2}{\varepsilon} \tag{3.24}$$

where k and ε are turbulent kinetic energy and the kinetic energy dissipation rate, respectively. The transport equations for k and ε are

$$u_i \frac{\partial k}{\partial x_i} = \frac{\partial}{\partial x_i}\left[\left(\nu + \frac{\nu_t}{\sigma_k}\right)\frac{\partial k}{\partial x_i}\right] + \nu_t\left(\frac{\partial u_i}{\partial x_j} + \frac{\partial u_j}{\partial x_i}\right)\frac{\partial u_i}{\partial x_j} - \varepsilon \tag{3.25}$$

$$u_i \frac{\partial \varepsilon}{\partial x_i} = \frac{\partial}{\partial x_j}\left[\left(\nu + \frac{\nu_t}{\sigma_\varepsilon}\right)\frac{\partial \varepsilon}{\partial x_i}\right] + C_{\varepsilon 1} P_k \frac{\varepsilon}{K} - C_{\varepsilon 2}\frac{\varepsilon^2}{k} \tag{3.26}$$

where C_ν, $C_{\varepsilon 1}$, and $C_{\varepsilon 2}$ are the standard model constants, and (σ_k, σ_ε) are the turbulent Prandtl numbers for k and ε, respectively. In (3.23), F_{ext} stands for the external forces applied to the wind, including those induced by the wind turbines. Wind turbine forces can be modeled through the actuator line model described in Section 3.3.1.2, or using the actuator disk model described in the following section.

3.3.2.2 The actuator disk model

In the actuator disk model, the wind turbine rotor is modeled as a disk with a diameter equal to the rotor diameter of the real wind turbine and a depth equal to the thickness of the blades. The F_{ext} is then defined as

$$F_{ext} = \frac{1}{2}\frac{\rho C_T U_0^2}{\Delta x} \tag{3.27}$$

where U_0 is the inlet velocity at hub height level, Δx stands for the control volume length and is equal to the actuator disk thickness, and C_T is the thrust coefficient defined as

$$C_T = \frac{T}{(1/2)\rho U_\infty^2 A_D} \tag{3.28}$$

in which U_∞ is free stream wind speed, $A_D = \pi D^2/4$ stands for the rotor swept area, and T is the thrust force and is a function of lift (C_L) and drag (C_D) coefficients obtained through airfoil lookup tables.

3.3.3 Stochastic models

The stochastic models are introduced to fill the gap between the accurate, however, computationally expensive CFD-based models and less accurate, however, computationally efficient analytical models. One of the most effective stochastic models is the wake model proposed by Doubrawa *et al.* [72] based on LESs of an offshore wind farm. The proposed stochastic model was found to successfully reproduce the mean characteristics of the original LES wake, including its area and stretching patterns, statistics of the mean azimuthal radius, the mean and standard deviation of the wake width and height, and the velocity deficit and meandering. In this model, the cross-section of the wake is defined as a series of wake radius versus azimuth $r_w(\theta)$, where θ is the azimuth angle with respect to the vertical direction and r_w is the distance between the center of the wake from the boundary of the wake at the azimuth angle of θ. The r_w is then decomposed into $\langle r_w \rangle$ and r'_w as $r_w = \langle r_w \rangle + r'_w$ in which $\langle r_w \rangle$ is the azimuthal mean radius and r'_w stands for radii perturbations. At each iteration, $\langle r_w \rangle$ is estimated using stochastic methods, and r'_w is obtained using spectral analysis. The azimuthal mean radius $\langle r_w \rangle$ is further decomposed into a constant temporal mean $\overline{\langle r_w \rangle}$ and a dynamic perturbation $\langle r_w \rangle'$ around the constant temporal mean. The values of $\overline{\langle r_w \rangle}$ are extracted upfront from the LES and will be provided by the user as initial conditions of the wake simulator. These values are given in [72] for different distances downstream of turbines. The perturbations $\langle r_w \rangle'$, however, are obtained at every time step through a first-order autoregressive model as

$$\langle r_w \rangle'_t = \rho_1 \langle r_w \rangle'_{t-1} + \varepsilon(t) \tag{3.29}$$

where $\rho_1 = 0.9$ is the first-order autocorrelation for the $\langle r_w \rangle'$ time series obtained from the LES data which was found to be approximately the same for different distances downstream, and $\varepsilon(t)$ are the random innovations in the form of white noise that make up the time series variability. These innovations are randomly sampled from a normal distribution of mean $\mu = 0$ and standard deviation $\sigma = 0.05R$ which were determined based on the original LES time series of wake radii. More information on stochastic wake models can be found in [73,74].

3.3.4 Linearized RANS models

3.3.4.1 Ainslie model

Ainslie wake model, proposed by Ainslie [75], is a two-dimensional model based on the assumptions that wake of a wind turbine is axisymmetric and pressure gradients are negligible in the wake region. The continuity and momentum equations in free stream direction and in cylindrical coordinates are as follows:

$$\frac{1}{r}\frac{\partial(rv)}{\partial r}+\frac{\partial u}{\partial x}=0 \tag{3.30}$$

$$u\frac{\partial u}{\partial x}+v\frac{\partial u}{\partial r}=-\frac{1}{r}\frac{\partial\left(r\overline{u'v'}\right)}{\partial r} \tag{3.31}$$

where

$$-\overline{u'v'}=\varepsilon(x)\frac{\partial u}{\partial r} \tag{3.32}$$

where ε is eddy viscosity and is assumed to be a function of distance downstream of the wind turbine. Ainslie decomposed the eddy viscosity term into the ambient eddy viscosity of the atmosphere and the eddy viscosity generated by the wake as

$$\varepsilon(x)=\varepsilon_a+\varepsilon_w(x)=\varepsilon_a+kb(U_\infty-u_c(x)) \tag{3.33}$$

in which k is constant and is empirically found to be 0.015, and b is the wake width and is defined as follows based on wind tunnel data

$$b=\sqrt{\frac{3.56C_T}{4(U_\infty-u_c)(2-(U_\infty-u_c))}} \tag{3.34}$$

where u_c is wind speed at the centerline of the wake. The Ainslie model is not valid in the near wake area (within 2D from downwind of the rotor), and the following Gaussian velocity profile is used as a boundary condition at $x=2\text{D}$,

$$1-\frac{u(r)}{U_\infty}=(U_\infty-u_c)\exp\left(-3.56\left(\frac{r}{b}\right)^2\right) \tag{3.35}$$

where u is the wind speed at radial distance r with respect to the center line and axial distance $x=2\text{D}$ downwind of the wind turbine. The initial wind speed deficit at the centerline of the wake is

$$U_\infty-u_c=C_T-0.05-(16C_T-0.5)\frac{I}{10} \tag{3.36}$$

where I is the ambient turbulence intensity.

3.3.4.2 Fuga model

Ott *et al.* [76] developed a linear RANS model called Fuga by employing a very simple closure instead of the one introduced through (3.24)–(3.26) in the $k-\varepsilon$ model. According to the Fuga model,

$$\mu_t=\rho ku^*z \tag{3.37}$$

where $k=0.4$ is the Von Karman constant, z is the height from the surface, and u^* is the shear velocity defined as $u^*=\sqrt{\tau/\rho}$ in which τ is the surface shear stress.

3.3.5 Empirical wake models

There are several wake loss models that are based on experimental data, such as the Ishihara model developed by Ishihara *et al.* [77] by using wind tunnel data for a scaled model and assuming a Gaussian velocity profile. The wind speed deficit in the Ishihara model is given by

$$U_\infty - u = \frac{\sqrt{C_T}U_\infty}{32}\left(\frac{1.666}{k_1}\right)^2\left(\frac{x}{D}\right)^{-p}\exp\left(-\frac{r^2}{D_w^2}\right) \tag{3.38}$$

where D_w is the wake diameter defined as

$$D_w = \frac{k_1 C_T^{0.25}}{0.833} D^{1-0.5p} x^{0.5p} \tag{3.39}$$

in which p is defined as

$$p = k_2(I_a + I_w) \tag{3.40}$$

where I_a and I_w are the ambient turbulence intensity and the turbulence intensity induced by the wind turbines, respectively. I_w is estimated as

$$I_w = \frac{k_3 C_T}{\max(I_a, 0.03)}\left(1 - \exp\left(\frac{-x^2}{25D^2}\right)\right), \tag{3.41}$$

and coefficients k_1, k_2, and k_3 are 0.27, 6, and 0.004, respectively.

3.3.6 Kinematic (analytical) models

The six most popular kinematic wake models, also called analytical wake models, which have been developed for wind energy applications are PARK (Jensen), Xie–Archer (XA), Bastankhah and Porté-Agel (BPA), Larsen, Frandsen, and GM. These models are respectively described in the following sections.

3.3.6.1 PARK

The PARK model, developed by Jensen [78,79], is underpinned by two major assumptions; first, the velocity deficit is conserved as the wake linearly expands downstream of the wind turbine, and second, the velocity deficit is only a function of the distance x downstream of the turbine. Accordingly,

$$\delta = \delta(x) = \frac{U_\infty - U(x)}{U_\infty}, \tag{3.42}$$

where x is the axial distance downwind of the turbine and is often expressed as multiples of the turbine diameter D, and $U(x)$ is the wind speed at distance x. In the PARK model, (3.42) is expressed as

$$\delta(x) = \frac{2a}{(1 + k_w(x/D))^2}, \tag{3.43}$$

where k_w is the wake decay coefficient, which is a dimensionless constant and its value depends on the surface roughness. Values of $k_w = 0.04$ and 0.078 are recommended for offshore and onshore conditions, respectively. In (3.43), the induction factor a is expressed as a function of thrust coefficient C_T as

$$a = 1 - \sqrt{1 - C_T}. \tag{3.44}$$

The diameter of the wake D_w is therefore

$$D_w = D_w(x) = D\left(1 + 2k_w \frac{x}{D}\right). \tag{3.45}$$

In the PARK model, the only relevant spatial variable is x, and hence, the wind speed and the wind speed deficit along y and z are uniform, which leads to an axis-symmetric conical-shaped wake.

3.3.6.2 Xie and Archer (XA) model

The XA wake loss model, developed by Xie and Archer [80], is the only wake loss model that truly depends on z and y as it predicts a wake that is not axis-symmetric or conical, but ellipsoidal, which is a more realistic approximation, in particular in the presence of wind shear [81]. The wind speed deficit in the XA model is defined as

$$\delta = \delta(x, y, z) = \delta_{hub} \exp\left\{-\left[\frac{(z-H)^2}{2\sigma_z^2} + \frac{y^2}{2\sigma_y^2}\right]\right\} \tag{3.46}$$

$$\delta_{hub} = \delta_{hub}(x) = 1 - \sqrt{1 - \frac{C_T}{8\left(\sigma_y \sigma_z / D^2\right)}} \tag{3.47}$$

$$\frac{\sigma_y}{D} = \frac{\sigma_y(x)}{D} = k_y \frac{x}{D} + \varepsilon; \quad \frac{\sigma_z}{D} = \frac{\sigma_z(x)}{D} = k_z \frac{x}{D} + \varepsilon, \tag{3.48}$$

where H is the hub height, $k_y = 0.025$ and $k_z = 0.0175$ are the growth rate of the wake in the y and z directions, obtained from a fit to LES results of a single turbine wake under neutral stability [80].

3.3.6.3 Bastankah and Porté-Agel (BPA) model

Although the BPA model has an explicit dependency on y and z, where y and z are the span-wise and vertical coordinates, respectively, the cross-section of the wake is always a circle. The wind speed deficit in the BPA model is given by the following equation:

$$\delta = \delta(x, y, z) = \delta_{hub} \exp\left\{-\frac{1}{2(k^*(x/D) + \varepsilon)^2}\left[\left(\frac{z-H}{D}\right)^2 + \left(\frac{y}{D}\right)^2\right]\right\} \tag{3.49}$$

$$\delta_{hub} = \delta_{hub}(x) = 1 - \sqrt{1 - \frac{C_T}{8(k^*(x/D) + \varepsilon)^2}} \tag{3.50}$$

where H is the hub height, $k^* = (\partial\sigma/\partial x)$ is the growth rate of the wake (which is not the same as $k_w = (\partial D_w/\partial x)$ in the previous models), σ is the standard deviation of the velocity deficit profile, and $\varepsilon = 0.25\sqrt{\beta}$. In the original study [82], k^* was found to vary between 0.030 and 0.055, from fitting LES results obtained with surface roughness z_0 between 0.5 and 0.00005.

3.3.6.4 Larsen model

Larsen developed an analytical wake loss model using similarity technique and assuming that the wake region behind a wind turbine can be described via Prandtl's turbulent boundary layer equations [83]. It was also assumed that flow is stationary, incompressible, and the wind shear is negligible. In the Larsen model, which was the recommended model by the European wind turbine standards II (EWTS II) for use in wake loading calculations [84], the wind speed deficit is a function of both axial distance x and radial distance r, while in the PARK model, the wake deficit is only a function of axial distance x. In the Larsen model, the wind speed deficit is calculated as

$$\delta=\delta(x,r)$$

$$= -\frac{1}{9}\left[C_T A (x+x_0)^{-2}\right]^{1/3}\left\{r^{3/2}\left[3c_1^2 C_T A(x+x_0)\right]^{-(1/2)} - \left(\frac{35}{2\pi}\right)^{3/10}\left(3c_1^2\right)^{-(1/5)}\right\}^2, \tag{3.51}$$

where C_T is the wind turbine thrust coefficient, x and r are the axial and radial distance of the wind turbine of interest from the upstream wind turbine, $A = \pi D^2/4$ is the swept area of the rotor, c_1 is a nondimensional mixing length defined as

$$c_1 = \left(\frac{D}{2}\right)^{-(1/2)}\left(C_T A x_0\right)^{-(5/6)}, \tag{3.52}$$

in which x_0 is a nondimensional reference distance defined as

$$x_0 = \frac{9.5D}{\left(D_{9.5}/D\right)^3} - 1, \tag{3.53}$$

where is a measure of the wake diameter at distance $9.5D$ given by the following equation:

$$D_{9.5} = D_{nb} + \min(H, D_{nb}), \tag{3.54}$$

where H is hub height and D_{nb} is a corrected wake diameter to take into account the blockage effect of the ground defined as

$$D_{nb} = \max[1.08D,\ 1.08D + 21.7D(I_a - 0.05)] \tag{3.55}$$

where I_a is the ambient turbulence intensity at hub height, assumed to be always greater than 5%. Similar to the PARK model, the wake diameter in the Larsen model is only a function of axial distance x as follows:

$$D_w = D_w(x) = 2\left(\frac{35}{2\pi}\right)^{1/5}\left(3c_1^2\right)^{1/5}(C_T A x)^{1/3}, \tag{3.56}$$

Similar to the PARK model, both wake diameter and wind speed deficit in the Larsen model are independent of free stream wind speed U_∞.

3.3.6.5 Frandsen model

Frandsen *et al.* [85] developed an analytical wake loss model by applying the momentum equation to a control volume and by assuming self-similarity. They also assumed that the velocity deficit is only a function of the distance x downstream of the turbine and wind speed has a constant profile, similar to that of PARK model. The wind speed deficit in Frandsen model is defined as

$$\delta = \delta(x) = \frac{1}{2}\left(1 \pm \sqrt{1 - 2\frac{A}{A_w(x)}C_T}\right), \tag{3.57}$$

in which A is the swept area of the rotor and $A_w(x) = \pi D_w^2(x)/4$ is the cross-section of the wake area at distance x downstream of the turbine, where $D_w(x)$ is the wake diameter and is defined as

$$D_w(x) = D\left(\beta^{k/2} + \alpha\frac{x}{D}\right)^{1/k}, \tag{3.58}$$

in which $\alpha = 0.7$, k is either 3 (Schlichting solution) or 2 (square root shape solution) [86], and β is the wake expansion parameter and is defined as

$$\beta = \frac{1 + \sqrt{1 - C_T}}{2\sqrt{1 - C_T}}. \tag{3.59}$$

It should be mentioned that the Frandsen model is recommended for both small and large regular wind farms with rectangular shapes and equal spacings between turbines. Similar to the PARK and Larsen models, the Frandsen model is also independent of free stream velocity U_∞.

3.3.6.6 Geometric model

The GM is a hybrid wake loss model that estimates the relative power generated by any downstream turbine with respect to the power generated by the front-row turbine [87]. The GM is considered as a hybrid model because it does not simulate the physical processes occurring in wakes but rather uses empirical coefficients, derived from a multilinear regression, to relate relative PP of a wind turbine sited in a wind farm to its geometric quantities, namely, blockage ratio BR and blockage distance BD.

The blockage ratio BR_i of a wind turbine i in a given wind direction is the fraction of the swept area of turbine i that is blocked by the swept area of any upstream turbine. Value of BR is always between 0 and 1. A blockage ratio of 0 in a given wind direction means that the turbine is not blocked at all and receives undisturbed wind in that direction, whereas a blockage ratio of 1 means that the turbine is completely blocked by the upstream wind turbines.

The blockage distance BD_i of wind turbine i, however, is a measure of the distance between the wind turbine of interest and the upstream blocking turbines. Hence, a larger blockage distance means a greater wind speed recovery, lower wake losses, and consequently more PP. Blockage ratio and blockage distance of a wind turbine for a given wind direction are calculated using the following equations:

$$BR_i = \frac{1}{A}\int_A \chi dA, \tag{3.60}$$

$$BD_i = \frac{1}{A}\int_A L\chi dA, \tag{3.61}$$

where $\chi = 1$ wherever the swept area of turbine i is blocked and zero otherwise, and L denotes the distance to the upstream blocking turbine. Once the two geometric properties are calculated for a given wind direction, then the relative power is obtained as follows:

$$P_i^{REL} = \frac{P_i}{P_{front}} = \begin{cases} \alpha + \beta BR_i + \gamma BD_i/L_\infty & BR_i \neq 0, \\ 1 & BR_i = 0, \end{cases} \tag{3.62}$$

where P_i and P_{front} are the power generated by turbine i and by the front turbine, and the fitting coefficients α, β, and γ depend on atmospheric stability. In the original paper [87], only values for neutral stability are presented.

3.3.6.7 Wake overlapping

When multiple wakes overlap, like in a wind farm, the wake overlapping is simply performed by taking the square root of the sum of the squared wind speed deficits induced by each individual wind turbine [79]. Hence, the total wind speed deficit at the location of turbine j is

$$\delta_{TOT} = \sqrt{\sum_{n=1}^{N} \delta_i^2\left(x_{ij}\right)}, \tag{3.63}$$

where N is the number of upstream wind turbines and x_{ij} is the distance between turbine j and the upstream turbine i. Depending on the employed model, each $\delta_i\left(x_{ij}\right)$ is calculated using one of the wind speed deficit equations, i.e., either (3.43), (3.46), (3.49), (3.51), or (3.57). Substituting the total wind speed deficit δ_{TOT} in $U_j = (1 - \delta_{TOT})U_\infty$ yields the wind speed experience by the wind turbine of interest U_j, and since wind power density is proportional to the cube of wind speed, relative power of turbine j is calculated as

$$P_{rel,j} = \left(\frac{U_j}{U_\infty}\right)^3 \tag{3.64}$$

where $P_{rel,j}$ is a measure of power produced by wind turbine j divided by the maximum power that is produced at the front row. The sum of squares (SS) method described in this section is the most popular wake overlapping technique used in industry. In addition to the SS technique, three other approaches have been introduced in literature, including the geometric superposition, the linear superposition, and the sum of energy deficits [88].

3.4 Search algorithms

A list of most popular search algorithms used for wind-farm-layout optimization analyses are presented in Table 3.4. According to literature, GA is the most popular search algorithm that has been used to perform wind-farm-layout optimization analysis. The general procedure of the GA is illustrated in Figure 3.3. First, the search process is *initialized* by creating random strings of 1 and 0, respectively, standing for places with turbine and without turbines. Then, the *selection*, which is to select and retain certain layouts that can generate higher AEPs according to a given selection probability, is conducted. During the *crossover*, selected layouts (parents) are combined to create new layouts (children). Then, parts of the layouts are randomly changed during the *mutation*. In the last step, layouts of the initial population are *replaced* with new layouts (children) that have been generated provided that the new layouts perform better in comparison with the layouts of the initial population. This process continues until the solution converges or the termination criteria are met.

A Greedy algorithm is a heuristic procedure that tries to find an optimal solution close to the global optimum by determining a locally optimal solution at each stage. For instance, one wind turbine is randomly located within the legal area of the wind farm, and then the location of the second turbine is determined so that the AEP of the combination of the two turbines is optimized. Then, the optimal location of the third turbine is determined so that the AEP of the combination of the three turbines is maximized. This process continues until all turbines are placed within the wind farm area.

Table 3.4 Popular search algorithms for wind-farm-layout optimization problem

Optimization technique	**Studies**
Genetic algorithms	[7,18,55,56,64,89] [12,25,39,63,90–92]
Greedy algorithms	[5,9,43,93,94]
Particle swarm optimization	[17,18,20,58,91,95–97]
Ant colony search algorithm	[98–100]
Mixed integer linear and quadratic optimization	[101–105]
Spread sheet	[106]
Simulated annealing	[43,60,107]
Definite point selection	[108]

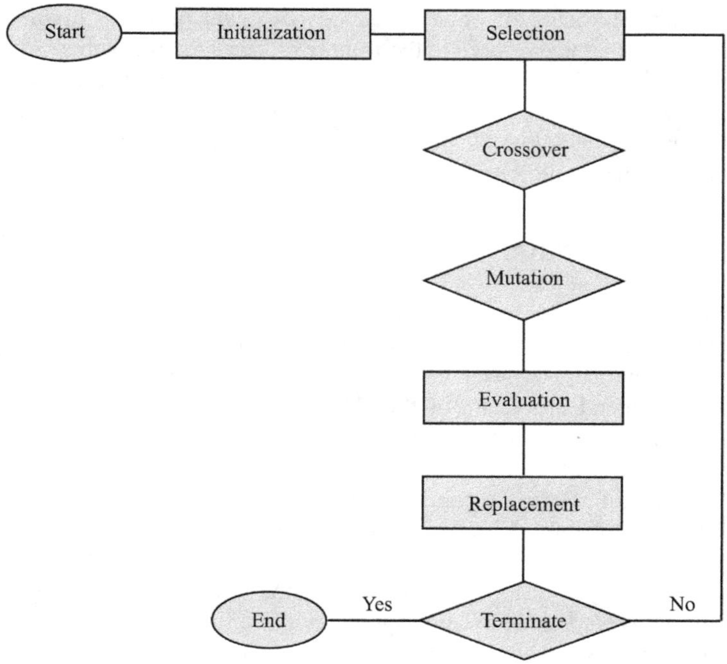

Figure 3.3 Genetic algorithm

Particle swarm optimization (hereafter PSO), developed by Kennedy and Eberhart [109], is a population-based stochastic optimization technique that iteratively improves a candidate solution. The PSO technique is somehow similar to the GA as the procedure is initialized using a population of random solutions. The major difference between the PSO and the GA is that in the PSO, a randomized velocity is also assigned to each potential solution based on which the potential solutions, which are called particles, evolve through the hyperspace.

In nature, ants initially move around randomly when they are searching for food, and once they find a food source, they leave a pheromone trail on the ground on their way transferring the food back to their colony. This pheromone trail helps other ants to not to move on random paths but instead to follow the trail to the food source. Based on this behavior, Dorigo [110] developed an optimization algorithm, namely, ant colony search algorithm (hereafter ACSA). The ACSA has been adapted for the wind-farm-layout optimization problem by several researchers of this field. For instance, Eroğlu and Seçkiner [99] developed an algorithm in which the contribution of each turbine to the total wake losses of the farm is assumed to be the pheromone. Accordingly, more ants are assigned to turbines with higher pheromone to improve their location. Ants move these turbines in random locations, and if any new location causes the total AEP of the wind farm to increase, then the previous layout will be replaced by the newly found more efficient layout. This procedure continues until a convergence occurs.

Simulated annealing (hereafter SA) is a probabilistic optimization technique which is more applicable to discrete spaces where determining an approximate global optimum is more preferred than a precise local optimum over the same amount of time. In fact, the SA is an approach that attempts to avoid entrapment in a poor local maximum by allowing an occasional downhill move. The acceptance of a downhill move depends on a control parameter, called the temperature, and on the magnitude of the variation. Rivas *et al.* [107] used the SA algorithm coupled with a local search module to perform a wind-farm-layout optimization analysis. The disadvantage of the employed local search was its likelihood of finding a local rather than a global optimum. The main idea behind the SA algorithm developed by Rivas *et al.* [107] is that the algorithm moves to a neighboring layout by removing a turbine, adding a turbine, or moving a turbine, and then, the AEP is calculated for the new layout. If the AEP of the new layout has increased in comparison to the previous layout, the new layout may be readily accepted; however, if the AEP has decreased, the new layout is accepted according to the probability calculated via the following equation:

$$P(\delta AEP) = \exp(-|\delta AEP|/t) \tag{3.65}$$

where δAEP is the variation of the AEP from the previous layout to the new layout, and t stands for the control parameter called temperature that is gradually cooled (decreased) to make the system converge.

In the definite point selection algorithm, developed by Shakoor *et al.* [108], the wake of each wind turbine is assumed to be a triangle with one vertex located upstream of the turbine and the other two vertices located downwind of the turbine [see Figure 3.4(a)]. The divergence angle of the wake triangle is calculated as $\gamma = 2\tan^{-1}(\alpha)$ where α is the entrainment constant and is defined as $\alpha = 0.5/\ln(z/z_0)$ in which z and z_0 are turbine hub height and surface roughness, respectively. Overlapping of the wake triangles associated with wind turbines of a wind farm forms a wake polygon with n vertices [see Figure 3.4(b)]. The shape and

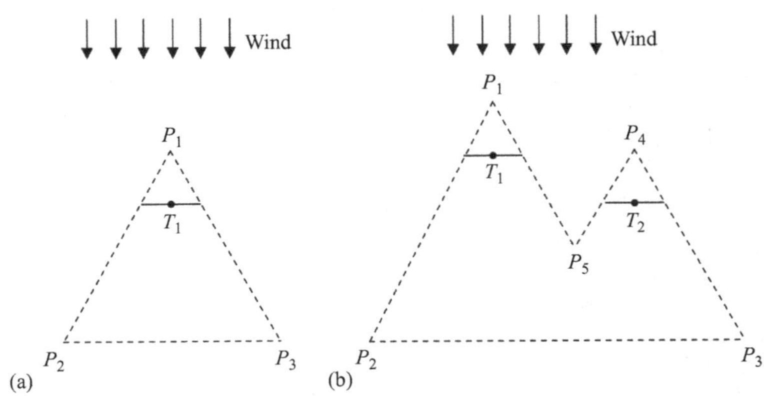

Figure 3.4 (a) Wake triangle and (b) wake polygon employed in definite point selection algorithm to select the optimal place for wind turbines

the area of this polygon varies with the wind direction; hence, there are A_i polygons where $i = 1:360$ stands for the wind direction. A point inside the wind farm area that does not fall inside any of the predetermined wake polygons is then selected as the best position for placing the next turbine inside the wind farm area.

3.5 Practice your knowledge

Increasing climate change concerns and extremely high economic, health, and social expenses caused by adverse environmental impacts of fossil fuels have pushed the energy industry towards sustainable energy resources, in particular, wind energy. Although wind energy provides approximately 8% of the United States generating capacity [111], which is more than any other renewable source, the global contribution of wind energy is yet too small, and for wind energy to play a more significant role in the market, the issues associated with the wind farm underperformance must be addressed. Solving the wind-farm-layout optimization problem is the most demanding task of wind farm design and development; hence, considerable research is actively conducted to develop more efficient solutions to this problem. Latest findings, references, and investigations on the major concepts associated with the wind-farm-layout optimization problem were reviewed and discussed in this chapter. That includes objective functions, optimization variables, wake loss models, and search algorithms. In this section, some specific cases are proposed to assist the readers to put the knowledge shared in this chapter into practice to some extent.

3.5.1 Case I: Shape of the wind farm

If the area of the two lands given in Figure 3.5(a) is equal ($A_1 = A_2$), and assuming that the wind frequency at the location of both sites is as presented in Figure 3.5(b), explain which area is more suitable for developing a wind farm.

3.5.2 Case II: Wake of wind turbines

A square land is available for developing a wind farm with five horizontal axis wind turbines (see Figure 3.6). If wind frequency at the location of this land is as presented in Figure 3.6(a), explain which of the three layouts proposed in Figure 3.6(b)–(d) is likely to be the most efficient one.

3.5.3 Case III: Wind speed deficit in wind farms

The layout of the Lillgrund wind farm, an offshore wind farm in Sweden with 48 SWT-2.3-93 wind turbines, is given in Figure 3.7. Assuming a southwesterly wind scenario along the direction of alignment of the wind turbines, draw a qualitative plot that illustrates the variation of the relative PP of wind turbines along columns C_1 and C_2. (Hint: Relative PP of a wind turbine at a given wind direction is defined as the ratio of power produced by that wind turbine to the power produced by the front row turbine. Hence, the relative PP of front row turbines is always 1, while the relative PPs of all downwind turbines are below 1.)

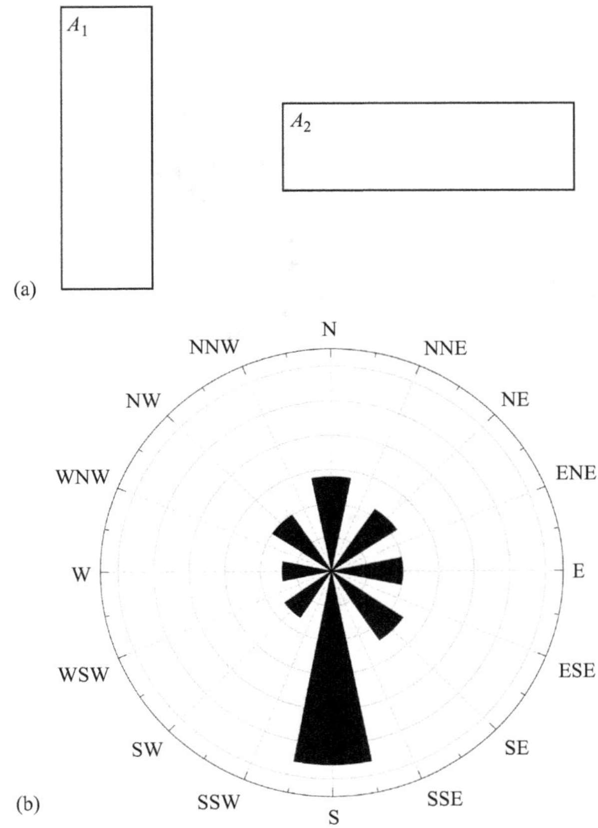

Figure 3.5 Case I: (a) two lands with equal areas, one oriented along the south-north direction and the other one oriented along the west-east direction and (b) wind rose

3.5.4 Case IV: Yaw angle of wind turbines

Nørrekær is an onshore wind farm with 13 Siemens SWT 2.3-93 wind turbines. The layout of this wind farm is illustrated in Figure 3.8. First, draw a qualitative plot that shows the variation of the relative PP of wind turbines along the column. How does this plot change if turbine 5 is yawed by 20 degrees as is illustrated in Figure 3.8(b). How would the plot change if turbine 5 was yawed in the opposite direction?

3.5.5 Case V: Variation of power production with wind direction

Plot the relative PP of turbine T_{22} in Figure 3.7 and turbine T_2 in Figure 3.8(a) versus wind direction for wind directions varying from 0° to 360° with respect to the North.

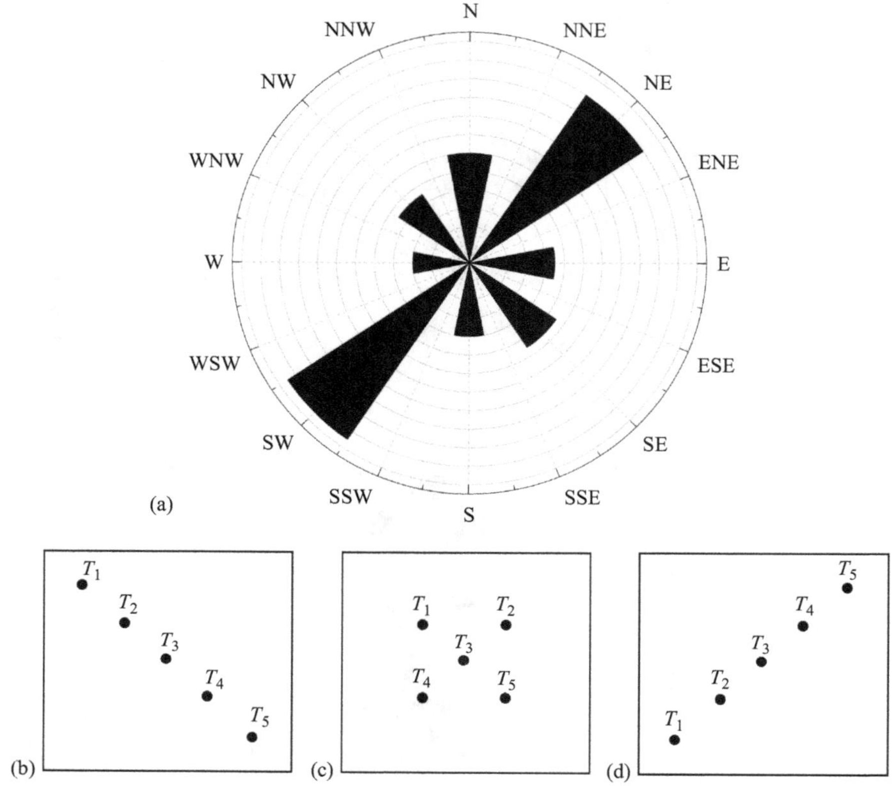

Figure 3.6 Case II: (a) wind rose, (b) layout 1, (c) layout 2, and (d) layout 3

3.5.6 Case VI: Surface roughness

How does the surface roughness affect the performance of a wind farm? Explain using the logarithmic boundary layer profile. A community wind farm and its associated wind rose is presented in Figure 3.9. Which turbine has the lowest production and needs to be relocated? Which turbine has the highest production?

3.5.7 Case VII: Inner turbines versus outer turbines

Figure 3.10(a) illustrates the wind frequency at the location of the area given in Figure 3.10(b). Among layouts proposed in Figure 3.10(c) and (d), which one is potentially a more efficient one?

3.5.8 Case VIII: Wind farm noise production

In Figure 3.11, coordinates of the house and four SWT-2.3-96 wind turbines are (1,337;292), (279;215), (395;821), (757;1,243), and (1,337;1,431) m, respectively. Assuming that the wind turbines are the only noise sources, estimate the total sound pressure level at the location of the house.

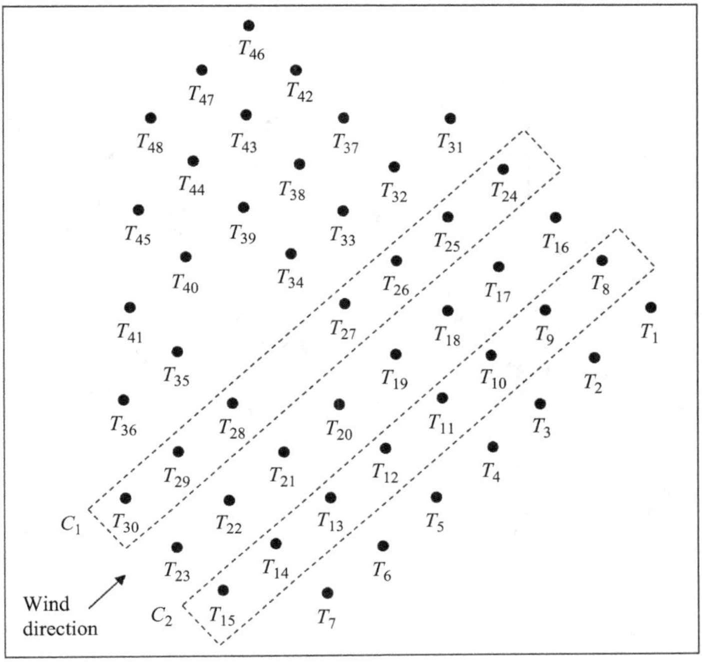

Figure 3.7 Cases III and XI

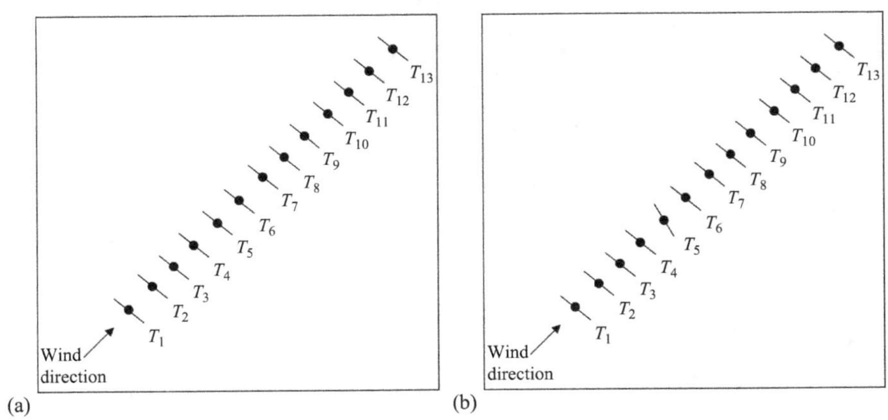

Figure 3.8 (a) Case IV: all turbines are perpendicular to the wind direction (i.e., yaw angle of all turbines is zero) and (b) case V: turbine 5 is misaligned (yawed), while all other turbines are perpendicular to the wind direction

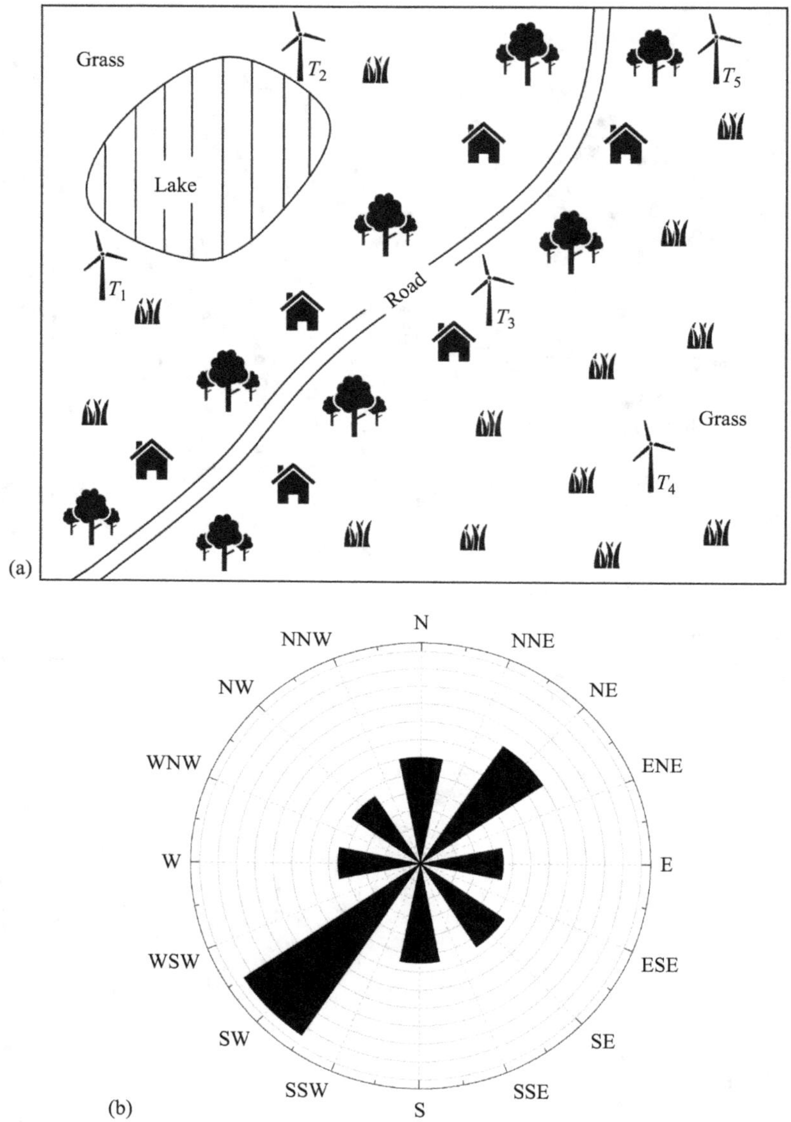

Figure 3.9 Case VI: (a) layout of the wind farm and (b) wind rose

3.5.9 Case IX: Hub height optimization

Assume both turbines shown in Figure 3.2 are SWT-2.3-93 manufactured by Siemens company. If $H_u = H_{\max} = 120$ m, then plot the relative power of the downwind turbine versus its hub height H_d, where $10 \text{ m} \leq H_d \leq H_{\max}$.

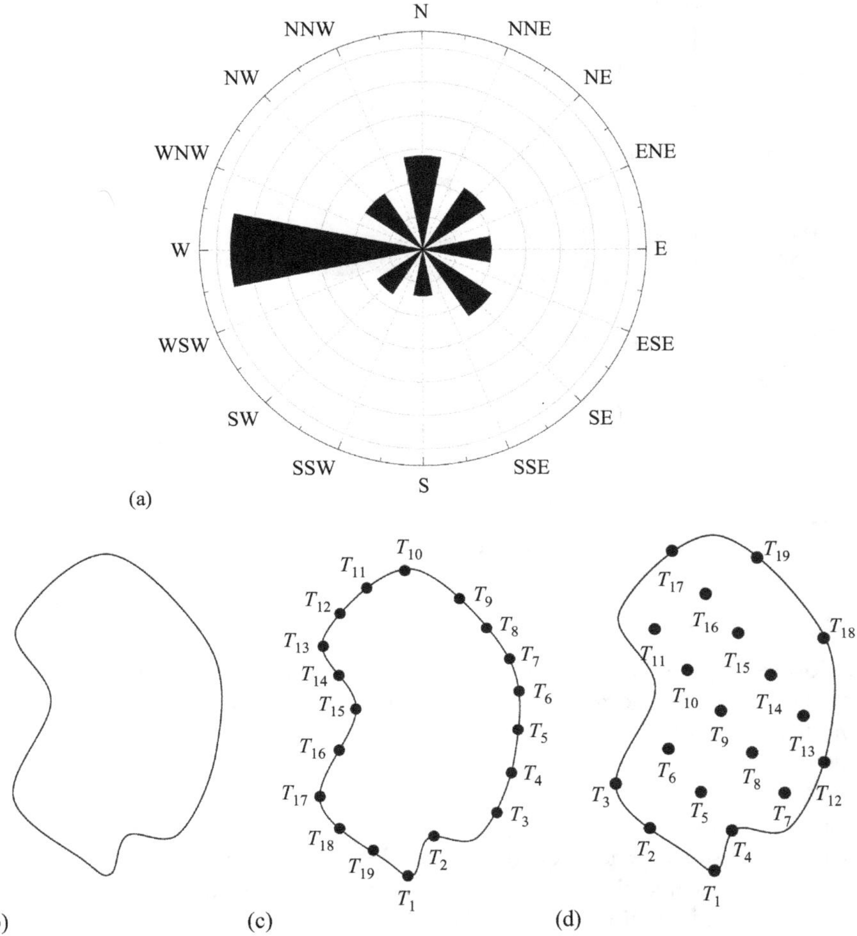

Figure 3.10 Case VII: (a) wind rose, (b) wind farm area, (c) layout 1, and (d) layout 2

3.5.10 *Case X: Fatigue loads*

How does the thrust coefficient of a wind turbine relate to the fatigue loads acting on that turbine?

3.5.11 *Case XI: Turbine type*

How would the AEP of the Lillgrund wind farm change if turbines T_i, where $i = 2, 4, 6, 8, 9, 10, 11, 12, 13, 14, 15, 24, 25, 26, 27, 28, 29, 30, 32, 34, 35, 37, 38, 39, 40, 41, 42, 44, 46, 47, 48$, were removed and the rest of them were replaced by SWT-8.0-154 turbines (see Figure 3.7). Conduct your calculations using the PARK model described in Section 3.3.6.1, and the wind rose given in Vasel-Be-Hagh and Archer [5].

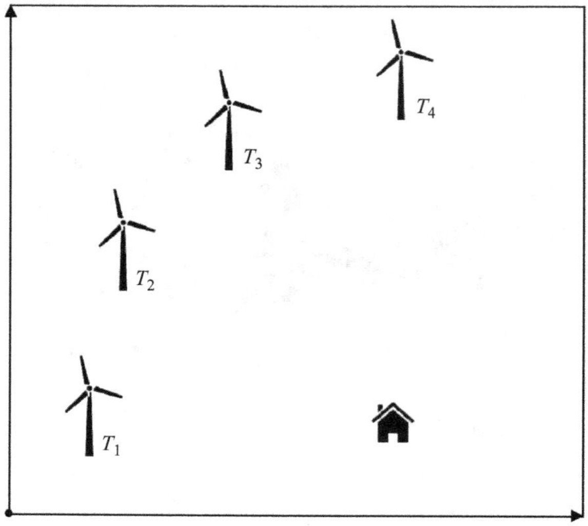

Figure 3.11 Case VIII

3.5.12 Case XII: Atmospheric stability

Atmospheric stability is a term used to qualitatively describe the potential for vertical motion in the atmosphere. Atmosphere is considered stable when it is stratified without any vertical motion. This condition usually exists at night time, when there is a negative heat flux at the surface and the air is cooled down from the bottom. On the other hand, when the air is heated up from the bottom, which usually happens during day time, strong vertical motions are generated and atmosphere is considered unstable. At sunset and sunrise, when the heat flux from the surface is approximately zero, atmosphere is considered neutral. Explain the effect of atmospheric stability on the wake of wind turbines and PP of wind farms assuming an equal geostrophic wind speed.

3.5.13 Case XIII: Wind farms and hurricanes

Discuss the possibility of using wind turbines to protect communities against hurricanes (see [112]).

References

[1] Mittal P, Mitra K, and Kulkarni K. Optimizing the number and locations of turbines in a wind farm addressing energy-noise trade-off: A hybrid approach. Energy Conversion and Management. 2017;132:147–160.

[2] Chen L, Harding C, Sharma A, *et al.* Modeling noise and lease soft costs improves wind farm design and cost-of-energy predictions. Renewable Energy. 2016;97:849–859.

[3] DuPont B, Cagan J, and Moriarty P. An advanced modeling system for optimization of wind farm layout and wind turbine sizing using a multi-level extended pattern search algorithm. Energy. 2016;106:802–814.

[4] Rahbari O, Vafaeipour M, Fazelpour F, *et al.* Towards realistic designs of wind farm layouts: Application of a novel placement selector approach. Energy Conversion and Management. 2014;81:242–254.

[5] Vasel-Be-Hagh AR, and Archer CL. Wind farm hub height optimization. Applied Energy. 2017;195:905–921.

[6] MirHassani SA, and Yarahmadi A. Wind farm layout optimization under uncertainty. Renewable Energy. 2017;107:288–297.

[7] Chen Y, Li H, Jin K, *et al.* Investigating the possibility of using different hub height wind turbines in a wind farm. International Journal of Sustainable Energy. 2017;36(2):142–150.

[8] Wang L, Tan ACC, Cholette M, *et al.* Comparison of the effectiveness of analytical wake models for wind farm with constant and variable hub heights. Energy Conversion and Management. 2016;124:189–202.

[9] Chen K, Song MX, Zhang X, *et al.* Wind turbine layout optimization with multiple hub height wind turbines using greedy algorithm. Renewable Energy. 2016;96:676–686.

[10] Chen Y, Li H, Jin K, *et al.* Wind farm layout optimization using genetic algorithm with different hub height wind turbines. Energy Conversion and Management. 2013;70:56–65.

[11] Vasel-Be-Hagh AR, and Archer CL. Wind farms with counter-rotating wind turbines. Sustainable Energy Technologies and Assessments. 2017;24: 19–30.

[12] Pillai AC, Chick J, Johanning L, *et al.* Optimisation of offshore wind farms using a genetic algorithm. International Journal of Offshore and Polar Engineering. 2016;26(3):225–234.

[13] Feng J, Shen WZ, and Xu C. Multi-objective random search algorithm for simultaneously optimizing wind farm layout and number of turbines. Journal of Physics: Conference Series. 2016;753(3):032011, 11 pages.

[14] Réthoré PE, Fuglsang P, Larsen GC, *et al.* TOPFARM: Multi-fidelity optimization of wind farms. Wind Energy. 2014;17(12):1797–1816.

[15] Van Eeckhout B, Van Hertem D, Reza M, *et al.* Economic comparison of VSC HVDC and HVAC as transmission system for a 300 MW offshore wind farm. European Transactions on Electrical Power. 2010;20(5): 661–671.

[16] Stevens RJAM, Hobbs BF, Ramos A, *et al.* Combining economic and fluid dynamic models to determine the optimal spacing in very large wind farms. Wind Energy. 2017;20(3):465–477.

[17] Hou P, Hu W, Soltani M, *et al.* Combined optimization for offshore wind turbine micro siting. Applied Energy. 2017;189:271–282.

[18] Amaral L, and Castro R. Offshore wind farm layout optimization regarding wake effects and electrical losses. Engineering Applications of Artificial Intelligence. 2017;60:26–34.

[19] Wędzik A, Siewierski T, and Szypowski M. A new method for simultaneous optimizing of wind farm's network layout and cable cross-sections by MILP optimization. Applied Energy. 2016;182:525–538.

[20] Hou P, Hu W, and Chen Z. Optimisation for offshore wind farm cable connection layout using adaptive particle swarm optimisation minimum spanning tree method. IET Renewable Power Generation. 2016;10(5):694–702.

[21] Ghaisas NS, and Archer CL. Geometry-based models for studying the effects of wind farm layout. Journal of Atmospheric and Oceanic Technology. 2016;33(3):481–501.

[22] Gonzalez-Rodriguez AG, Burgos-Payan M, Riquelme-Santos J, *et al.* Reducing computational effort in the calculation of annual energy produced in wind farms. Renewable and Sustainable Energy Reviews. 2015;43:656–665.

[23] Padrón AS, Stanley APJ, Thomas JJ, *et al.* Polynomial chaos for the computation of annual energy production in wind farm layout optimization. Journal of Physics: Conference Series. 2016;753(3):032021, 10 pages.

[24] Pookpunt S, and Ongsakul W. Design of optimal wind farm configuration using a binary particle swarm optimization at Huasai district, Southern Thailand. Energy Conversion and Management. 2016;108:160–180.

[25] Mittal P, Kulkarni K, and Mitra K. A novel hybrid optimization methodology to optimize the total number and placement of wind turbines. Renewable Energy. 2016;86:133–147.

[26] Brusca S, Lanzafame R, and Messina M. Wind turbine placement optimization by means of the Monte Carlo simulation method. Modelling and Simulation in Engineering. 2014;2014:760934, 8 pages.

[27] van Dijk MT, van Wingerden JW, Ashuri T, *et al.* Wind farm multi-objective wake redirection for optimizing power production and loads. Energy. 2017;121:561–569.

[28] van Dijk MT, van Wingerden JW, Ashuri T, *et al.* Yaw-misalignment and its impact on wind turbine loads and wind farm power output. Journal of Physics: Conference Series. 2016;753(6):062013, 13 pages.

[29] Park J, and Law KH. Bayesian ascent: A data-driven optimization scheme for real-time control with application to wind farm power maximization. IEEE Transactions on Control Systems Technology. 2016;24(5):1655–1668.

[30] Park J, and Law KH. Cooperative wind turbine control for maximizing wind farm power using sequential convex programming. Energy Conversion and Management. 2015;101:295–316.

[31] Abdulrahman M, and Wood D. Some effects of efficiency on wind turbine interference and maximum power production. Wind Engineering. 2015;39(5):495–506.

[32] Song M, Chen K, Zhang X, *et al.* Optimization of wind turbine micro-siting for reducing the sensitivity of power generation to wind direction. Renewable Energy. 2016;85:57–65.

[33] Wang L, Tan ACC, Cholette ME, *et al.* Optimization of wind farm layout with complex land divisions. Renewable Energy. 2017;105:30–40.

[34] Rabiee A, and Mohseni-Bonab SM. Maximizing hosting capacity of renewable energy sources in distribution networks: A multi-objective and scenario-based approach. Energy. 2017;120:417–430.

[35] Gao X, Yang H, and Lu L. Study on offshore wind power potential and wind farm optimization in Hong Kong. Applied Energy. 2014;130:519–531.

[36] Chowdhury S, Zhang J, Messac A, *et al.* Optimizing the arrangement and the selection of turbines for wind farms subject to varying wind conditions. Renewable Energy. 2013;52:273–282.

[37] Hou P, Hu W, Zhang B, *et al.* Optimised power dispatch strategy for offshore wind farms. IET Renewable Power Generation. 2016;10(3):399–409.

[38] Ashuri T, Ponnurangam C, Zhang J, *et al.* Integrated layout and support structure optimization for offshore wind farm design. Journal of Physics: Conference Series. 2016;753(9):092011, 12 pages.

[39] Gao X, Yang H, and Lu L. Investigation into the optimal wind turbine layout patterns for a Hong Kong offshore wind farm. Energy. 2014;73:430–442.

[40] Chen L, and MacDonald E. A system-level cost-of-energy wind farm layout optimization with landowner modeling. Energy Conversion and Management. 2014;77:484–494.

[41] Ashuri T, Zaaijer MB, Martins JRRA, *et al.* Multidisciplinary design optimization of offshore wind turbines for minimum levelized cost of energy. Renewable Energy. 2014;68:893–905.

[42] Rezaei Mirghaed M, and Roshandel R. Site specific optimization of wind turbines energy cost: Iterative approach. Energy Conversion and Management. 2013;73:167–175. Cited By 21.

[43] Elkinton CN, Manwell JF, and McGowan JG. Algorithms for offshore wind farm layout optimization. Wind Engineering. 2008;32(1):67–83.

[44] Lackner MA, Elkinton CN. An analytical framework for offshore wind farm layout optimization. Wind Engineering. 2007;31(1):17–31.

[45] Li W, Özcan E, and John R. Multi-objective evolutionary algorithms and hyper-heuristics for wind farm layout optimisation. Renewable Energy. 2017;105:473–482.

[46] Chowdhury S, Mehmani A, Zhang J, *et al.* Market suitability and performance tradeoffs offered by commercialwind turbines across differingwind regimes. Energies. 2016;9(5):352, 32 pages.

[47] Shamshirband S, Petković D, Ćojbašić Ž, *et al.* Adaptive neuro-fuzzy optimization of wind farm project net profit. Energy Conversion and Management. 2014;80:229–237.

[48] González JS, Rodríguez TG, Mora JC, *et al.* Overall design optimization of wind farms. Renewable Energy. 2011;36(7):1973–1982.

[49] Castro Mora J, Calero Barón JM, Riquelme Santos JM, *et al.* An evolutive algorithm for wind farm optimal design. Neurocomputing. 2007;70(16–18): 2651–2658.

[50] Attias K, and Ladany SP. Optimal economic layout of turbines on windfarms. Wind Engineering. 2006;30(2):141–151.

[51] Kwong WY, Zhang PY, Romero D, *et al.* Multi-objective wind farm layout optimization considering energy generation and noise propagation with NSGA-II. Journal of Mechanical Design, Transactions of the ASME. 2014;136(9).

[52] Yamani Douzi Sorkhabi S, Romero DA, Yan GK, *et al.* The impact of land use constraints in multi-objective energy-noise wind farm layout optimization. Renewable Energy. 2016;85:359–370.

[53] Rodrigues SS, and Marta AC. On addressing noise constraints in the design of wind turbine blades. Structural and Multidisciplinary Optimization. 2014; 50(3):489–503.

[54] Muljadi E, and Butterfield CP. Pitch-controlled variable-speed wind turbine generation. IAS Annual Meeting (IEEE Industry Applications Society). 1999; p.323–330.

[55] Guirguis D, Romero DA, and Amon CH. Toward efficient optimization of wind farm layouts: Utilizing exact gradient information. Applied Energy. 2016;179:110–123.

[56] Parada L, Herrera C, Flores P, *et al.* Wind farm layout optimization using a Gaussian-based wake model. Renewable Energy. 2017;107:531–541.

[57] Kuo JYJ, Romero DA, Beck JC, *et al.* Wind farm layout optimization on complex terrains – Integrating a CFD wake model with mixed-integer programming. Applied Energy. 2016;178:404–414.

[58] Yang H, Xie K, Tai HM, *et al.* Wind farm layout optimization and its application to power system reliability analysis. IEEE Transactions on Power Systems. 2016;31(3):2135–2143.

[59] Quan N, and Kim HM. A mixed integer linear programing formulation for unrestricted wind farm layout optimization. Journal of Mechanical Design, Transactions of the ASME. 2016;138(6).

[60] Hernandez EA, Uddameri V, and Singaraju S. Combined optimization of a wind farm and a well field for wind-enabled groundwater production. Environmental Earth Sciences. 2014;71(6):2687–2699.

[61] Eroğlu Y, and Seçkiner SU. Wind farm layout optimization using particle filtering approach. Renewable Energy. 2013;58:95–107.

[62] Bansal JC, and Farswan P. Wind farm layout using biogeography based optimization. Renewable Energy. 2017;107:386–402. Cited By 0.

[63] Chen Y, Li H, He B, *et al.* Multi-objective genetic algorithm based innovative wind farm layout optimization method. Energy Conversion and Management. 2015;105:1318–1327.

[64] Abdulrahman M, and Wood D. Investigating the Power-COE trade-off for wind farm layout optimization considering commercial turbine selection and hub height variation. Renewable Energy. 2017;102:267–278.

[65] Chowdhury S, Zhang J, Tong W, *et al.* Modeling the influence of land-shape on the energy production potential of a wind farm site. Journal of Energy Resources Technology, Transactions of the ASME. 2014;136(1): 011203, 10 pages.

[66] Tong W, Chowdhury S, and Messac A. A consolidated visualization of wind farm energy production potential and optimal land shapes under different land area and nameplate capacity decisions; 2014, 10th AIAA Multidisciplinary Design Optimization Conference, AIAA SciTech Forum, (AIAA 2014-0998), https://doi.org/10.2514/6.2014-0998.

[67] Wang L, Tan ACC, Gu Y, *et al.* A new constraint handling method for wind farm layout optimization with lands owned by different owners. Renewable Energy. 2015;83:151–161.

[68] Smagorinsky J. General circulation experiments with the primitive equations. Monthly Weather Review. 1963;91(3):99–164.

[69] Moeng CH. A large eddy simulation model for the study of planetary boundary-layer turbulence. Journal of the Atmospheric Sciences. 1984; 41(13):2052–2062.

[70] Sørensen JN, and Shen WZ. Numerical modeling of wind turbine wakes. Journal of Fluids Engineering, Transactions of the ASME. 2002;124(2):393–399.

[71] Martinez-Tossas LA, Churchfield MJ, and Leonardi S. Large eddy simulations of the flow past wind turbines: Actuator line and disk modeling. Wind Energy. 2015;18(6):1047–1060.

[72] Doubrawa P, Barthelmie RJ, Wang H, *et al.* A stochastic wind turbine wake model based on new metrics for wake characterization. Wind Energy. 2017;20(3):449–463.

[73] Bastine D, Vollmer L, Wächter M, *et al.* Stochastic wake modeling based on POD analysis. Wind Energy Science Discussions. 2016;2016:1–36.

[74] Ali N, Kadum HF, and Cal RB. Focused-based multifractal analysis of the wake in a wind turbine array utilizing proper orthogonal decomposition. Journal of Renewable and Sustainable Energy. 2016;8(6): 063306, 19 pages.

[75] Ainslie JF. Calculating the flowfield in the wake of wind turbines. Journal of Wind Engineering and Industrial Aerodynamics. 1988;27:213–224.

[76] Ott S, Berg J, and Nielsen M. Linearised CFD models for wakes. Danmarks Tekniske Universitet, Risø Nationallaboratoriet for Bæredygtig Energi; Denmark: Forskningscenter Risoe; 2011.

[77] Ishihara T, Yamaguchi A, and Fujino Y. Development of a new wake model based on a wind tunnel experiment. Global Wind; 2004.

[78] Jensen NO. A note on wind generator interaction. Denmark: Risø National Laboratory; 1983. Risø-M-2411.

[79] Katic I, Højstrup J, and Jensen NO. A simple model for cluster efficiency. European Wind Energy Association Conference and Exhibition. 1986; p. 407–410.

[80] Xie S, and Archer C. Self-similarity and turbulence characteristics of wind turbine wakes via large-eddy simulation. Wind Energy. 2015;18(10):1815–1838.

[81] Xie S, Archer CL, Ghaisas N, *et al.* Benefits of collocating vertical-axis and horizontal-axis wind turbines in large wind farms. Wind Energy. 2017; 20(1):45–62.

[82] Bastankhah M, and Porté-Agel F. A new analytical model for wind-turbine wakes. Renewable Energy. 2014;70:116–123.

[83] Larsen GC. A simple wake calculation procedure. Denmark: Risø National Laboratory; 1988. Risø-M-2760.

[84] Pierik JTG, Dekker JWM, Braam H, *et al.* In: Petersen EL, Hjuler Jensen P, Rave K, *et al.*, editors. European wind turbine standards II (EWTS-II). London: James and James Science Publishers; 1999. p. 568–571.

[85] Frandsen S, Barthelmie R, Pryor S, *et al.* Analytical modelling of wind speed deficit in large offshore wind farms. Wind Energy. 2006;9(1–2):39–53.

[86] Barthelmie R, Larsen G, Frandsen S, *et al.* Comparison of wake model simulations with offshore wind turbine wake profiles measured by sodar. Journal of Atmospheric and Oceanic Technology. 2006;23(7):888–901.

[87] Ghaisas N, and Archer CL. Geometry-based models for studying the effect of wind farm layout. Journal of Atmospheric and Oceanic Technology. 2016;23(3):481–501.

[88] Kuo JYJ, Romero DA, and Amon CH. A mechanistic semi-empirical wake interaction model for wind farm layout optimization. Energy. 2015;93:2157–2165.

[89] Zhang YH, Gong YJ, Gu TL, *et al.* Flexible genetic algorithm: A simple and generic approach to node placement problems. Applied Soft Computing Journal. 2017;52:457–470.

[90] DuPont B, and Cagan J. A hybrid extended pattern search/genetic algorithm for multi-stage wind farm optimization. Optimization and Engineering. 2016;17(1):77–103.

[91] Long H, and Zhang Z. A two-echelon wind farm layout planning model. IEEE Transactions on Sustainable Energy. 2015;6(3):863–871.

[92] Feng J, and Shen WZ. Solving the wind farm layout optimization problem using random search algorithm. Renewable Energy. 2015;78:182–192.

[93] Chen K, Song MX, and Zhang X. The iteration method for tower height matching in wind farm design. Journal of Wind Engineering and Industrial Aerodynamics. 2014;132:37–48.

[94] Chen K, Song MX, and Zhang X. The investigation of tower height matching optimization for wind turbine positioning in the wind farm. Journal of Wind Engineering and Industrial Aerodynamics. 2013;114:83–95.

[95] Zhang B, Song B, Mao Z, *et al.* A novel wake energy reuse method to optimize the layout for Savonius-type vertical axis wind turbines. Energy. 2017;121:341–355.

[96] Hou P, Hu W, Chen C, *et al.* Optimization of offshore wind farm layout in restricted zones. Energy. 2016;113:487–496.

[97] Tong W, Chowdhury S, and Messac A. A multi-objective mixed-discrete particle swarm optimization with multi-domain diversity preservation. Structural and Multidisciplinary Optimization. 2016;53(3):471–488.

[98] Wu YK, Lee CY, Chen CR, *et al.* Optimization of the wind turbine layout and transmission system planning for a large-scale offshore wind farm by

AI technology. IEEE Transactions on Industry Applications. 2014;50(3): 2071–2080.

[99] Eroĝlu Y, and Seçkiner SU. Design of wind farm layout using ant colony algorithm. Renewable Energy. 2012;44:53–62.

[100] Zeng J, and Zhang H. Wind speed forecasting model study based on least squares support vector machine and Ant Colony optimization. Taiyangneng Xuebao/Acta Energiae Solaris Sinica. 2011;32(3):296–300.

[101] Turner SDO, Romero DA, Zhang PY, *et al.* A new mathematical programming approach to optimize wind farm layouts. Renewable Energy. 2014;63:674–680.

[102] Quan N, and Kim H. A tight upper bound for grid-based wind farm layout optimization. ASME 2016 International Design Engineering Technical Conferences and Computers and Information in Engineering Conference, Volume 2A: 42nd Design Automation Conference, Charlotte, NC, USA, August 21–24, 2016; 2016.

[103] Irawan CA, Song X, Jones D, *et al.* Layout optimisation for an installation port of an offshore wind farm. European Journal of Operational Research. 2017;259(1):67–83.

[104] Wędzik A, Siewierski T, and Szypowski M. A new method for simultaneous optimizing of wind farm's network layout and cable cross-sections by MILP optimization. Applied Energy. 2016;182:525–538.

[105] Archer R, Nates G, Donovan S, *et al.* Wind turbine interference in a wind farm layout optimization mixed integer linear programming model. Wind Engineering. 2011;35(2):165–175.

[106] Massan SUR, Wagan AI, Shaikh MM, *et al.* Wind turbine micrositing by using the firefly algorithm. Applied Soft Computing Journal. 2015;27: 450–456.

[107] Rivas RA, Clausen J, Hansen KS, *et al.* Solving the turbine positioning problem for large offshore wind farms by simulated annealing. Wind Engineering. 2009;33(3):287–298.

[108] Shakoor R, Hassan MY, Raheem A, *et al.* Wind farm layout optimization using area dimensions and definite point selection techniques. Renewable Energy. 2016;88:154–163.

[109] Kennedy J, and Eberhart R. Particle swarm optimization. Proceedings of IEEE International Conference on Neural Networks. vol. 4; 1995. p. 1942–1948.

[110] Dorigo M, Di and Caro G. Ant colony optimization: A new meta-heuristic. Proceedings of the 1999 IEEE Congress on Evolutionary Computation. vol. 2; 1999. p. 1470–1477.

[111] Today in energy [Governmental page]. U.S. Department of Energy, Washington DC: U.S. Energy Information Administration; 2017. Available from: https://www.eia.gov/todayinenergy/detail.php?id=31032.

[112] Jacobson MZ, Archer CL, and Kempton W. Taming hurricanes with arrays of offshore wind turbines. Nature Climate Change. 2014;4(3):195–200.

Chapter 4

Financing for community wind and solar project development

Lindsay Miller[1] *and Rupp Carriveau*[1]

Abstract

Community energy, and in particular, community wind and solar, has experienced significant growth in recent years. The main challenge to further expansion and implementation of community renewable energy projects is the high capital costs associated with project development. This chapter will present case studies of community wind and solar financing and include innovative mechanisms such as lease financing, sales of renewable energy credits, crowdfunding, and unconventional loan strategies. Through compilation of these cases, the objective is to provide prospective small renewable energy developers options as to how they can go about financing their projects and ultimately empower communities to independently produce energy while reducing greenhouse gases.

Community energy, in particular, community wind and solar, has experienced significant growth in recent years. Driving this trend is the desire for communities to become energy self-sufficient and more environmentally conscious. Benefits of community wind and solar extend to include increasing community capacity, empowering residents, and promoting energy efficiency and conservation. The main challenge for further expansion and implementation of community renewable energy projects is the high capital costs associated with project development. With initial costs representing approximately 70 per cent of the life cycle costs of the project [1], it can take several years to recoup this investment and can impede the ability of local developers to raise enough capital.

Small utility-scale wind power projects have provided proving grounds for new technology and innovative financing structures. Historically, financing of community wind projects has generated innovative structures that were later adopted by the broader wind market and are now popular mainstream financing structures, such as the special allocation partnership flip structure [2]. More

[1]Turbulence and Energy Laboratory, Department of Engineering, University of Windsor, Canada

recently, community wind projects have been financed with creative structures that capitalize on incentives and make use of public market capital.

Community solar is a popular community energy choice due to the scalability of projects. Small communities and neighbourhoods can pursue solar energy more easily by selecting the number of panels required to suit their needs. Community solar has relied on three main financing structures: utility-sponsored, special-purpose entity, and non-profit, to implement projects of various scales and make use of available incentives.

Despite many community wind and solar projects achieving financing success, there are challenges in replicating the elements of some of these deals. Some of the cases only work in the exact circumstance presented due to specific incentive availability or legalities in different areas. Also, since these proposed structures are new to the industry, transaction costs may be high due to the learning process of executing these deals and development of the financing package could be lengthy for the same reason.

This chapter will present case studies of community wind and solar financing and include innovative mechanisms such as lease financing, sales of renewable energy credits, crowdfunding, and unconventional loan strategies. Through compilation of these cases, the objective is to provide prospective small renewable energy developers options as to how they can go about financing their projects and ultimately empower communities to independently produce energy while reducing greenhouse gases. These examples serve as encouraging case studies for other community projects looking for ways to raise capital and make use of programmes and incentives to piece together a financial package. Furthermore, the cases presented have the potential to extend beyond community projects to commercially renewable energy project finance.

4.1 Introduction

Community energy, in particular, community wind and solar, has experienced significant growth in recent years. Driving this trend is the desire for communities to become energy self-sufficient and more environmentally conscious. Benefits of community wind and solar extend to include increasing community capacity, empowering residents, and promoting energy efficiency and conservation. The main challenge for further expansion and implementation of community energy projects is the high capital costs associated with project development. With initial costs representing approximately 70 per cent of the life cycle costs of the project [1], it can take several years to recoup this investment and can impede the ability of developers to raise enough capital.

Initial costs of project development, referred to as installed capital costs, include the purchase, transportation, and installation of equipment, design and engineering costs, preliminary data collection, purchasing of monitoring equipment and software, project management, contingency, and others. These costs are typically expressed as $/kW installed. Once operating, the project will also incur operation and maintenance (O&M) costs such as materials and wages for scheduled and unscheduled maintenance and operational expenses such as fuel costs, taxes, land leases, and administrative expenses [2]. The installed capital costs and the O&M are two components of the life cycle cost of energy (LCOE), a metric that is used to express the cost of energy as a ratio of the present value of total costs divided by the energy produced over the lifetime of the project (expressed as $/kW h). The LCOE allows for equivalent cost comparisons between projects of different sizes and generation types. For renewable projects, with no cost for fuel, the capital costs far outweigh the O&M costs, and therefore, upfront financing can pose significant barrier to community energy projects.

Community energy projects generally involve a collective pooling of resources to implement a project and therefore share in the benefits of clean energy. The community wind sector differs from traditional commercial wind in scale and ownership structure and has been defined to 'consist of relatively small utility-scale wind power projects that sell power on the wholesale market and that are developed and owned primarily by local investors' [3]. This definition has largely been adopted in the literature [4] and will be applied in this chapter. Community solar has been defined as 'a solar photovoltaic (PV) project with multiple individual owners living in geographic proximity to the solar project, sharing in the costs and benefits of ownership' [5].

Interest in community-owned energy in North America has been spurred in part by the success that European countries such as Sweden, Denmark, and Germany have realized with similar projects for decades [6]. Furthermore, government incentives have been available recently which has allowed for more projects to be financially viable. Even with incentives, however, there exists regulatory, market, and financial challenges to community energy project development.

Small utility-scale wind power projects and community solar projects have provided a proving grounds for new technology and innovative financing structures. Financing structures that were initially developed for community wind projects have proven to be successful in the broader markets. The best example of this is the special allocation partnership flip structure that was originally developed by community wind projects in Minnesota, and therefore often referred to as the 'Minnesota flip' model [3]. This model is now widely used in commercial wind projects. The smaller scale nature of community energy projects are more suited for experimentation with innovative financial structures, and the commercial sectors have benefitted from the outcomes of this.

More recently, community energy projects have been financed with creative structures that capitalize on incentives and make use of public market capital. Examples include the 25.3-MW Ridgewind project in Minnesota [3], the first sale/leaseback financing of a wind project and the Crow Lake wind project in South Dakota [7], and Fort Dix solar project in New Jersey [8] which were financed using crowdfunding models, allowing residents to invest in the project. Despite many community projects achieving financing success, there are challenges in replicating the elements of some of these deals. Some of the cases only work in the exact circumstance presented due to specific incentive availability or legalities in different areas. Also, since these proposed structures are new to the industry, transaction costs may be high due to the learning process of executing these deals and development of the financing package could be lengthy for the same reason.

This chapter provides a review of financial mechanisms used to finance community wind and solar, both commonly used and first-of-a-kind structures, for the purpose of providing blueprints to future developers and to assess the barriers that may exist to further community wind development. Innovative mechanisms such as lease financing, sales of renewable energy credits, crowdfunding, and unconventional loan strategies will be explored. Through compilation of these cases, the objective is to provide prospective small energy developers options as to how they can go about financing their projects and ultimately empower communities to independently produce energy while reducing greenhouse gases. These examples serve as encouraging case studies for other community projects looking for ways to raise capital and make use of programmes and incentives to piece together a financial package. Furthermore, the cases presented have the potential to extend beyond community energy to commercial renewable energy project finance. Through the dissemination of case studies, best practices can be applied to future projects.

4.1.1 Community wind and solar – defined

Community wind has been defined in the literature as 'locally owned, utility-scale wind development that is interconnected to the grid on either the customer or utility side of the meter' [9]. In general, community wind has traditionally involved construction of commercial-sized turbines in rural areas to provide power to the community, to sell the power externally, or a combination of the two [10]. In contrast to commercial wind projects where local involvement is limited, community wind is typically locally owned and operated.

Community solar is similar in concept and has been defined as 'solar PV projects with multiple individual owners living in geographic proximity to the solar project, and sharing the costs and benefits of ownership of the solar project' [5]. Community solar projects are also typically owned and operated, and the scalability of solar is a major advantage while considering community energy options. The smaller scale nature of community solar may allow for projects to be more easily realized. The National Renewable Energy Laboratory reports that only 22–27 per cent of residential rooftop area is suitable for hosting on-site PV systems [11]. Community solar projects provide residents with options to participate in solar energy without necessarily installing panels on their property.

Community projects can have various degrees of ownership from 100 per cent community-owned to co-ownership arrangements where the community owns a small part of a larger project [12]. Different models of community projects include cooperatives, community charities, development trusts, and shares owned by a local community organization [12].

Community energy generally brings to mind energy independence, although globally, there are relatively few examples of communities that are entirely energy independent [10]. Some states have specific qualifications in order for a project to be deemed 'community energy'. For example, Minnesota has requirements that Community-Based Energy Development (C-Bed) projects have no single beneficiary owning more than 15 per cent of the project unless the project includes only one or two turbines or is owned by a public entity [5]. Additionally, 51 per cent of the revenue from the project must flow to local owners and other local entities, and the project must have the support of the county board in which it is located.

4.2 Benefits of community wind and solar

Community energy provides power for the community, potential income for the community, promotes social cohesion, and provides local jobs. Community renewable energy can also make claims to the environmental benefits associated with carbon-free energy sources, such as reduced CO_2 emissions, and has been suggested to be a practical bottom-up approach to carbon reduction [13]. Community renewable energy investments also contribute to energy security by reducing dependence on foreign energy imports. Individuals within the community benefit from direct ownership economically, from the business venture itself, and also potentially through lowered energy costs. These benefits are depicted in Figure 4.1.

Previous studies have demonstrated significantly higher local economic benefits when a project is community owned instead of corporate owned [15]. A US-based study demonstrates that local ownership increased economic benefits by 50–240 per cent, and the British government states that 12–13 times as much community value is reinvested into communities compared with commercial structures [16].

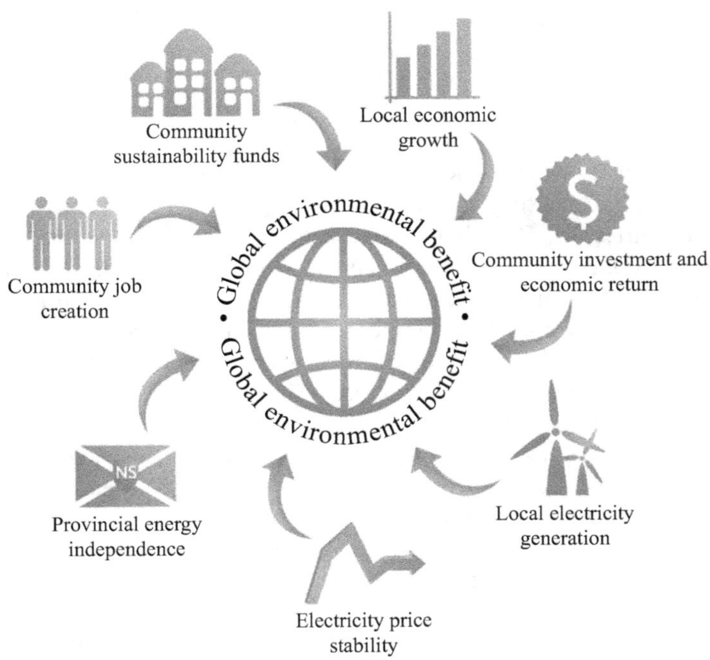

Figure 4.1 Benefits of community wind [14]

Another benefit is the associated high social acceptance from the community members. While communities generally appreciate that wind and solar development brings economic benefits to the community, they can struggle with trusting the intentions of the stakeholders who may be far removed from the community [17]. On the contrast, when projects have a high level of local involvement, more people are likely to recognize the positive value of renewable energy [18]. Community wind has been credited with improved social acceptance of projects through transparent local ownership and the creation of equitable outcomes amongst community members [19]. Recent research has also demonstrated that community funding structures generate significantly higher socio-economic impacts than other funding structures with greater opportunities for equitable distribution of benefits [20]. Maruyama *et al.* [21] explain how the movement towards a sustainable society is fuelled in part by economic, environmental, and social aspects but also by consciousness, participation, compassion, and cooperation – all of which are elements of renewable community energy.

Although there are many aspects involved in the successful development of community energy projects, this chapter focuses on financing. Details on some other planning aspects such as identifying community members and developing community risk, architectural, and sustainability plans can be found in [10]. Aspects such as whether there is a convenient grid interconnect, resource assessment, zoning laws, federal restrictions, and community acceptance should all be carefully considered prior to initiating the funding process. Furthermore, the choice of technology should be carefully considered, an example of which can be found in [22].

4.3 Lessons from overseas

Globally, community wind has been responsible for large amounts of installed wind power capacity, accounting for the majority of installed capacity in several European countries [9]. Community wind and solar projects have played a large role in Denmark, Germany, Sweden, and the Netherlands since the early 1990s, and more recently, also in the United Kingdom. The fundamental contributor to the success of community energy in these countries is the support mechanisms at the national level. In contrast to North America, where incentives have largely been tax-based, requiring that a project owner have significant tax liability to capitalize, in Europe, 'feed-in' laws have been the choice of support. These feed-in laws require utilities to purchase renewable power at premium prices, thereby creating a stable and profitable market. Furthermore, in Denmark, Sweden, and Germany, there are standard interconnection rules, removing uncertainty in who will bear these costs [9]. Some countries also do not tax community renewable energy investors on income earned from their projects and do tax CO_2 emissions, further making community energy an appealing alternative.

In Denmark, community wind ownership takes the form of a wind partnership. Simply put, individuals pool their savings to invest in a turbine and then sell the energy generated back to the utility at an attractive rate. The partnership members received the full carbon tax and a partial refund of the energy tax [6]. Denmark also has appealing options for financing along tax advantages and tradable green certificates. Recently, Denmark also set a world crowdfunding record, raising 1.3 million in 13 h from 1,700 Dutch households to buy shares in a wind turbine [23].

Germany has been implementing community energy since the 1990s. An example of this is the Wildpoldsried village where solar and wind initiatives began in 1997. Today, the village has nine community buildings with solar panels and seven installed wind turbines and produce over 300 per cent more energy than it needs [24]. Selling this energy back to the national grid generates revenue for the members. This village has been recognized as best practice model for community energy [24].

The most common community wind ownership structure in Germany is a limited partnership (LP) with a limited liability company as a general partner [25]. In this model, the developer incorporates his business as a limited liability company and for each individual project, forms a LP with the company as the general partner and individual investors as limited partners. Revenues are distributed proportionately to investments. Overall, from a financing perspective, the success in these countries can be attributed to the proper use of tax exemptions, feed-in tariffs, and capital investment subsidies.

Research on community energy in the United Kingdom can also provide some insight. Stamford [25] detailed the barriers associated with the United Kingdom falling behind the countries mentioned above. First, the programmes that were in place to support renewables were administratively burdensome and therefore not well suited to smaller projects. Additionally, at the time, the United Kingdom did not offer attractive tax incentives or capital investment subsidies for renewable energy projects. Toke [26] also identified a lack of confidence and knowledge amongst farmers in the United Kingdom as a barrier. To address these obstacles, and in the absence of incentives and subsidies, Kellett [13] suggested one possibility to increase community-based renewables in the United Kingdom. Through a case study of community-based energy in South Yorkshire, United Kingdom, the authors demonstrated that community-based energy service companies could assist communities by accelerating the transition to energy, leading to reduced fuel poverty and maximum energy purchasing power. More recently, crowdfunding is also being encouraged as means of raising capital for community energy projects in the United Kingdom.

Renewable energy networking sites such as EnergyShare are offering to match funds to crowdfunders of wind or solar projects. Another company, Abundance Energy, is also encouraging individuals and businesses to invest small amounts into wind turbines in their communities. Community solar had some success in the United Kingdom with projects being implemented in fuel-deprived areas. A project in Newport, South Wales was successful in providing free PV panels to 74 households in fuel poverty. The project was financed through community share offers and loans [27].

The countries mentioned above are not the only ones who have successfully implemented community renewable energy projects. Crowdfunding is also being pursued in Australia by local residents looking to realize community solar installation projects. The Philippines has also successfully installed small wind turbines to supply energy in remote communities [28].

In summary, the success realized in these countries is largely attributed to financial and fiscal incentives that are available for community project development along with standard rules for items such as interconnection costs. Furthermore, the sustained success is largely due to the interest of the community members in supporting renewable energy as well as the many blueprints that are available for implementation due to the many successful cases available. Although North America does not have the same incentives and nor the same level of implementation, this has not prevented creative owners from devising innovative ways to capture the available incentives and still maintain local ownership and operation.

4.4 Overview of available incentives and credits in North America

Different tax incentives apply to different projects based on location, project size, and other qualifications. This section will discuss some of the more widely applicable incentives that have been credited for recent wind development in North America.

In the United States, there have been two tax credits that have been credited with allowing for the expansion and development of renewable energy, the production tax credit (PTC), currently worth \$23/MW h, and the investment tax credit (ITC) which is an upfront credit (30 per cent) against the capital expense of the project [29]. Cash benefits also include revenues from the sale of renewable energy certificates (RECs). RECs are tradable energy commodities representing proof of 1 MW h of renewably generated electricity, which are certified and tracked by third parties [30]. The holder or purchaser of RECs own the rights to the environmental benefits of the renewable energy generated. A REC can be sold separately from the physical electricity generated. Although there are challenges surrounding RECs, such as their value being widely variable by location and the possibility of the benefits of renewables being double-counted, RECs have been a major contributor to the success of many renewable deals by acting as a payment that can be put towards the projects costs. RECs represent electricity produced using environment-friendly processes that are marked by credits which can be sold to a third party.

Renewable portfolio standards (RPS) will also influence the availability of incentives, and ultimately, the economics of a project. An RPS is a regulatory mandate to increase renewable energy production. These standards can also be stated specifically in relation to community energy. For example, within Oregon's overall RPS of 50 per cent renewables by 2040, it is also stipulated that 8 per cent come from small-scale and community-based renewable energy projects [31].

New Market Tax Credits (NMTCs) are another option for some community energy developers. NMTCs were established in 2000, however, have only recently been used in renewable project financing [3]. The NMTC program provides an ITC of 39 per cent over 7 years against federal income taxes in exchange for making a Qualified Equity Investment in a Community Development Entity. Communities must meet certain low-income characteristics. NMTCs can also be used along with PTCs for qualifying communities.

US States are also actively developing programmes to promote community energy, particularly in rural Midwestern states where community wind is recognized as a means to supplement farmer income. Policies and incentives are also tracked in a database 'Database of State Incentives for Renewables and Efficiency' [32] where developers can search for incentives and policies by state and by state and policy type. The International Energy Agency also maintains a database of renewable energy policies [33].

Canada has a similar tax incentive for renewable energy developments to make these projects more fiscally attractive. The accelerated Capital Cost Allowance allows for capital costs of eligible equipment to be written off at 50 per cent per year on a declining basis [34]. In addition, developers can choose to have certain expenses that are incurred during new energy start-up projects be fully deducted in the year that they are incurred, carried forward, or transferred to investors. Funding in Canada is available from various sources such as the Green Municipal Fund [35] which developers can apply for to fund feasibility studies and some project costs. Provinces have also developed their own programmes such as the Prince Edward Island Renewable Energy Act [36] and the Ontario Feed-In Tariff Programme [37]. The microFIT program in Ontario supports renewable energy projects with generating capacities less than 10 kW by standardizing prices and contracts. In Alberta, RECs are being credited with promoting renewable energy development. This system grants one REC for 1 MW h of green renewable energy. Bundling the sale of future RECs into financing packages has allowed for projects to become economically feasible. The Government of the Northwest Territories recently launched the Community Renewable Energy Program to assist community-based development of alternative energy systems. The programme will fund up to one half of the project cost, up to $50,000 per year.

Incentives and grants are available in North America; however, they often require complex arrangements to make full use of their value. Also, the incentives and grants have qualifications that can be challenging to navigate and are subject to renewals or changes in structures.

4.5 Historical financing models

There have been several financing structures that have been developed and have had widespread use in the last decade. These will be described briefly below.

4.5.1 Municipal wind

Municipal wind refers to projects that are developed and installed on public property by a municipal utility, school, or other small jurisdiction and are generally financed through the sale of tax-exempt municipal bonds. Power can be sold to the local utility or used to displace other generation. An economic advantage of this structure is that the internal rate of return for municipal projects in generally lower than for private equity-driven ownerships; however, a major drawback is the inability of non-profit municipal entities to collect the PTC [15].

4.5.2 Cooperatives (wind and solar)

Community wind and solar cooperatives involve members pooling their resources into an energy project for the purpose of personal consumption of the energy generated. Energy can be delivered to the members either through the cooperative or through an agreement with the local utility. Most cooperatives are non-profit and offer annual dividends to their members proportional to consumption.

4.5.3 Private placements (wind and solar)

A private placement is a transaction in which a company raises money directly from private investors. The private placement structure involves the project sponsor selling equity shares of the project to local investors. Cash and tax benefits are distributed to owners in proportion to their investment. Typically in this model, about half of the project's construction cost is financed through the private placement of equity with the remainder being financed through a bank loan. Although this model has been successful in some instances (as will be discussed below in the MinWind case studies), it faces the challenge of finding enough investors with sufficient passive income to adequately capitalize on the tax credits [3].

4.5.4 Private equity (wind and solar)

A private equity fund is a collective investment scheme used for making investments in various equity securities. A private equity investment will generally be made by private equity or venture capital firm. Several global equity firms are now focused in renewable energy such as JCM Capital, Hudson Clean Energy Partners, Riverstone, Global Energy Efficiency, and Renewable Energy Fund. The advantage of these funds is the ability to access a large, diverse group of investors. These funds have typically been used to invest in large-scale renewable projects; however, in theory, they can also be used to fund community projects and an example of this is discussed below.

4.5.5 Multiple local owner (wind and solar)

As the name implies, this model consists of multiple owners who solicit investments from the local community through the formation of an LLC. Investors purchase shares and the LLC obtains debt from a local financial institution. The power from the project is sold to the utility through a power purchase agreement (PPA) and the income and incentives or credits from the project are split amongst the investors, proportional to their initial investment [15]. In theory, this is a simple and straight forward financing mechanism; however, in reality, there exists barriers to widespread implementation. The major barrier to this structure is that the shares being offered are likely to be considered 'securities' which would need to be registered at a federal level, a costly process. Exemptions from this are possible in some jurisdictions; however, the exemption process itself can be costly and lengthy. Additionally, the PTC, which is often relied upon to make any wind project profitable, has qualifications, one of which being the requirement that investors having passive income, which can be challenging to meet with this model.

4.5.6 Flip structures (wind)

The most common financing mechanisms for community wind have been partnership flip structures and private placements [3]. The objective of flip structures is to bring in a tax-motivated equity partner to capitalize on incentives such as the PTC that might otherwise not achieve their full value in other financing mechanisms. Flip structures involve a partnership between the local project sponsor and a tax equity investor. In the most common flip structure, the 'Minnesota-style flip', the investor provides the majority of the equity for the project (~99 per cent) and for this, is initially allocated 99 per cent of the cash and tax benefits generated [3]. Benefits can include revenue from the sale of power and from RECs as well as receipt of grants or tax benefits, for example the ITC or PTC, if applicable. At some point, typically 10 years after commercial operation, when the PTC expires, the ownership 'flips' to 1 per cent corporate and 99 per cent local. After this point, the local partner essentially owns a debt-free wind project, having only contributed 1 per cent of the capital, which should continue to be profitable for at least another 10 years.

There are also hybrids or variation of the types of structures discussed above. The 'Wisconsin-style flip structure' has been referred to as a hybrid between multiple local owner and the Minnesota-style flip [15]. In this model, multiple local investors provide debt financing to the wind project and pool capital to cover 20 per cent of the capital costs. This mechanism is structured more like a loan, and income is in the form of interest payments. The local investors buy out the corporate investor's stake in the project at the end of the initial 10 years, similar to the 'flip' in the Minnesota-style flip structure.

4.5.7 On-site projects, behind the meter (wind and solar)

On-site projects differ from those discussed above as they intend to provide power directly to the farm or community rather than selling to a utility or corporation. In this model, the end-use electricity user finances and interconnects one or more wind turbines on their side of the meter to supply on-site power. One major financial barrier to this model is that the electricity savings that are achieved actually reduce the amount of utility expenses that a farm or business can deduct as an expense, thereby substantially reducing the value of the electricity savings [15]. Some institutions are tax-exempt, for example schools, and are therefore better candidates for on-site projects. Another barrier is that these projects do not qualify for some incentives, such as the PTC, since the power is not sold to a third party. Furthermore, demand and standby charges may apply depending on the jurisdiction.

4.5.8 Utility-sponsored model (wind and solar)

The utility-sponsored model involves the utility owning or operating the project with voluntary ratepayer participation. One advantage of this model is that the utilities have the financial and physical infrastructure in place to handle and implement projects quickly [11]. Community members can then participate by paying an up-front or ongoing payment and, in exchange, receive payment or credit on their bills proportional to their contribution.

4.5.9 Special-purpose entity (wind and solar)

This model involves community members forming a business enterprise for the purpose of developing a community shared solar project [11]. These projects are owned by the participating community members. It is up to the member to raise capital and negotiate agreements between land owners and utility companies.

4.5.10 Non-profit model (solar)

The non-profit involves a charitable non-profit organization administering a community solar project on behalf of its members [11]. The donors will pay for the construction of the project but do not receive direct benefits. The benefits are typically meant to be credited to schools, churches, and low-income communities. Donors can receive indirect benefits such as tax benefits. Although it is possible for this model to be applied to community wind as well, it so far has been successful in community solar due to the smaller scale (lower investment) nature.

4.6 Innovative financing models – case studies of community wind

Developers have made use of creative financing and fundraising to bring wind power to communities. The North-American-based cases discussed below have unique attributes that are worth highlighting. Case 1 is unique in the multiple avenues that were explored to piece together the final financial package and also in the variation of the flip model that was eventually applied. Case 2 discusses how the MinWind projects were financed with reliance on local debt and equity and without reliance on the PTC. Case 3 provides an example of the first sale/leaseback financing of a wind project. Case 4 discusses details on how Iowa is powering some of their schools with wind. Case 5 discusses the application of crowdfunding to community wind with a case study from South Dakota. Case 6 highlights how business models are being developed for the purpose of encouraging community investment in utility-scale wind. In Canada, the COMFIT program in Nova Scotia is highlighted, and Case 7 discusses the financing mechanism of this programme as well as how nearby projects can benefit from some cost sharing. Case 8 focuses on the financing model of a public–private partnership that was applied in Alberta. Case 9 demonstrates how financing in the absence of incentives is possible when back by long-term PPAs. Lastly, Case 10 highlights Canada's only fully community-owned wind turbine, the WindShare in Toronto, Ontario.

4.6.1 Cases from the United States

4.6.1.1 Case 1: PaTu Wind Farm, LLC

The developers of this 9-MW project formed the Oregon Trail Wind Farm, LLC and partnered with MMA Renewable Ventures to establish a partnership formation of the PaTu Wind Farm, LLC [4]. The original financing plan was to use a 'flip' structure for financing and also take advantage of the PTC. However, unfortunately in 2008, when the project was being developed, there was uncertainty around the availability of the PTC which resulted in one of the primary tax equity investors pulling out of the project. Additionally, the financial crisis of 2008 resulted in MMA Renewable Ventures being sold to another company. The Oregon Trail Wind Farm, LLC bought back their assets from the new company, abandoned the 'flip' structure, and forged forward with innovative financing to complete their development [4].

Initially, partial funding was obtained through Oregon's Energy Loan Program which provided a \$12 million (~40 per cent of the projected \$23 million project cost), 20-year term loan at an attractive interest rate. Additionally, the Oregon Trail Wind Farm, LLC, was able to retain a pre-certified application for Oregon's Business Energy Tax Credit (BETC) which provided a 50-per-cent state ITC which they could choose to take as an up-front, lump sum, discounted 'pass through' cash payment upon commencing operation [3]. The 'pass through' option allowed the owners to transfer their tax credits to a pass through partner in return for a discounted, lump sum, cash payment. The project was able to secure a 20-year fixed-PPA with Portland General Electric for energy only, with PaTu retaining the RECs. The benefits of the tax credit and RECs, however, would not be realized until operation, which left PaTu with substantial equity and debt financing still to fulfil [3].

The project acquired \$5.685 million in equity financing from Vert Investment group and two wealthy families [3,4]. A unique flip-style structure was applied here. The equity partners start with a 99-per-cent interest in gross income and a 100-per-cent interest in cash distributions. Two flips then occur – one down to 90 per cent once the investors have earned back their principal, and another down to 50 per cent once they have achieved their internal rate of return. The Oregon Trail Wind Farm, LLC has the option to buy out the investors after year 5 of the project. To complete the funding requirements, a construction loan for \$16.5 million came from CoBank, a cooperative bank part of the Farm Credit System. CoBank's comfort providing this loan came from knowledge of the multiple sources of income that would flow into the projects once operations began.

4.6.1.2 Case 2: Projects resulting from MinWind

Minnesota has been leading the way for community wind in the United States. In addition to having good wind resources and motivated local developers, there have been several state-based programmes that can be credited for the success of community wind in this state. From a financing perspective, Minnesota has a small wind tariff and standardized PPA, a 10-year production incentive, as well as a Renewable Development Fund [9].

MinWind Energy is regarded as a pioneer of community wind. The company is based in Southwest Minnesota and has relied solely on local debt and equity to build several community wind projects. MinWind projects are primarily farmer owned and rely on local investors along with financing from a local bank. After debt repayment, all project returns flow to local entities. The MinWind projects consist of nine farmer-owned wind projects near the town of Luverne in southwestern Minnesota. The nine projects are organized as LLCs and have the stipulations that all shareholders are Minnesota residents (85 per cent must be from rural communities) and that ownership is limited to 15 per cent per project for each investor [9]. The MinWind projects also were able to take advantage of renewable energy grants from the US Department of Agriculture (USDA). A unique element of the MinWind projects was that they were not reliant on the PTC for financial viability. Long-term PPAs for the projects were negotiated with Xcel Energy, another critical factor in the success of these projects.

4.6.1.3 Case 3: Ridgewind, Pipestone, Minnesota

The Ridgewind project is a C-Bed project consisting of 11 Siemens 2.3-MW turbines along the Buffalo Ridge. The project involves 17 landowners, renters, and tenant farmers. The 25.3-MW project, developed by the Project Resources Corporation (PRC), is an example of the first sale/leaseback financing of a wind project [3]. In the past, the PRC relied on flip structures to finance their projects. However, flip structures often rely on foreign investors and they were looking for a way to stick with local investors and keep landowners more connected with the project.

The power from the project along with the RECs was sold to Xcel Energy under a 20-year contract. Union bank provided US$51 million in construction financing which was repaid after the commencement of commercial operations to a Union bank affiliate [3]. Ridgewind then leased the project back for a 20-year term during which they will manage and operate the project. At the end of the lease term, Ridgewind will have the opportunity to buy back the project's hard assets. Once the project is operating, the PRC expanded community ownership by implementing its Minnesota Windshare Program [38]. This is another unique element of this deal – rather than raising capital from local investors prior to construction (and having these investors exposed to construction risk), the offering to local investors was made once the project is operational.

The lease financing structure applied here is attractive due to the relative simplicity of the deal owing to their being one entity for both construction and permanent financing; however, there are also drawbacks due to issues of performance risk. In this particular case, the fact that Ridgewind used Siemens, a 'tier 1' turbine supplier, likely increased Union Banks confidence in the performance of the turbines, thereby increasing their confidence that the lease payments would be made on time.

4.6.1.4 Case Study 4: Iowa – powering schools with wind

Iowa has been a leader in powering schools with wind energy. A 2006 report details how ten schools in Iowa developed and financed wind energy [39]. The majority of these developments relied on no interest loans from the Iowa Energy Center's (IEC) Alternative Energy Revolving Loan Program (AERLP). In addition to AERLP loans, financing for these school-based wind projects came from Energy Bank approved loans, grants from the US Department of Energy, Funds also came in the form of donations and fundraisers, and in one case, turbines were gifted to the school by a local family. The unique element in most of these deals was the availability of the AERLP, a state specific programme.

4.6.1.5 Case 5: Crow Lake Wind Project, South Dakota

A very unique case can be illustrated with a portion of the funding for Basin Electric Power Cooperative's Crow Lake Wind Project located just east of Chamberlain, South Dakota [7]. The entire 162-MW, US$363 million project consists of 108 GE 1.5-MW turbines and represents the first-of-its-kind community wind investment partnership. Seven of these turbines (10.5 MW) are owned by local community investors. In order to raise funds to be able to add more wind power to their portfolio, the cooperative formed the South Dakota Wind Partners (SDWP) with the South Dakota Corn Utilization Council, the South Dakota Farmers Union, and the South Dakota Farm Bureau. The process began with the partnership contributing $80,000 in seed capital and was followed by a 'crowd-funding' model with the SDWP offering the South Dakota residents an opportunity to invest in the project, resulting in US$16 million in investment and equity, funded mostly by residents, towards the US$21.5 million portion of the project. Shares were sold for US$15,000 and offered a 7-per-cent return, with additional accelerated depreciation benefits for farmers leveraging assets. Additional financial support for the project came from Chevrolet, who purchased RECs from the partnership as part of its Chevy Carbon Reduction initiative and was also supported by a $6.7 million grant through the 1603 Program, a tax incentive administered by the Treasury Department. This innovative, community supported project, representing the largest wind project in the US owned by a co-op, earned the title of Wind Cooperative of the Year Award.

4.6.1.6 Case 6: Hale Community Wind, Texas

Tri Global Energy developed a proprietary business plan, the Wind Force Plan, with the objective to achieve utility-scale wind while encouraging direct equity participation from landowners and community members [40]. Although the mechanism of this model is proprietary, the objective is to allow local landowners and community investors the opportunity to partner with and have a substantial ownership in wind projects in their community. The model has been applied in Texas and is responsible for several wind projects, most notably the Hale Community Wind Farm in Hale County, Texas.

The Hale Community Wind Project in Texas, developed by Tri Global Energy and later sold to NextEra Energy, is the largest community-sponsored wind development in the world. With the potential to generate 1,200 MW of capacity upon project completion, this project encompasses 122,312 leased acres with more than 350 landowners [40]. Hale Community Energy, LLC, is joint venture comprised of Hale County Wind Farm, LLC, Cotton Wind Farms, LLC, Lakeview Wind Farm, LLC, and East Mound Renewable Energy Project, LLC, for which Tri Global Energy, LLC is project developer. Although the exact financing details are part of the proprietary Wind Force Plan, the success of this project has been attributed to the US$7.3 million that was put into the project by landowners and local investors. The project began construction in December 2014, in order to take advantage of appropriate federal tax credits, and is expected to be completed by November 2016. Hale Community Energy is unique in that it is able to supply two major national grids, the Energy Reliability Council of Texas and the Southwest Power Pool.

Smaller US community wind projects have also contributed unique ideas to the growing options for financing mechanisms. Innovative, yet complex, financial structures such as the inverted lease structure and use of NMTCs used to finance the 6-MW Coastal Energy Project in Grayland, Washington [3] and the unique combination financing of the 4.5-MW Fox Islands project off the coast of Maine serve as interesting examples of creative financing to push projects into financial reality. The specifics of these deals are not discussed in detail here due to the limited potential replicability of these project specific transactions.

4.6.2 Cases from Canada

4.6.2.1 Case 7: Millbrook Community Wind and Truro Heights Community Wind Projects, Nova Scotia

Some provinces are developing their own programmes to encourage community wind development. In Nova Scotia, the Community Feed-In Tariff (COMFIT) program, which began issuing approvals in 2011, offered a guaranteed rate per kilowatt hour for local renewable energy that enters the province's electrical grid [41]. Millbrook Community Wind and Truro Heights Community Wind are two success stories of the COMFIT program. The 6-MW Millbrook project operates on a site in a community of Millbrook, Nova Scotia and is co-located with the 4-MW Truro Heights project. The Eskasoni and Millbrook First Nations native communities teamed up on these projects and developed the Truro-Millbrook Wind LP [14]. The partnership will also involve Juwi Wind Canada, the Canadian arm of a German energy developer who will provide the development capital, and Community Wind Farms Inc. The costs of the project were lowered by partnering together and sharing some of the costs including road construction and environmental assessments. Furthermore, the projects were economically feasible due to the feed-in tariff programme provided by COMFIT which both projects received approval for. Over 60 projects, totalling 200 MW, were developed and commissioned through the COMFIT program, including the 10-MW Chebucto Pockwock Community Wind project in Halifax, Nova Scotia [42], and the 4-MW Whynotts Wind Farm close to Bridgewater, Nova Scotia [43]. The programme is no longer accepting new applications due to the upward pressure on power rates that was resulting due to the number of renewable projects coming onto the grid. Even though the programme is currently not available, the concept deserves attention as it resulted in several successful projects and has elements of replicability. These serve as examples of how government funding mechanisms can drive projects into the economically feasible and ultimately result in increased renewable energy generation.

4.6.2.2 Case 8: Box Springs Wind Farm, Alberta

The 6-MW Box Springs Wind project in Western Canada was financed through a public–private partnership between the city of Alberta and the WindRiver Power

Corporation, which means that the $12 million project costs were not paid for by tax dollars [44]. The land is owned by the city of Medicine Hat, and Wind Power has a long-term lease with the city to place the turbines on the land. The city will benefit through land lease payments and taxes. The project will operate under a long-term PPA between Medicine Hat and Box Springs Wind Corp., a subsidiary of WindRiver Power Corporation. Medicine Hat will receive all of the energy generated by the project along with the associated carbon credits. When the lease expires, the turbines can either be torn down, at WindRiver's cost, or the city can take them over, which they hope to do.

4.6.2.3 Case 9: Frampton Community Wind, Quebec

Even in the absence of grants and incentives, community wind projects can be financed relying solely on loans that are backed with a long-term PPA. The Frampton Wind Farm in Quebec, Canada is a community wind project partnership between the Frampton Municipality (33.3 per cent) and Boralex, Inc. (66.7 per cent) [45]. The 24-MW wind farm was financed with a C$73.5 million construction loan, to be converted to a term loan at the start of commercial operations, and short-term bridge financing totalling C$7.9 million. A 20-year PPA with Hydro-Quebec provides security to lenders, in this case, the National Bank of Canada.

4.6.2.4 Case 10: WindShare, Ontario

Canada's only fully community-owned wind project is the WindShare turbine, located at the Exhibition Place in Toronto, Ontario. The Toronto Renewable Energy Co-op (TREC) initiated the wind turbine project and partnered with Toronto Hydro to build the first urban-sited turbine in North America and the first community-owned wind power project in Ontario [46]. WindShare has the capacity to power up to 250 homes and is structured as a community-based cooperative. Community members invested in WindShare out of concerns for air quality and climate change and as a means to promote wind energy. The energy produced from the turbine enters the Ontario electricity grid. For their investment, members receive an annual dividend payment for the power that the turbine produces. TREC is another resource for developers in Ontario, providing services for community energy developers and tracking community investment opportunities.

4.6.2.5 Case 11: Gunn's Hill Wind Farm and the Oxford Renewable Energy Co-operative

The Oxford Renewable Energy Co-operative (OCEC) was established for the purpose of investing in Gunn's Hill Wind farm in Ontario which is currently being constructed [47]. This project serves as a successful case study of both the FIT program in Ontario and the co-operative financing structure. Residents of Oxford County, and all of Ontario, are invited to become members of this co-operative that is responsible for partial financing of the 18-MW farm. The OCEC was conceived as a partnership between Prowind Canada, Ontario Sustainability Services, and IPC Energy and was incorporated as a 'For-Profit Co-operative with Share Capital'. OCEC is a limited partner in the Gunn's Hill Wind Farm LP [47]. Prowind Canada acts as the general partner and a third limited partner is the Six Nations of the Grand River. The general partner (Prowind) will secure 75 per cent of the project costs from Senior Lenders as a long-term loan and 25 per cent equity investment from the limited partners. As a limited partner, OCEC is aiming to raise up to 49 per cent of the equity portion and a minimum of 10 per cent. OCEC has developed a simple investment structure consisting of one class of shares (dividends expected in the range of 10 per cent annually) and one class of bonds (10-year maturity, fixed rate of 5.5 per cent) [47]. This project also benefitted from securing a 20-year FIT contract for $0.148/kW h.

4.6.3 Discussion on replicability and challenges

To date, there isn't a one-size-fits all financing package for wind project development. Each project will have unique requirements and access to incentives and programmes. Community wind projects have typically relied on some type of flip structure, which is desirable due to the ability of such a structure to capitalize on tax incentives and also allows for ownership to flip back to the community at a future date. Flip structures have been tailored to meet the needs of the project, such as in the case of the PaTu Wind Farm. Flip structures have been a popular financing mechanism for both community as well as commercial wind projects and have proven to be replicable to several types of wind projects.

Projects that rely on a co-op structure also have elements of replicability. Co-ops allow for members to invest in their community, which acts as a means of raising capital, and members receive dividends. This structure has also realized success both globally, and more recently in North America as discussed in the WindShare and Oxford Community Energy Co-op cases, and should be considered as a viable structure for future developments.

It may also be of value for developers to consider cost sharing with nearby projects when possible. Piggybacking on other projects can help ease some of the financial burdens on development as was demonstrated by the Millbrook and Truro Heights case studies as well as the Crow Lake Wind project.

Some the cases presented below relied on complex structures consisting of multiple financing mechanisms. These may be less replicable due to possible high transaction costs and the time required to piece these packages together. Furthermore, some cases relied on wealthy families, such as the PaTu case, or donors, such as in the Iowa school cases, to overcome high capital costs which cannot be counted on when considering possible financial mechanisms.

The sale/leaseback structure in the Ridgewind project can possibly be replicated and is desirable due to the simplicity of the being one entity for both construction and permanent financing; however, there were some unique attributes of the project that lead to success such as the project relying on tier 1 suppliers in order to increase the lender's confidence in turbine performance.

Crowdfunding has the potential to fund community wind projects and has been proven to work overseas as well as in a US-based project. The crowdfunding model is simple and can provide intrinsic value to the members who are contributing to renewable energy in their community. Companies are also realizing the benefits of and interests in community wind and are developing specific business models, such as the Wind Force Plan, to act as a financing model geared towards community involvement in wind project development.

The states that are leading the way for community wind typically have some type of community wind collaborative to provide development services to interested owners. Most of the cases depended on a federal, state, or provincial programme and, therefore, are only replicable where such programmes exist. Feed-in tariffs, tax incentives, RECs, and renewable energy loans were major drivers of financial success. For example, there are several programmes in Minnesota which have been credited for the success of the MinWind as well as other community wind projects in Minnesota. The MinWind projects each received renewable energy grants from the USDA. Similarly, the COMFIT program provided support for many successful community wind projects in Nova Scotia including the Millbrook and Truro farms projects highlighted here.

Many factors will affect replicability including the availability of incentives and the ability to use them, grant and policy programmes, jurisdictions, regulations, project size, developer objectives, and preference for individual or co-operative ownership. Developers will have to carefully assess these factors when selecting a financing structure that works for their project.

Nearly all recent community wind literatures identify common barriers to further community wind. These include market incentives and barriers, developer incentives, and cost-effectiveness [12]. Where sales of securities are involved, high administrative costs may apply. Identifying and engaging equity partners, finding a willing power purchaser, and negotiating a PPA can also pose challenges. Furthermore, when incentives are available, their use may be limited due to inabilities to utilize tax credits and successful programmes being cancelled due to upward pressure on rates, as in the case of COMFIT in Nova Scotia. Grid connection can also be a barrier to further project development. Long project application and development lead times present a challenge to raising local capital.

4.7 Innovative financing models – case studies of community solar

Historically, of the financing models discussed above, three of them have been the main project models for community solar: utility-sponsored, special-purpose entity, and non-profit, each with unique costs and benefits associated with them. Examples of each of these types along with some innovative models will be presented in the case studies below. Cases 1 and 6 are variations of the utility-sponsored which involves the utility owning or operating the project with voluntary ratepayer participation. Cases 2 and 3 are examples of the special-purpose entity model, with Case 3 having a unique REC component. Case 4 is a non-profit example involving charitable donations from the community towards a community solar project. Case 5 highlights one project of many that have been developed through the Clean

Energy Collective, an example of collective ownership through a utility. Case 8 is a unique example of a large-scale project funded entirely through government grants, and Case 9 highlights SolarShare – Canada's leading renewable energy co-op.

4.7.1 Cases from the United States

4.7.1.1 Case 1: Sacramento Municipal Utility District – SolarShares Program

An example of the utility-sponsored model is the SolarShares Program offered by the Sacramento Municipal Utility District (SMUD). SMUD has been a pioneer in community solar, building the first utility-scale solar array in the United States in 1984, generating enough electricity to power 2,200 single-family homes [48]. Through its SolarShares Program, SMUD provides customers the option to purchase solar energy directly from SMUD and achieve net metering benefits that are similar to behind-the-meter PV projects. SMUD purchases the output of local, community-scale PV systems under 20-year PPAs and then resells the energy to local participating customers. Subscribers to the SolarShares Program pay a flat monthly fee, based on the share size they select (from 0.5 to 4 kW), for a portion of solar power that is produced at a local solar farm. Their monthly bills are then offset by a credit that will vary each month depending on the solar energy being generated.

The programme began in 2008 with a 1-MW system constructed by enXco who handled the capital financing of the project. EnXco is entitled to the ITC generated by the project and SMUD purchases 100 per cent of the output and retains the RECs [48]. SMUD is planning for expansion up to 25 MW. Many similar utility-sponsored community solar programmes are available with the same premise: the utility funds the project development and then sells shares in the programme to recover the upfront costs and provide residents with solar energy.

4.7.1.2 Case 2: University Park Solar, Maryland

The residents of a Maryland community formed a for-profit private membership LLC to develop a 22.7-kW solar power generation site in their community [49]. Solar panels were installed on the host site, University Park Church of the Brethren by Standard Solar Inc., who also maintains the panels. Members could buy into the projects starting at $2,000.00 and membership proceeds funded the purchase and installation of panels. Members sell the power generated to the church at rates slightly lower than the utility rate and excess power is sold back to the grid at wholesale prices. Revenue leftover from the sale of energy and RECs, after O&M costs, are divided amongst the members with members expecting to receive an approximate 8 per cent return on investment over 20 years [49].

4.7.1.3 Case 3: Boardman Hill Solar Farm (BHSF)

This 150-kW AC project is one of the largest single-phase PV arrays in Vermont and is entirely owned by 30 Vermont families and small businesses who purchased shares in the project [50]. This project has been recognized for achieving low costs without selling their RECs. Additionally, the $512,000 project was financed without any third party. This project was carried out through the formation of a non-profit BHSF LLC which enabled the community to take responsibility for the project, including financial and administrative matters. The $512,000, or $2.73 per watt, did not account for site land lease, budgeted at 5 per cent of the project's output, or taxes and insurance. To keep prices low, interested participants saw value in selling the RECs that the project would generate. However, the only available land for the project was from an organic farmer who stipulated that if the project was on his property, that it would be truly green renewable and therefore, that the RECs were retired and not sold. The members of the LLC came up with an innovative solution. They contracted with the land owner to pay the land lease through energy credits to the landowner's electricity bill from the output of panels bought by the solar farm. The landowner still receives the 5 per cent over the lifetime of the project and the costs for the members decreased enough to compensate for not selling the RECs. This case study is an important contribution to community renewable financing as it demonstrates the ability of a project to be economically feasible without selling RECs. The financial success of the project can be attributed to the member-managed, non-profit LLC and that the panels were financed entirely by panel owners. The simple approach of this project has been accepted by the Institute for Energy and the Environment at Vermont Law School and Vermont Natural Resources Council for replication and expansion to other communities [50].

4.7.1.4 Case 4: Solar for Sakai, Washington (Non-profit)

An example of a non-profit model is the 5.1-kW Solar for Sakai project on Bainbridge Island, Washington. The project was funded by a $25,000 grant from Puget Sound Energy's Solar4R Schools programme and from 26 private, tax-deductible donations totalling approximately $30,000 [51]. The local contributing citizens do not receive payment for the energy produced by the system but do qualify for a tax deduction as their contributions were to a non-profit entity.

4.7.1.5 Case 5: Clean Energy Collective (CEC)

The Clean Energy Collective (CEC) has pioneered an innovative community solar concept. The CEC model finances solar power through collective ownership by participating utility customers. The CEC was the first to propose the 'community-owned solar garden' concept. Their first project was a 77.7-kW system in Colorado, developed by the CEC and collectively owned by 18–20 customers of Holy Cross Energy, the local electric cooperative [11]. The cost of this system was $466,000. Capital was raised by selling individual panels at a cost of $725. The owners purchased as little as one and up to 80 of the 340 panels in total. Owners received 11 cents/kW h produced by their share of the panels under a PPA negotiated with Holy Cross. Since then, the CEC has expanded and currently has 56 projects across 15 states, totalling 51 MW. The community solar arrays are centralized solar facilities that are owned by community members who receive credits on their electricity bill for the power produced, capitalizing on the power of buying in bulk.

Similar volume purchasing programmes have been successful such as Solarize Portland, which ran from 2009 to 2013, bringing solar energy to over 1,000 Portland homes, and has now been revived and expanded to Connecticut [52]. This programme works by bringing down the prices by buying in bulk from pre-selected installers to lower overall costs. Additionally, the collective knowledge and support helps to streamline the process for interested community members. Financing of the panels is still paid for by individual community members, but the cost of financing is brought down making solar more accessible in the community.

4.7.1.6 Case 6: Solar Pioneers II, Oregon (Utility-Sponsored)

The 63-kW community solar project was partially financed through a shared ownership programme through Ashland's Municipal Utility. Members made upfront purchases in 1/4, 1/2, or full solar panel increments, at a price of $825 per panel, and in return receive payment for the corresponding energy and RECs produced over 20 years [51]. The cost of unsold shares was absorbed by the utility. Ashland was approved to use Clean Energy Renewable Bond (CREB) financing and also took advantage of the Oregon BETC pass-through option. Specifically, the Bank of the Cascades purchased the 35-per-cent tax credit (over 5 years) in return for a one-time payment to Ashland equal to 25.5 per cent of the system cost, which lowered the installed cost from $420,000 to $312,900 [51].

4.7.1.7 Case 7: Fort Dix, New Jersey (Crowdfunding)

The Fort Dix Solar project is an example of a successful project that was financed through crowdfunding. In this case, the community turned to a company, Mosaic, to coordinate the crowdfunding of the project. The project is a 12.3 MW solar project that powers a military housing project in Fort Dix, New Jersey [8]. For a 1-per-cent fee, Mosaic allows people to invest in the project for as little as $25 and coordinates the investors and payments. Mosaic has been responsible for 21 community solar energy projects and has been referred to as a pioneer in crowdfunding renewable energy [8]. The system, using 55,189 panels, is expected to output 13.82 GW h annually.

4.7.1.8 Case 8: Elsie Whitlow School, Washington, DC [Property Assessed Clean Energy Program (PACE)]

Property Assessed Clean Energy (PACE) programmes offer long-term private financing for renewable energy projects. PACE programs are available in 19 states, with residential programmes currently limited to California, Florida, and Missouri [53]. The programme pays for 100 per cent of a project's costs and is repaid over a period of up to 20 years as payments added to the property's tax bill. PACE can also be combined with utility, local, and federal incentive programmes. An example of successful implementation of this programme is the 35-kW rooftop solar PV project at Elsie Whitlow Stokes Community Freedom Public Charter School [53]. The capital for the project was provided by Greenworks Lending, a Connecticut-based

PACE lender and will be repaid at a fixed rate over a 20-year term. The project also received an incentive from the DC Sustainable Energy Utility and also qualifies for Solar Renewable Energy Credits. The project expects to realize $9,800 in annual savings, taking into account their annuals PACE payments.

The PACE program has been credited for several other successful projects, such as the 56.4-kW solar PV project at Temple Israel in Cortland, New York and another solar project at Ivy Knoll Senior Retirement Community in Covington, Kentucky [53].

4.7.2 Cases from Canada

4.7.2.1 Case 8: Drake Landing, Alberta

The Drake Landing community solar project in the town of Okotoks, Alberta, is a first of its kind planned neighbourhood that uses solar energy to meet over 90 per cent of the heating needs of the participating residents [54]. The 52 house sub-division has its space and water heating needs met by 800 garage mounted panels throughout the community which generate 1.5 MW of thermal power during a typical summer day. Short-term thermal storage and borehole thermal storage facilitate collection and storage in the summer for use in space heating in the winter. Funding for this innovative project was provided by three subsidization sources: The Government of Alberta (C$1,125,000), the Government of Canada (C$2,600,000), and the Federation of Canadian Municipalities (C$2,900,000) [50]. Since the costs of this project were fully funded, there was no need to obtain financing or collect memberships.

There are a number of additional examples of successful community solar programmes that are similar to those described here. Community solar is currently more widespread than community wind due to the scalability of solar projects to meet the needs of the community members.

4.7.2.2 Case 9: SolarShare Canada

SolarShare is Canada's leading renewable energy co-op with over 1,275 members [55]. The co-op develops commercial-scale solar energy installations and allows community members to purchase shares in these projects. The projects range from 10 to 600 kW arrays with a combined portfolio of over 6 MW of installed capacity. The Goodmark project near Toronto is an example of a successful project. The 133 kW project was developed by RESCo Energy and cost a total of C$690,611. Annual revenue from the project is C$94,841 and members can expect a 5–6 per cent return on their investment [55].

4.7.2.3 Case 10: Solar Portfolio in Ontario, Canada

Nautilus Solar Energy LLC completed a 4.1-MW solar portfolio project, consisting of 14 rooftop installations, in Ontario in 2015, with plans to expand to 24 MW [56]. This case study highlights the involvement of a private equity firm in a community energy project. Private equity financing was secured from North Sky Clean Tech

Alliance Fund LP and NewWorld Environmental Infrastructure LP. Community investment partners included Green Energy Cooperative of Ontario Inc., and Eagle Lake First Nation. The projects secured 20-year power purchase contracts under the Ontario Power Authority's feed-in-tariff programme. This project represents one of the largest financings of a distributed solar portfolio.

4.7.3 Discussion on replicability and challenges

The solar projects have good replicability due to their more widespread use and scalability. The three main financing models are, in general, replicable; however, some of the case studies did rely on incentives or one time grants to achieve success. The utility-sponsored model has been successfully executed in many different areas. This model is favoured from a replicability perspective because the utilities generally have similar infrastructure and programme management and the model is simple in that the utility purchases the power, retains any RECs, and then resells to the community. Since the utility has the capital funds and a tax appetite to take advantage of available credits, there is no requirement for complex capital funding arrangements. Communities have also achieved success replicating the co-operative model.

The projects that relied on incentives and subsidies to attain affordability are less replicable. For example, the Solar Pioneers II Program in Oregon made use of the CREB and the Oregon BETC which are limited to the location of the project. Similarly, the Drake Landing project was fully subsidized by three separate government sources, which cannot be expected to be available for future project financing. In summary, when evaluating replicability, the projects that did not rely on local incentives and subsidies are more likely to act as models for community energy financing.

Future community solar projects are can turn to crowdfunding for upfront costs, or more specifically can arrange financing through start-up companies who coordinate community crowdfunding and investment in projects. Mosaic, the company who helped enable the 12.3 MW solar project in Fort Dix, New Jersey is one such company, and others such as Village Power, a Palo Alto, California-based platform has also entered the renewable energy finance scene, allowing community organizations to fund and manage solar projects through investments from the local community. Private equity firms that, once focused on large-scale projects, are also now involved in community energy finance. In addition to the case presented, an additional 100 MW community-solar portfolio project focused in the Minnesota community solar sector and funded by North Sky Capital, a private equity firm, was recently announced [57]. The growth of renewable energy fund companies is likely to continue as more communities realize the environmental and economic benefits of renewable energy but lack funds and expertise to execute the deals. Involving a private equity firm with experience in energy financing also assists communities with the challenge of navigating energy transactions and markets as these firms will develop expertise as they execute more deals.

Another simple option is for communities considering solar projects to start work with their local utility. The utilities generally already have the legal, financial, and physical infrastructure in place and are willing to work with the community to provide an option for shared solar power. Taken together, these cases demonstrate the community solar is an economically feasible option across North America and regardless of the scale of the project.

4.8 Summary and conclusions

A summary of the case studies and project details that are discussed in this chapter is presented in Table 4.1.

Community energy is expanding in North America yet still faces challenges due to high capital costs, more so with wind than solar, associated with project development. The case studies presented demonstrate that community energy is feasible in North America and has many economic, environmental, and social benefits. Community energy projects are expected to continue to expand in North America. Particularly in remote off-grid communities where diesel generation and freight costs are high, renewable energy technologies will make economic sense [58]. Society engagement in carbon reduction initiatives and instability in energy markets could further drive interest in community renewable energy. Through incentives and creative financing, it is possible to increase community wind and solar in North America.

Developers can look to case studies, such as these, as resources when trying to piece together financing to make their projects possible. Although the models presented here will provide examples for future developers, models alone will not be enough. Looking at examples of where community energy has been successful, legal and financial government drivers are what fuelled community ownership elsewhere and are what will be necessary to push community wind and solar forward in North America.

These case studies highlight success stories. There are also numerous cases of failed community wind projects, many of dedicated significant resources and time to a project to have it halted. For example, SaskWind spent 4 years working on a Swift Current Community Renewables project consisting of 25 MW of wind generation and 10 MW of solar generation to be 100 per cent community-owned – the first of this scale in North America, to have SaskPower decide not to consider the project at the time [59]. Even with proven technology and complete financing packages, projects may not be realized.

Economics is an important consideration; however, there are several others. Selecting the best resources for a community energy project is an important first step. Geographic and climatic considerations will drive the decision between wind and solar as will the scale of the project, the availability of incentives and grants, and compatibility of the technology with the built environment in the community [22]. Furthermore, successful projects require innovation and community support.

The cases that were selected for this chapter were intended to capture a sample of projects spanning various geographies, scales, and financing structures. In addition to the demonstrated case studies, many models of how community wind

Table 4.1 Summary of case study project details

Project name	Project size	Date of operation	Location	Financing structure(s)	Loans/incentives/ programs
			Wind		
PaTu Wind Farm, LLC	9 MW	1 December 2010	Wasco, Sherman County, Oregon	Unique flip structure	Oregon's Energy Loan Program Oregon's Business Energy Tax Credit
MinWind	MinWind 1 and 2: 1.9 MW MinWind 3–9: 1.65 MW	2002–2004	Minnesota	Local debt and equity	Renewable energy grants from the USDA
Ridgewind	25.3 MW	December 2010	Minnesota	Sale/leaseback	Sales of RECs Minnesota WindShare Program
Iowa, Powering Schools	50 kW–1.65 MW	1991–2004	Iowa	Grants, donations, fundraising	Iowa Energy Center's (IEC) Alternative Energy Revolving Loan Program (AERLP) Grants from the US Department of Energy
Crow Lake Wind	10.5 MW	February 2011	South Dakota	Cooperative Crowdfunding	Section 1603 Grant Program Sales of RECs

Hale Community Wind	1,200 MW (potential)	November 2016 (expected)	Texas	Through the Wind Force Plan (exact details are proprietary) Landowner and local investors	Tax credits
Millbrook and Truro Heights Community Wind	6 and 4 MW	2014	Nova Scotia	LLC Partnerships	COMFIT
Box Springs Wind	6 MW	2012	Alberta	Public–private partnership	
Frampton Community Wind	24 MW	December 2015	Quebec	Construction loan and bridge financing backed by 20-year PPA with Hydro-Quebec	
WindShare	750 kW	February 2002	Ontario	Community-based cooperative	
Gunn's Hill	18 MW	Under construction	Ontario	Limited partnership/ cooperative	FIT Program
			Solar		
Sacramento Municipal Utility District SolarShares	1 MW (plans for expansion to 25 MW)	2008	Sacramento	Solar Utility Model	ITC, RECs

(Continues)

Table 4.1 (Continued)

Project name	Project size	Date of operation	Location	Financing structure(s)	Loans/incentives/ programs
University Park	22.7 kW	July 2010	Maryland	Community for-profit LLC	RECs
Boardman Hill Solar Farm	150 kW	October 2015	Vermont	Community LLC with REC retirement	ITC
Solar for Sakai	5.1 kW	2008	Washington	Non-profit	Grant, private donations
Clean Energy Collective	77 kW	2010	Colorado	Collective ownership through utility	Bulk purchasing
Solar Pioneers II	63 kW	2007	Oregon	Shared ownership through utility, REC payments	Clean Energy Renewable Bond, Business Energy Tax Credit
Fort Dix Solar	12.3 MW	2009	New Jersey	Crowdfunding	ITC
Elsie Whitlow School	35 kW	2016	Washington, DC	PACE Program	Incentives SRECs
Drake Landing	1.5 MW	2007	Alberta		Fully funded through subsidies: Government of Canada, Government of Alberta, Federation of Canadian Municipalities
SolarShare Canada	10–600 kW projects, totalling 6 MW	2011	Across Canada	Community-based cooperatives	FIT Program
Nautilus Solar Portfolio	14 projects, totalling 4.1 MW	2015	Ontario	Private equity	FIT Program

can potentially be implemented have been proposed [10]. By simply make developers aware of the potential mechanisms available, more innovative deals can be possible and more communities can reap the benefits of clean, reliable energy.

References

[1] American Wind Energy Association. *The Cost of Wind Energy in the United States* [online] 2016. Available from http://www.awea.org/Resources/Content.aspx?ItemNumber=5547 [Accessed 28 July 2016].

[2] Miller L., Carriveau R., Harper S., and Singh S. 'Evaluating the link between LCOE and PPA elements and structure for wind energy'. *Energy Strategy Reviews*. 2017;16:33–42.

[3] Bollinger M. *Community Wind: Once Again Pushing the Envelope of Project Finance*. LBNL-4193E, 2011.

[4] Yin Y. 'An analysis of empirical cases of community wind in Oregon'. *Renewable and Sustainable Energy Reviews*. 2013;17:54–73.

[5] Farrell J. *Community Solar Power – Obstacles and Opportunities* [online]. 2010. Available from https://ilsr.org/community-solar-power-obstacles-and-opportunities/ [Accessed 7 January 2017].

[6] Bolinger M. *Community Wind Ownership Schemes in Europe and Their Relevance to the United States*. LBNL-48357, 2001.

[7] U.S. Department of Energy. *Want to Finance a Wind Farm Project in Your Community? Try Crowdfunding* [online]. 2013. Available from https://energy.gov/eere/articles/want-finance-wind-farm-project-your-community-try-crowdfunding [Accessed 12 December 2016].

[8] Gilpin L. *How Crowdfunding Solar Power is Democratizing the Way We Finance Clean Energy* [online]. 2014. Available from https://www.techrepublic.com/article/how-crowdfunding-solar-power-is-democratizing-the-way-we-finance-clean-energy/ [Accessed 8 February 2017].

[9] Bolinger M.A. 'Making European-style community wind work in the US'. *Renewable and Sustainable Energy Reviews*. 2005;9:556–75.

[10] Bral D., Wang C., and Yeh C.-P. 'Community Wind Power – A Tipping Point Strategy for Driving Socio-Economic Revitalization in Detroit and Southeast Michigan, Wind Farm – Technical Regulations, Potential Estimation and Siting Assessment' in Suivre G.O. (ed.). *Wind Farm – Technical Regulations, Potential Estimation and Siting Assessment*. Rijeka: InTech; 2011, pp. 66–91.

[11] Coughlin J., Grove J., Irvine L., *et al. A Guide to Community Shared Solar: Utility, Private, and Non-Profit Project Development*. NREL Report DOE/GO-102012-3569, 2012.

[12] Walker G. 'What are the barriers and incentives for community-owned means of energy production and use?' *Energy Policy*. 2009;36:4401–5.

[13] Kellett J. 'Community-based energy policy: A practical approach to carbon reduction'. *Journal of Environmental Planning and Management*. 2007; 50(3):381–96.

[14] Alberstat J. *Eskasoni, Millbrook Join Forces on Wind Project* [online]. 2012. Available from http://thechronicleherald.ca/business/150515-eskasoni-millbrook-join-forces-on-wind-projects [Accessed 10 November 2016].

[15] Kildegaard A., and Myers-Kuykindall J. *Community vs. Corporate Wind: Does it Matter Who Develops the Wind in Big Stone County, MN*? IREE Report Grant No. SGP4c, 2004.

[16] Lipp J., and Dolter B. *The Power of Community* [online]. 2015. Available from http://www.trec.on.ca/report/the-power-of-community/ [Accessed 3 November 2016].

[17] Wustenhagen R., Wolsink M., and Burer M.J. 'Social acceptance of renewable energy innovation: An introduction to the concept'. *Energy Policy.* 2007; 35(5):2683–91.

[18] Walker G., and Devine-Wight P. 'Community renewable energy: What should it mean?' *Energy Policy.* 2008;36:497–500.

[19] Warren C.R., and McFadyen M. 'Does community ownership affect public attitudes to wind energy? A case study from south-west Scotland'. *Land Use Policy*. 2010;27(2):204–13.

[20] Beery J.A., and Day J.E. 'Community investment in wind farms: Funding structure effects in wind energy infrastructure development'. *Environmental Science and Technology*. 2015;49:2648−55.

[21] Maruyama Y., Niskikido M., and Iida T. 'The rise of community wind power in Japan: Enhanced acceptance through social innovation'. *Energy Policy.* 2007;35:2761–9.

[22] Ferrer-Marti L., Garwood A., Chiroque J., *et al.* 'Evaluating and comparing three community small-scale wind electrification projects'. *Renewable and Sustainable Energy Reviews*. 2012;16:5379–90.

[23] Bayar T. *Dutch Wind Turbine Purchase Sets World Crowdfunding Record* [online]. 2013. Available from http://www.renewableenergyworld.com/articles/2013/09/dutch-wind-turbine-purchase-sets-world-crowdfunding-record.html [Accessed 13 November 2016].

[24] Singh T. *German Village Produce 321 Per Cent More Energy than it Needs!* [online]. 2011. Available from http://inhabitat.com/german-village-produces-321-more-energy-than-it-needs/ [Accessed 1 February 2017].

[25] Stamford M. *Community ownership: The best was forward for UK windpower*? Master's thesis. University of East Anglia. 2004.

[26] Toke D. 'Community wind power in Europe and the UK'. *Wind Engineering*. 2003;29(3):301–8.

[27] Generation Community. *Newport Solar Project* [online]. 2014. Available from http://www.gen-community.co.uk/newport-solar-pv/ [Accessed 2 February 2017].

[28] Corbyn A. 'Small wind-turbine community-based renewable energy systems in the Philippines'. *Wind Engineering*. 2007;31(5):353–61.

[29] Bolinger M., Wiser R., Cory K., and James T. *PTC, ITC, or Cash Grant? An Analysis of the Choice Facing Renewable Power Projects in the United States*. NREL/TP-6A2-45359, 2009.

[30] U.S. EPA. *Green Power Partnership – Renewable Energy Credits* [online]. 2017. Available from https://www.epa.gov/greenpower/renewable-energy-certificates-recs [Accessed 27 July 2017].
[31] Community Renewable Energy Association (CREA). *8% Mandate for Community Based Renewable Energy* [online]. 2017. Available from https://www.community-renewables.org/oregon-s-rps-and-8 [Accessed 27 July 2017].
[32] U.S. Department of Energy. *Database of State Incentives for Renewables and Efficiency* [online]. 2016. Available from http://www.dsireusa.org/ [Accessed 20 October 2016].
[33] International Energy Agency. *IEA/IRENA Joint Policies and Measures Database* [online]. 2016. Available from http://www.iea.org/policiesandmeasures/renewableenergy/ [Accessed 20 October 2016].
[34] Natural Resources Canada. *Tax Savings for Industry* [online]. 2016. Available from http://www.nrcan.gc.ca/energy/efficiency/industry/financial-assistance/5147 [Accessed 15 November 2016].
[35] Federation of Canadian Municipalities. *Green Municipal Fund* [online]. 2016. Available from http://www.fcm.ca/home/programs/green-municipal-fund.htm [Accessed 1 November 2016].
[36] Prince Edward Island. *Renewable Energy Act* [online]. 2016. Available from https://www.princeedwardisland.ca/sites/default/files/legislation/r-12-1.pdf [Accessed 2 November 2016].
[37] Independent Electricity System Operator. *Feed in Tariff Program* [online]. 2016. Available from http://fit.powerauthority.on.ca/ [Accessed 2 November 2016].
[38] PRC Wind. *Minnesota WindShare* [online]. 2013. Available from http://www.prcwind.com/ [Accessed 3 December 2016].
[39] Galluzza T., and Osterberg D. *Wind Power and Iowa Schools* [online]. 2006. Available from http://www.iowapolicyproject.org/2006docs/060307-WindySchools.pdf [Accessed 15 November 2016].
[40] Tri Global Energy. *Unique Development Model: The Wind Force Plan* [online]. 2016. Available from https://www.triglobalenergy.com/tge_wind_projects [Accessed 12 November 2016].
[41] Nova Scotia Department of Energy, COMFIT [online]. 2016. Available from http://energy.novascotia.ca/renewables/programs-and-projects/comfit [Accessed 13 November 2016].
[42] Chebuckto Pockwork Community Wind [online]. 2013. Available from http://www.pockwockwindfarm.ca/ [Accessed 16 November 2016].
[43] Whynotts Community Wind [online]. 2014. Available from http://www.whynottswindfarm.ca/ [Accessed 16 November 2016].
[44] Dvorak P. *Gamesa to Supply 6 MW for Community Wind Farm in Alberta* [online]. 2013. Available from http://www.windpowerengineering.com/construction/gamesa-supply-6-mw-community-wind-farm-alberta/ [Accessed 10 November 2016].
[45] Boralex Inc. *Boralex: Financing of the Frampton Community Wind Project in Quebec* [online]. 2015. Available from http://www.newswire.ca/news-releases/boralex-financing-of-the-frampton-community-wind-project-in-quebec-520640332.html [Accessed 18 November 2016].

[46] Windshare. *Windshare Co-operative* [online]. 2016. Available from http://www.windshare.ca/ [Accessed 14 November 2016].

[47] Oxford Community Energy Co-operative Inc. *Executive Summary* [online]. 2014. Available from https://oxford-cec.ca/Resources/Documents/20140520%20OCEC%20investment%20summary%201.3.pdf [Accessed 5 November 2016].

[48] Sacramento Municipal Utility District. *SolarShares* [online]. 2016. Available from https://www.smud.org/en/about-smud/environment/renewable-energy/solar.htm [Accessed 16 January 2017].

[49] University Park Solar. *A For-Profit Private Membership for Solar Energy* [online]. 2015. Available from http://universityparksolar.com/index.htm [Accessed 16 January 2017].

[50] Vermont Energy and Climate Action Network. *Boardman Hill Solar Farm Case Study* [online]. 2015. Available from http://www.vecan.net/wp-content/uploads/BHSF-CASE-STUDY-FINAL-1.compressed.pdf [Accessed 11 January 2017].

[51] Bonneville Environmental Foundation. *The Northwest Community Solar Guide* [online]. 2013. Available from http://www.nwseed.org/wp-content/uploads/2013/05/NW-Community-Solar-Guide.pdf [Accessed 10 December 2016].

[52] Irvine L., Sawyer A., and Grove J. The Solarize Guidebook: A Community Guide to Collective Purchasing of Residential PV Systems. NREL Report DOE/GO-102012-3578, 2012.

[53] PACENation. *Building the Clean Energy Economy* [online]. 2017. Available from http://pacenation.us/ [Accessed 26 July 2017].

[54] Drake Landing Solar Community. *How it Works* [online]. 2016. Available from http://www.dlsc.ca/how.htm [Accessed 10 January 2017].

[55] SolarShares. *Solar Projects* [online]. 2016. Available from https://www.solarbonds.ca/all-projects/all-projects [Accessed 12 February 2017].

[56] Beetz B. *Ontario: Nautilus Solar Completes 14 Rooftop Solar Projects* [online]. 2015. Available from https://www.pv-magazine.com/2015/06/17/ontario-nautilus-solar-completes-14-rooftop-solar-projects_100019854/ [Accessed 11 July 2017].

[57] Ryan C. *US Solar Secures Financing for 100MW Community Solar Portfolio* [online]. 2017. Available from https://www.pv-tech.org/news/us-solar-secures-financing-for-100mw-community-solar-portfolio [Accessed 11 July 2017].

[58] Thompson S., and Duggirala B. 'The feasibility of renewable energies at an off-grid community in Canada'. *Renewable and Sustainable Energy Reviews*. 2009;13:2740–5.

[59] SaskWind. *Swift Current Community Renewables* [online]. 2016. Available from https://www.saskwind.ca/swift-current/ [Accessed 15 November 2016].

Chapter 5

Community-level solar thermal systems

Vishal Bhalla[1], Vikrant Khullar[2] and Himanshu Tyagi[1]

Abstract

This chapter provides a brief background in the various conventional methods (surface-based absorption of solar energy) available for harnessing solar thermal energy at a community level. It describes some of the main differences between the typical non-concentrating as well as concentrating type solar collectors, and highlights some of their main attributes. It also presents in detail the results of a novel technique for harnessing solar energy. In this technique, the solar energy is directly absorbed by the fluid (volumetrically) using nanoparticle-laden fluids (hence, it is categorized as volumetric-based absorption of solar energy). Such a collector is analysed in detail using a numerical model. The results of the numerical model are then discussed, which simulates the requirements of hot water for a typical community consists of about 10 households (40 persons). Two of the main performance evaluation parameters – collector efficiency and the fluid outlet temperature – have been extensively studied, and the influence of various design and operational parameters (particle volume fraction, mass flow rate, solar irradiation, collector height, collector length) on these two have been studied in detail. Moreover, the variation of spectral intensity, energy generation rate and spatial temperature distribution within the collector has been quantified. The calculations also show the dimensions of the desired solar collector in order to meet the daily hot water requirements (100 kg/person) for this community.

5.1 Introduction

Solar energy has been used in various forms in communities within India (and across the world) for a long time. Right from the antiquity where the sun was used for day lighting, navigational purposes to middle ages where applications such as drying, disinfection, curing evolved, till the present time where lot solar-driven processes are taking place in various segments of the society. For example, in

[1]Department of Mechanical Engineering, Indian Institute of Technology Ropar, India
[2]Mechanical Engineering Department, Thapar University, India

current times within the rural agrarian societies, the processing of food product (lentil seeds) and agricultural by-products (manure-cakes) is highly reliant on solar energy. Similarly, with the advent of solar thermal energy (STE)-based systems (water and air solar collectors), lot of residential and industrial demand for hot water is being met through solar energy. Even production of electricity has been taking place in large scale using STE-based systems [parabolic trough collectors (PTC) and central tower receivers] in addition to the more prevalent solar PV (photovoltaic).

When the use of solar energy is compared to the overall use of energy from all available sources an interesting picture emerges. Even more so, when this comparison is carried out over the last three to four decades. Energy has always been seen as driver of economical grow, and even more so since the advent of the industrial revolution. In the last two to three centuries, availability of energy sources has been seen as drivers of economical growth, and various developed countries have been heavily utilizing the main source of energy – fossil fuels. These fossil fuels mainly consist of coal (28.6 per cent share), petroleum products (31.3 per cent share) and natural gas (21.2 per cent share) [1]. When the three of these are taken together, they have been responsible for approximately 81.1 per cent [1] of the net global energy utilized since the beginning of the industrial revolution till the present times. When such figures are compared for individual sectors, this share may become even larger – for instance, in transportation sector, fossil fuels have been responsible for about 95 per cent share [2], in electricity production about 66.7 per cent share [1]. These figures of course vary from country to country and also with time. There are a few exceptions which stand out, for example France has relied heavily on nuclear energy for generation of electricity, Scandinavian countries have invested a lot in wind power, Iceland in geothermal energy and recently lot of countries (Spain, India, USA, China, etc.) have invested heavily in solar energy.

With the past record of such high proportion of fossil fuel usage, there are certain challenges faced globally. Among them few are fluctuations in price of fossil fuel products (two extreme examples occurred in the 1970s and 2000s), geostrategic and security related issues, and most predominantly, the impact on the environment by the overwhelming use of fossil fuels. It is estimated that about 1,120 billion tonnes of carbon dioxide has been pumped into the atmosphere due to human activity during the whole of recoded history, and at present, every year an estimated 200 million tonnes is further being added to the atmosphere [3]. It is expected that due to this additional carbon dioxide the average temperature of the planet could rise about 0.9 °C [4], which may have huge impact on marine life, affect crop yield, change rainfall pattern, etc. These figures are cause of alarm and necessary action is being planned at multiple levels (international, national, state and province, community and personal) to slowly reduce the reliance on fossil fuels by harnessing alternative forms of energy, primarily – solar, wind, hydro, nuclear, geothermal, biomass, tidal, etc. In addition, several steps are being undertaken to reduce the global energy footprint by increasing efficiencies of equipment, processes (e.g. making buildings and appliances more efficient), as well as use of smart materials. In this context, the utilization of solar energy provides an opportunity to utilize a clean and sustainable form of energy.

This chapter deals with the potential application of various STE technologies at community level within India. The focus of the chapter remains predominately on technical issues such as sizing, design, optimization of operational parameters in order to achieve high efficiency, etc. It begins by giving an overview about the global energy requirement and the unique opportunity to harness solar energy. It then talks about one of the main techniques for harnessing solar energy for thermal applications – flat-plate collector (FPC). Next, a very detailed description has been provided for the novel technique of absorbing the sunlight using nanoparticles (volumetric-based absorption of solar energy). Such a collector has been analysed using a numerical model. The results of this numerical model have been provided which simulate the hot water requirement for a small community in India consisting of about ten households. Further, the detailed analysis of the performance of such a collector has been presented, including the influence of various operations and design parameters (particle volume fraction, mass flow rate, solar irradiation, collector height, collector length) on the overall performance of the collector.

5.2 Solar energy

Solar energy is available at almost all locations in the world. It is estimated that every day on average about 5.6 million exajoules of sunlight strikes the surface of the earth. This value is approximately about 1,000 times the current global energy consumption (570 EJ) [5]. A large portion of this sunlight gets reflected back into space, while some portion is absorbed by the land and water bodies. Some portion of sunlight is also absorbed by the plants which convert that energy into biomass.

At any given terrestrial location on the earth the value of the global horizontal solar irradiation (defined as the amount of solar flux incident on a horizontal surface from all directions, and has the units of W/m^2) has pattern of variation. On a given day, it varies from hour to hour, due to the relative motion of the sun and earth (caused by the rotation of the earth), beginning at 0 W/m^2 close to dawn, then reaching a peak value during the solar noon and again reaching 0 W/m^2 around dusk time. If on a particular day the sunlight is obstructed by local weather conditions (clouds, fog, dust storm, etc.), the value of the solar irradiation would be affected accordingly. Unexpected global events such as volcanic eruptions and solar/lunar eclipses also affect the amount of sunlight reaching the ground. A diurnal variation of solar irradiation further varies from region to region, depending on the latitude of the location (e.g. variation between the equatorial region vs. the polar region on the same day). Moreover, the pattern of solar movement (tracked by solar-based angles such as zenith, azimuth, altitude, etc.) varies from month to month. In the northern hemisphere, the sun reaches a higher altitude angle during solar noon from the months of April to September compared to the rest of the months; and the reverse happens in the southern hemisphere. All these variations in solar position (on an hourly as well as seasonal basis), coupled with the intermittency issues, along with the demand-side variation creates a very complicated picture of the optimum utilization of solar energy resources.

Various types of solar collectors are available for efficient conversion of the solar irradiation into the thermal energy of the working fluid. Their classification depends on various factors, predominant amongst which is the requirement of the outlet temperature.

1. Non-concentrated type collectors

   For low-temperature applications (50–90 °C) [6]:

   (i) FPCs: These may also have some design variations
      (a) unglazed
      (b) single glazed
      (c) double glazed
   (ii) Evacuated-tube collectors

2. Concentrated type collectors

   For mid-temperature applications (150–200 °C) [6]:

   (i) PTC

   For high-temperature applications (200–400 °C) [6]:

   (ii) Central tower receivers
   (iii) Paraboloid dishes

Generally speaking, there are significant differences between the non-concentrated and concentrated type collectors [7]. The main differences between the non-concentrated and concentrated type collectors have also been highlighted in Table 5.1.

As the focus of this chapter is on providing hot water supply to a community consisting of about ten households, in subsequent sections, attention has been paid to only non-concentrated type solar collectors which produce the output fluid at temperatures (<95 °C) which are of the current interest. The next section provides details of the conventional FPC such as it is constructional, operational and design related features. After that, the subsequent section discusses the results of a novel

Table 5.1 The main differences between the non-concentrated and concentrated type collectors [7]

Attributes	**Non-concentrating type solar collectors**	**Concentrating type solar collectors**
Examples	Flat-plate collectors, evacuated tube collectors	PTC, central tower receivers, paraboloid dishes
Solar irradiation	Both direct and diffuse	Only direct
Tracking	No tracking required	1-axis or 2-axis tracking required
Temperature range	Low (50–90 °C)	Medium to high (150–300 °C)
Typical applications	Water and air heating, drying, passive building heating	Steam generation, industrial processing, electric power generation

type of solar collector which uses nanoparticles to directly absorb the incident sunlight (volumetric-based absorption of solar energy).

5.3 Flat-plate collector

FPCs are very useful for use by individuals as well as communities to obtain low to medium temperature range of water (50–80 °C) [8]. Since it does not require any tracking mechanism its operation is relatively very simple and can provide up to about 20–50 l of heated water per day (per square meter of roof space), depending on the levels of incident solar irradiation.

5.3.1 Construction and operation of a flat-plate collector

Figure 5.1 details the constructional and thermal resistance circuit details of two most popular designs of FPC. In both these designs, the solar selective plate is essentially responsible for STE collection. However, they differ from each other in the manner the absorber plate is attached to the array of working fluid carrying tubes/conduits. This aspect is critical as it dictates the efficiency with which the absorbed solar energy is transferred to the tubes/conduits (housing the working fluid). The latter of the two designs [see Figure 5.1(b)] is more robust as in this design the absorber plate and tubes form an integral component. Opposed to this, in the former design [see Figure 5.1(a)], the efficiency shall depend on the thermal contact resistance of the bonding material between the absorber plate and the tubes.

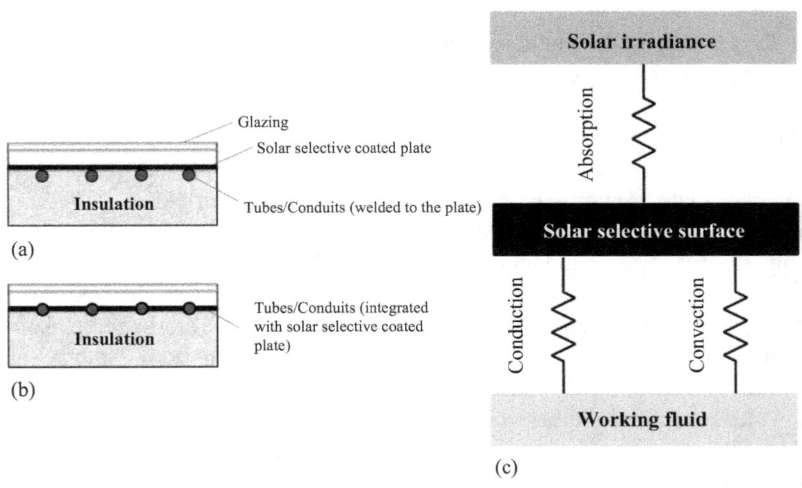

Figure 5.1 Schematic of a solar flat-plate collector designs: (a) heat transfer fluid (HTF) carrying tubes/conduits welded/soldered to the absorber plate, (b) HTF carrying tubes/conduits integral to the absorber plate, and (c) the thermal resistance circuit for solar flat-plate collector design

Table 5.2 Components of a flat-plate collector

Component	Typical materials used	Desirable attributes
Solar selective absorber plate	Black-chrome-coated copper/aluminium sheets	High solar weighted absorptivity and low infrared radiation (IR) emittance and high thermal conductivity
Tubes/conduits	Black-chrome-coated copper/aluminium	High solar weighted absorptivity and low infrared radiation (IR) emittance and high thermal conductivity
Glass glazing	Low iron borosilicate glass	High solar irradiance transmittance and not allowing the long wavelength IR to escape the collector
Heat transfer fluid (HTF)	Water, water + antifreeze solutions	Non-reactive, non-acidic, non-corrosive
Insulation	Fibreglass	Should retain its low thermal conductivity characteristics even at elevated temperatures
Outer cover/ casing	–	Houses all the components

It is imperative to ensure good thermal conductance of the bonding material for optimum performance of the collector. Quantitatively, thermal conductance of the bonding material is given by the following equation:

$$C_b = \frac{k_b b}{\gamma}, \tag{5.1}$$

where k_b is the thermal conductivity of the bond material and b and γ are the width and thickness of the bond, respectively. Efficiency factor of the collector increases with increase in bond material conductance, for instance, it has been reported by Whillier and Saluja [9] that as the bond thermal conductance is increased from 1.73 to 17.3 W/m K, the efficiency factor increases approximately by 33 per cent. Once the absorber plate converts the solar radiant energy into thermal energy, this thermal energy is transferred to the tubes (through conduction) and subsequently to the working fluid (though conduction and convection). To ensure low radiative and convective losses glass glazing(s) is employed. Air gap between the absorber plate and the glass cover ensures low convective losses. Furthermore, the glass cover being highly transparent to incident sunlight and opaque to IR radiations ensures low radiative losses. At the back, insulation is provided to reduce thermal (conduction) losses from the tubes/conduits.

The FPC consists of various components such as the outer casing, absorbing surface, transparent glass cover, etc. All the components are described in detail, along with the materials that they are generally made of, as well as the attributes desired from them, in Table 5.2.

Now that we know the basic construction and operation of a FPC, let us get into more detail of the heart of the FPC i.e. the absorber plate. Generally, the plate is constructed from a metal (copper, aluminium, etc.) which is coated on one-side with

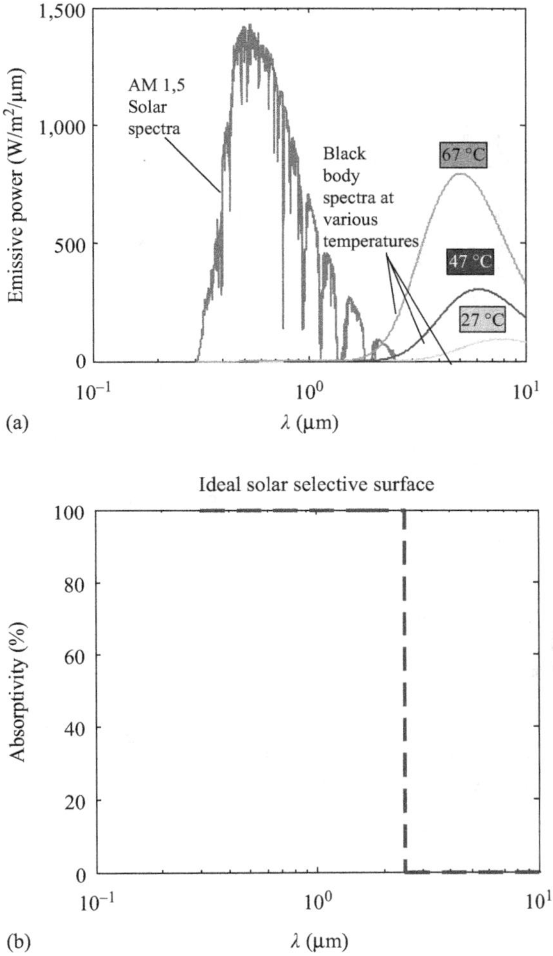

Figure 5.2 Wavelength dependant parameters: (a) solar irradiation and black body spectra and (b) spectral absorptivity, α_λ

a selective-coating. The purpose of this coating is to absorb the maximum amount of solar irradiation, and at the same time should have minimum amount of emission losses. This solar selectivity can be achieved by having a wavelength-dependent property (absorptivity as a function of wavelength). Figure 5.2(a) shows the spectral solar irradiance (AM 1.5) and the black body emission spectra at typical collector temperatures. Furthermore, Figure 5.2(b) shows the spectral absorptivity of an ideal solar selective surface which has a sharp cut-off wavelength.

5.3.2 Design and operational parameters

Various design and operational parameters that affect the performance of the solar collector are summarized in Tables 5.3 and 5.4.

Table 5.3 Various design parameters affecting the performance of the flat plate solar collector

Design parameters	Effect on			
	Optical efficiency	**Thermal efficiency**	**Outlet temperatures**	**Cost/maintenance**
Location and orientation	Optimum tilt angle (= to the latitude of that location) angle is imperative to ensure high optical and hence thermal efficiency (and outlet temperatures)			–
Collector area	–	May increase if proper insulation mechanisms are devised	Increase owing to increase in the energy gain	Increases with size
Glazing (glass cover) optical properties	Optimum optical properties of the glazing are imperative to ensure high optical and hence thermal efficiency (and outlet temperatures)			Needs to be cleaned and monitored for cracks if any due to thermal cycles or natural calamities
Number of glass covers	Decreases with increase in number of glass covers			–
Gap between the glazing and the absorber plate	–	Not affected if the gap is varied under suitable limits (20–40 mm). Convective losses shall increase at higher gaps and conduction losses may increase at lower gap heights		–
Optical properties of absorber plate	Increases with increase in absorptivity of the absorber plate	Increases with increase in absorptivity of the absorber plate and decrease in emissivity of the absorber plate		Solar selective paints and coatings may lose their desired properties with time and thermal cycles
Thermal contact between absorber plate and conduit	–	Most critical to ensure efficient heat transfer from the absorber plate to the tubes/conduits housing the working fluid		Needs to be periodically monitored and repaired (if perforated or broken)
Working fluid	–	High thermal conductivity and ability to wet the surface shall enhance the thermal efficiency	Low specific heat shall result in higher outlet temperatures	Nature of the working fluid shall dictate the overall life of the collector. The extent of corrosion and erosion shall depend on the acidity/nature of the working fluid

Table 5.4 Various operating parameters affecting the performance of the flat plate solar collector

Operating parameters	**Effect on**			
	Optical efficiency	**Thermal efficiency**	**Outlet temperatures**	**Cost/maintenance**
Solar irradiance	–	Increases as the energy gain increases for the same collector area	Decrease with increase in mass flow rate	–
Mass flow rate	–	Increases with increase in mass flow rate	Decrease with increase in mass flow rate	–
Inlet temperature	–	Decreases	–	Cost as well maintenance cost may increase owing to usage of better
Dust and dirt	Decreases with increase in dust layer thickness	Decreases	Decrease	Needs to be cleaned periodically to ensure consistent optimum performance
Wind speed	–	Decreases owing to increase in convective losses	Decrease owing to increase in convective losses	High wind speed may damage the collector

It is apparent from the critical review that there are numerous designs and operation parameters that need to be taken care in order to ensure optimum operation and life of the FPCs. Furthermore, the mechanism of transfer of the absorbed energy to the working fluid is not a very efficient process and also the existing FPC designs are metal components intensive resulting in corrosion and erosion problems. Therefore, in order to overcome these problems researchers have proposed numerous improvements vis-à-vis to the existing designs. In this respect, volumetric absorption-based FPCs promise higher thermal efficiencies and lower maintenance costs. Such collectors are discussed in detail in the next section, where a detailed analysis has been carried out on them using a numerical model, to simulate the process of producing hot water for a small community of about ten households in India.

5.4 Community-level volumetric absorption-based solar collectors (using nanofluids)

This section presents the results of the numerical model of a volumetric absorption-based solar collector that is assumed to meet the demands of a typical community in India (consisting of about ten households). The section is divided into two parts. In the first part, the background details of the volumetric absorption-based solar collectors are provided. Here, various boundary conditions and assumptions have been laid out. In the second part, the full details of the results which are obtained by varying the important parameters (such as particle volume fraction, mass flow rate, incident radiation intensity) along with their effects on the performance of the solar collector (collector efficiency, fluid outlet temperature), have been provided.

5.4.1 Numerical model of the volumetric absorption-based solar collector

Figure 5.3 shows the schematic of a typical closed-loop community-level solar collector system. The system consists of the solar collector which absorbs the solar energy and converts it into thermal energy, which is carried away by the fluid flowing through the collector. This fluid once it gets heated passes through the heat exchanger which also acts like a partial storage device where the hot fluid can be stored up to a limit (depending on the requirements of the community). This enables the system to utilize maximum potential of solar energy both during the peak sunshine hours as well as during the periods when the sunlight is not available. On the other side of the heat exchanger, the cold water from the supply tank flows this cold water, gets heated once it flows through the heat exchanger and after becoming sufficiently warm, it exits and is supplied to various places within the community.

Such a system is able to provide reliable and continuous supply of warm water to the community with very little issues related to maintenance as well as very less reliance for external power input (except for operating the pump). To make such a

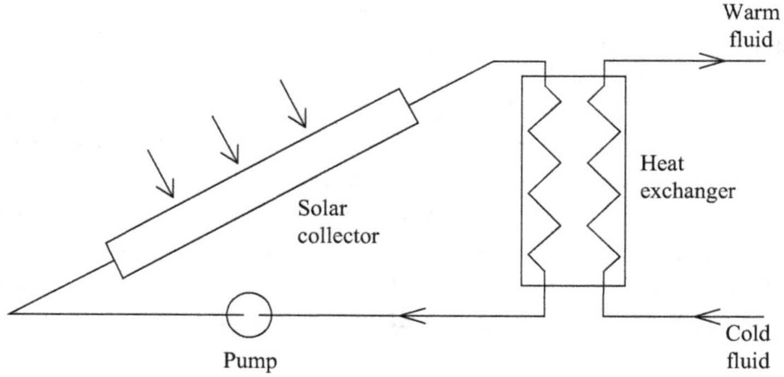

Figure 5.3 Schematic of a closed-loop community-level solar collector system

system truly stand-alone, a small PV panel (along with a battery backup) can also be used, so that the electricity needed to operate the pump can be harnessed from sunlight directly.

Figure 5.4 shows the schematic and the thermal resistance circuit of the volumetric absorption-based solar collector. Such a collector is able to absorb the incident solar irradiation directly within the (volume of) fluid itself. This is carried out by utilizing sufficient quantity of nanoparticles which are mixed within the fluid itself to enhance its radiative properties. Some of the working fluids (base fluids) used in solar collectors are water, oil, molten salts [10]. The main criterion for choosing a particular fluid for a given solar collector is its boiling point. For relatively low temperature (say <95 °C), which are usually required in community-scale solar collectors, water is a good candidate [11]. Water has very good thermal (thermal conductivity, thermal diffusivity, specific heat) and physical properties (density). However, within the solar spectrum (which is between 0.2 and 3 μm), water is mostly transparent [10]. As a result, it cannot be directly used as a medium for absorbing sunlight. Therefore, sufficient quantities of nanoparticles need to be added to the base fluid to enable the fluid to significantly absorb the solar energy. As compared to larger particles, the nanoparticles have very high absorption cross-section per unit particle volume [12]. Furthermore, careful selection of nanoparticle material, shape and size could ensure high solar weighted absorptivity at low-volume fractions of nanoparticles.

Generally, the nanoparticles that are considered good candidates for such applications are – graphite, aluminium, copper [13–16]. Nanoparticles are made of such materials exhibit good absorption and scattering coefficients (K_a and K_s, respectively), within the solar spectrum [17]. In addition to the material, other parameters also influence the radiative properties of the nanoparticle, mainly particle diameter (D) and particle volume fraction (f_v). These values are typically expected to be within $D = 5$ nm to $D = 80$ nm, and $f_v = 0.001$ per cent to $f_v =$ 0.5 per cent [15].

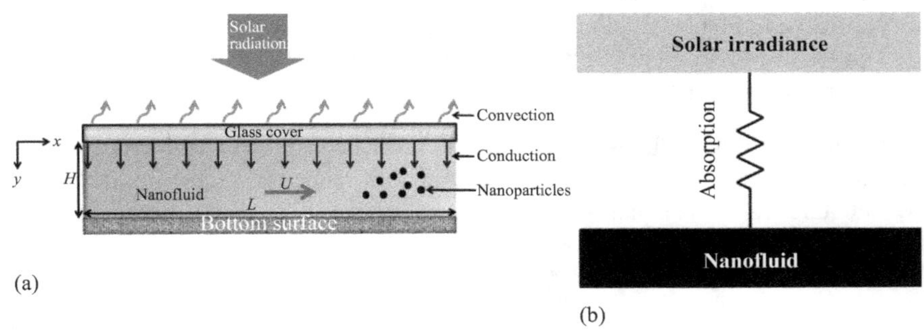

Figure 5.4 (a) Schematic of volumetric absorption (using nanofluids)-based solar collector and (b) thermal resistance circuit for volumetric absorption-based solar collector

The schematic of the volumetric absorption-based solar collector discussed in this study is shown in Figure 5.4. The nanofluid enters the collector from the left side and flows across the collector, and finally exits from the right side. As shown in the schematic, the problem is assumed to be 2-dimensional (in x and y directions). The fluid is contained between two surfaces. On the top it is covered by a transparent glass (having an overall transmissivity of $\tau = 0.9$, within the solar spectrum). At the bottom it is contained by an opaque surface which does not allow any of the irradiation to pass through it. The top glass cover is exposed to the ambient conditions (surroundings at a temperature of $T_{surr} = 35$ °C, and a convective heat loss coefficient of $h_{conv} = 5$ W/m^2 K). The solar irradiation incident on the top surface is assumed to be normally incident, at a flux of 1,000 W/m^2 [15]. In order to solve the numerical analysis, following assumptions have been considered: (1) the top surface is exposed to the ambient atmosphere so losses from top surface are due to convection only, (2) bottom surface is adiabatic and transparent i.e. no heat flux can pass through it. (3) Emission losses from the top surface are negligible, (4) scattering by the nanoparticles is an independent scattering because the volume fraction of the nanoparticles is very small. (5) Spectrally distributed solar irradiance is incident normally to the solar collector with an intensity of 1,000 W/m^2 (no atmospheric absorption).

In the present model, the following parameters have been assumed:

Basefluid	Water
Type of nanoparticles	Graphite
Incident flux (normal) on the top surface, q_o	1,000 W/m^2
Transmissivity of the glass cover, τ	0.9
Overall convective heat loss coefficient, h_{conv}	5 W/m^2 K
Surrounding temperature, T_{surr}	25 °C
Fluid inlet temperature, T_{in}	35 °C
Fluid inlet velocity, U	1 cm/s
Height of the collector, H	1.2 cm

The model has been simulated for the following requirements of a community:

Total number of households in the community, 10
Number of persons residing in each household, 4
Hot water requirement per person per day, 100 kg [18]
Desired hot water temperature, 60 °C

This model is numerically solved using the 2-D heat transfer and 1-D radiative transport equations in MATLAB®, using finite difference technique. For describing 1-D radiative transport equation, the following equation has been used:

$$I_\lambda = I_{o,\lambda} e^{-K_{a\lambda,water} + (-K_{e\lambda,nanoparticles})y} \tau \tag{5.2}$$

where I_λ is the intensity of light propagating inside the nanoparticles-laden fluid, I_o is the intensity of the incident light, $K_{a\lambda}$ is the spectral absorption coefficient, $K_{e\lambda}$ is the spectral extinction coefficient, y is depth of solar collector, τ is the transmittance of the glass.

After finding intensity distribution in the solar collector, radiative heat flux (q_r) has been evaluated which is defined as

$$q_r = \int_\lambda \int_\phi I_\lambda d\phi d\lambda \tag{5.3}$$

For solving 2-D heat transfer equation, the following equation has been used:

$$\rho C_p \mathrm{u} \cdot \nabla T = \nabla \cdot (k \cdot \nabla T) - \nabla q_r \tag{5.4}$$

where ρ is the density of the nanofluid, C_p is the specific heat of the nanofluid, k is the thermal conductivity of the nanofluid, q_r is the divergence of radiative heat flux, which results in volumetric heating due to incident solar radiation.

The details of the equations and the numerical methods have been discussed elsewhere and those details are not the focus of the current study [15,16]. The results obtained by solving those governing equations are presented here. The calculations performed by the model gives the following results:

Design fluid mass flow rate, $m = 0.185$ kg/s
Length of the collector, $L = 16.6$ m
Width of the collector, $W = 1.85$ m
Total area of the collector, $A = 30.7$ m^2

5.4.2 Parameters influencing the performance of the solar collector

Figure 5.5 shows the spectral intensity variations within the solar collector at various depths of the solar collector. As can be seen from this figure, the incident solar energy on the top surface of the fluid corresponds to the solar blackbody radiation ($T_{sun} =$ 5,800 K). The significant portion of solar spectrum (99 per cent) lies within the

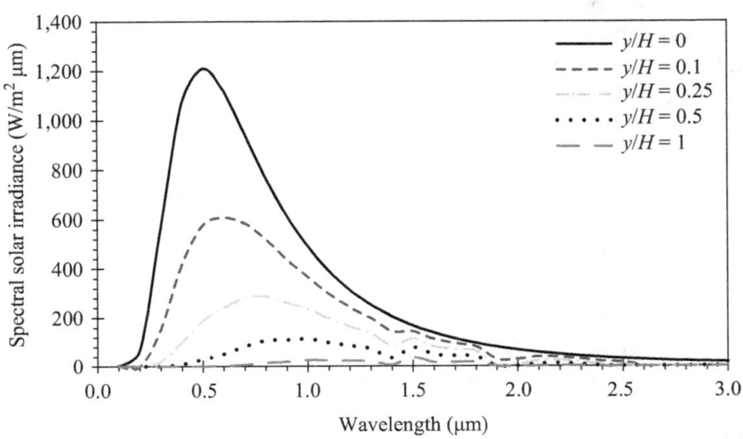

Figure 5.5 Spectral solar irradiation at various heights of the collector

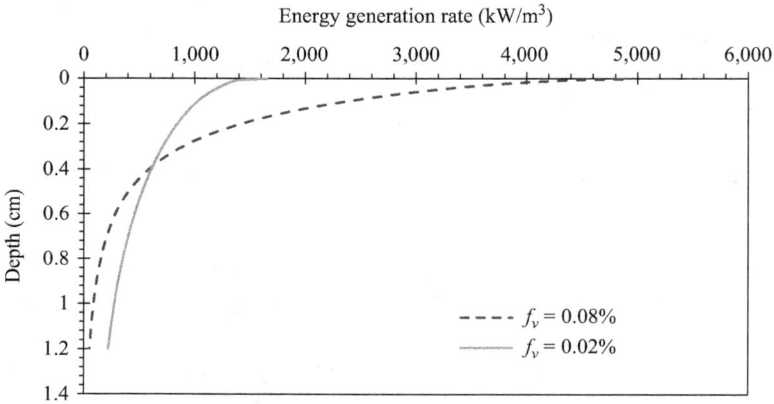

Figure 5.6 Energy generation rate as a function of collector depth

wavelengths of 0.2 and 3 μm, and the peak of the solar irradiation is about 0.5 μm. Once the irradiation passes through the glass cover ($\tau = 0.9$), and interacts with the top layer of the fluid, its flux is about 900 W/m^2. As the irradiation passes through the fluid, it is gradually absorbed. The absorption is also accompanied by scattering, and the sum together is denoted by extinction. Hence the extinction coefficient (K_e) can be expressed as the sum of absorption and scattering coefficients, $K_e = K_a + K_s$ [19]. All these quantities have a spectral dependence, due to which the attenuation of the intensity takes places somewhat non-uniformly over various wavelengths. For example, at a wavelength of about 1.5 μm, the attenuation is relatively the least, where at around 0.5 μm, the attenuation in intensity is very high.

Figure 5.6 shows the variation of energy generation rate along the depth of the solar collector, at two different particle volume fractions. For both cases, it can be seen that the energy generation rate is very high near the top surface and it decreased very rapidly along the depth of the collector. This is because the radiation intensity gets attenuated exponentially, as per Beer's Law. Hence the top portion of the fluid absorbs maximum radiation (and as a result get heated the most). The remaining radiation reaches the lower portion and is subsequently absorbed there.

The temperature profile within the collector has been shown in Figure 5.7. This figure shows the temperature variation within the collector at two different axial locations, first at $x/L = 0.5$, and second at $x/L = 1$. When the fluid enters the solar collector it has a uniform inlet temperature ($T_{in} = 35$ °C). As the fluid flows within the collector (i.e. moves along the x direction), it starts to get heated by the solar irradiation. The more time it spends within the collector, the more time it spends receiving the incident solar irradiation and consequently its temperature starts to increase accordingly. This process will continue even for longer collectors, and the fluid will continue to get warmer and warmer as it flows. Eventually, it will reach a point where the losses from the fluid to the ambient (convective and radiative

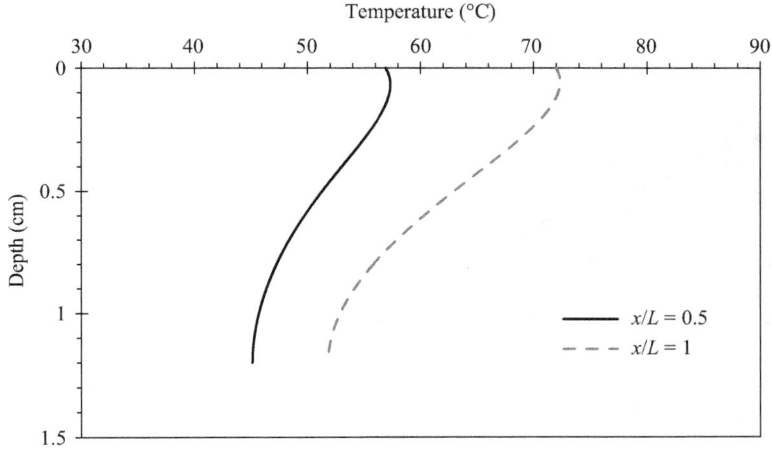

Figure 5.7 Temperature profile within the collector

losses) will become equal to the incident energy, at which point its temperature would stop increasing any further. In this figure, only two locations $x/L = 0.5$ and $x/L = 1$ have been considered just to show the transition of temperature profile along the length of the collector. These two locations show that the temperature before $x/L = 0.5$ will be less and after $x/L = 0.5$ the temperature will be more.

Also, from the same curves (Figure 5.7), another trend can be seen, regarding the variation of temperature along the depth of the collector. For instance, at both these axial locations, the temperatures are much higher near the top surface (near $y/H \sim 0.2$) and then decline at lower locations. The lowest temperatures are observed at the bottom surface (near $y/H = 1$). The temperature profile shows a slight bend near the top (i.e. near $y/H \sim 0.2$). This is because there are some convective losses from the top surface. As a result, the highest temperature for a given axial location, does not occur at the top ($y/H = 0$), instead it occurs at a location slightly below the top surface (say near $y/H \sim 0.2$).

Figure 5.8 shows the variations in mean fluid temperature as a function of the distance along the collector. The mean fluid temperature is obtained by averaging the fluid temperature along the depth (y direction) for each axial location. This figure shows curves for three different cases depending on the amount of nanoparticles present in the fluid – base fluid (pure water), nanofluid having low-volume fraction of nanoparticles and finally nanofluid having high-volume fraction of nanoparticles. In all the three cases a similar trend is observed – the mean fluid temperature experiences a steady increase along with the distance travelling by the fluid within the solar collector. These curves are a direct outcome of the temperature profiles shown in the previous figure (Figure 5.7), where the spatial distribution of the temperature was shown.

The variation of the fluid outlet temperature as a function of the mass flow rate has been shown in Figure 5.9. This figure shows three different curves

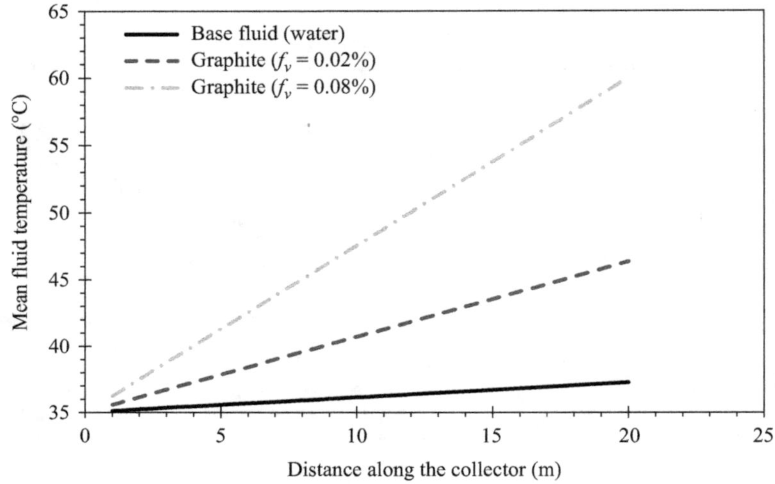

Figure 5.8 Variation of mean fluid temperature as a function of distance along the collector

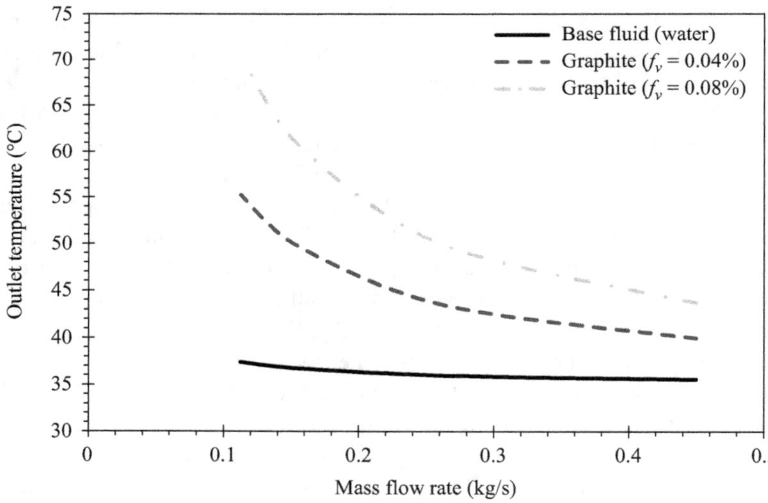

Figure 5.9 Solar collector outlet temperature as a function of the fluid mass flow rate

corresponding to the following three cases – base fluid (pure water), nanofluid having low-volume fraction of nanoparticles and finally nanofluid having high-volume fraction of nanoparticles. For each case it is seen that the outlet temperature shows a gradual dip with increase in the mass flow rate. These results have been obtained by keeping all the other parameters same and just by varying the mass

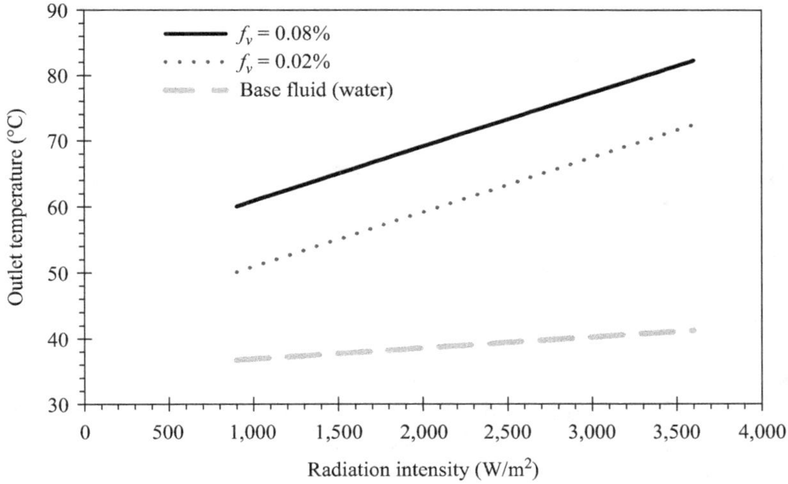

Figure 5.10 Solar collector outlet temperature as a function of the incident irradiation

flow rate. As a consequence, an increase in the fluid flow rate leads to a reduction in the time for which the fluid is able to absorb sunlight, leading to a natural decrease in the outlet temperature. For instance, in an extreme example, if the flow rate of the fluid through the collector were to be very high, it would be expected that the outlet temperature of this fluid would be very close to the inlet temperature. If the curves shown in this figures were to be extrapolated, a similar trend would be observed.

The outlet temperature of the fluid also depends on the incident radiation intensity. The radiation intensity incident on a particular solar collector can vary in real life conditions for a variety of reasons. First, during the natural course of the day the solar radiation intensity changes from a low value in the morning to a highest value near noon and again reduces towards the evening. Moreover, the sunlight can experience transients due to weather conditions (cloudy, rainy, foggy, etc.). And finally due to human intervention also the radiation intensity can be controlled, say by use of reflector mirrors (to increase its amount) or by partial shading (to reduce its amount), or by changing the tilt (to either increase or decrease its amount). Figure 5.10 shows a variation in the fluid outlet temperature with respect to the incident solar radiation. All the three curves in this figure show an increase in the fluid outlet temperature as the radiation intensity increases.

Figure 5.11 shows the variation in the efficiency of the solar collector. The efficiency of the collector is a very important parameter as it is a direct indicator of the performance of a particular collector. Generally, two types of efficiencies are defined for solar collectors – optical efficiency and thermal efficiency. Optical efficiency is defined as the ratio of the rate of energy absorbed by the collector to the rate of energy incident on the top plane of the collector. Optical efficiency can

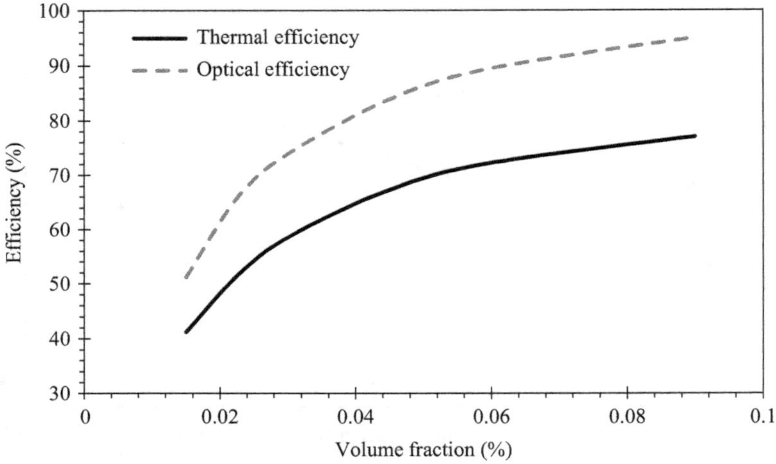

Figure 5.11 Efficiency of the solar collector as a function of nanoparticle volume fraction

theoretically be equal to one provided there are not transmission losses from the glass cover and no reflection losses from the collector [16]. Thermal efficiency is defined as the rate of thermal gain by the fluid to the rate of energy incident on the top plane of the collector. The factors which lead to lowering of thermal efficiency are primarily the losses from the collector – optical losses (reflection and transmission), convective and conduction losses, and finally the radiative emission losses from the collector.

The efficiency of the solar collector has been plotted as a function of particle volume fraction and is shown in Figure 5.11. Two types of efficiencies are shown in the figure. First, the optical efficiency and, second, the thermal efficiency have been shown. The value of thermal efficiency is lower than that of the optical efficiency, as it takes into account the various losses such as convective and radiative losses. As can be seen from the figure, the particle volume fraction has a large impact on the overall efficiency of the solar collector. At low values of particle volume fraction, the efficiency is low as not all the radiative intensity is absorbed by the collector. When the particle volume fraction increases larger portion of radiative intensity is absorbed by the fluid, and hence the overall efficiency is high. In the present configuration, the upper limit of nanoparticle volume fraction is 0.09 per cent. After this value the effect of adding more nanoparticles is not observable. This limiting value of particle volume fraction depends on couple of factors such as – optical properties of the nanoparticle, size of nanoparticle, optical depth of the collector.

Figure 5.12 shows the variation in efficiency as a function of radiation intensity. Three curves are shown which are at different volume fractions. The efficiency shows a slight increase with radiation intensity. The highest efficiency was

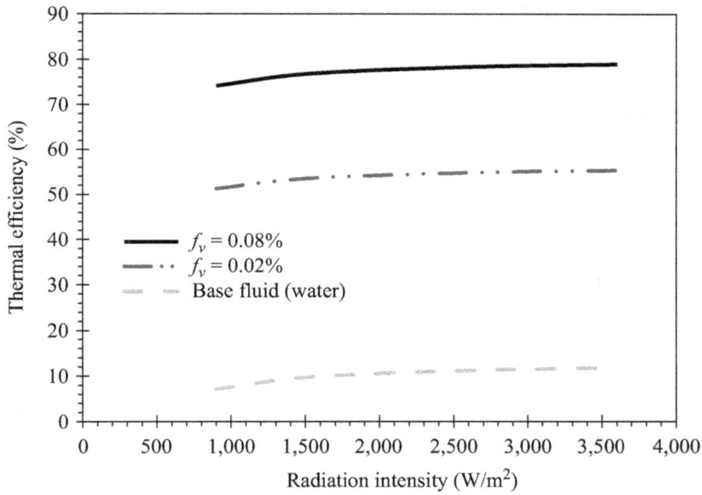

Figure 5.12 Solar collector efficiency as a function of the incident irradiation

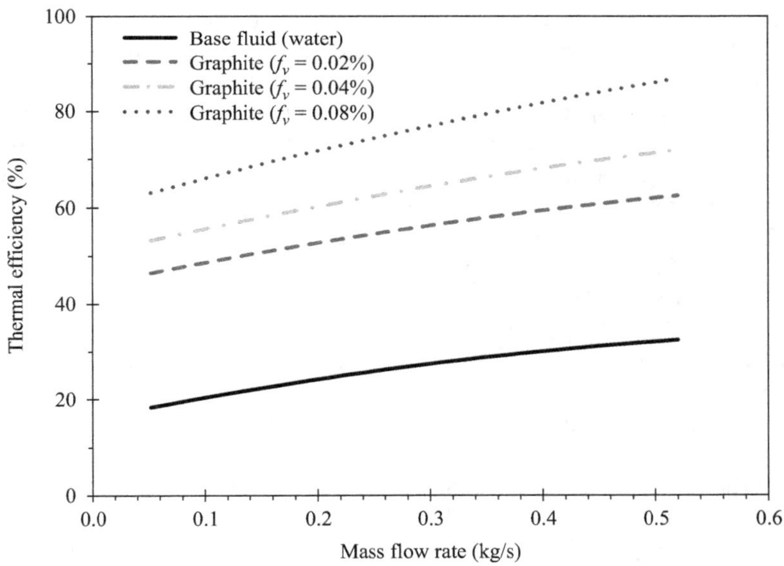

Figure 5.13 Variation of thermal efficiency as a function of mass flow rate

observed for the highest volume fraction, and the lowest efficiency was observed for pure water (base fluid).

As can be seen in Figure 5.13, the thermal efficiency of the solar collector is plotted as a function of the mass flow rate. These curves show the effect of adding the nanoparticles on the overall performance of the solar collector.

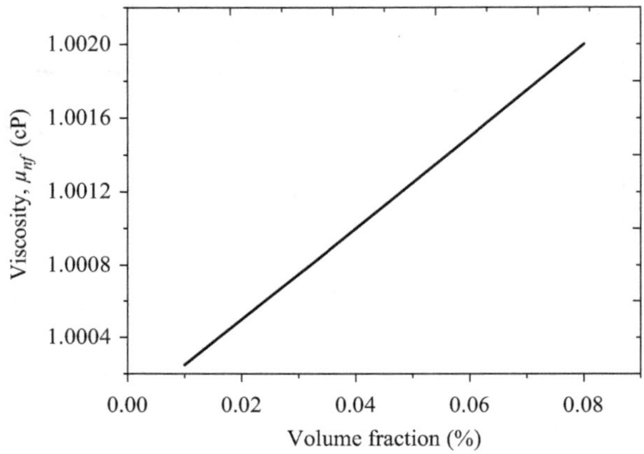

Figure 5.14 Effect of volume fraction on the viscosity of the nanofluid

As can be seen in the figure, the thermal efficiency of the collector increases with the mass flow rate (keeping all other operational parameters the same). As the solar irradiation amount is held constant across each case, once the mass flow rate of the collector increases (due to increase in demand) the outlet temperature of the collector sees a corresponding drop in the temperature. And once the outlet temperature reduces the effective temperature difference between the collector and the ambient experiences a drop as well. This leads to lowering of the overall losses from the collector and as a result, the efficiency of the collector increases.

Moreover, the influence of the nanoparticles is also clearly visible in this graph. As the amount of nanoparticles increases in the solar collector (from the lowest value of $f_v = 0.01$ per cent, to a medium value of $f_v = 0.05$ per cent, to the highest value of $f_v = 0.1$ per cent), the efficiency of the collector increases. This is primarily due to an increase in the absorption caused by the nanoparticles present in the collectors.

This figure shows that volumetric absorption-based solar collectors (using nanofluids) can provide high efficiencies during operation and are quite suitable for meeting the requirements of the community. The overall size of the collector has been easily evaluated by knowing the typical requirements of the community, as well as by knowing the seasonal solar irradiation data for that particular location (mainly dependent on latitude, longitude and the climatic conditions).

In this numerical analysis, very low volume fractions of the nanoparticles have been considered which are dispersed in water. In order to analysis the effect of volume fraction on the viscosity of water, Einstein model has been considered which can be applied for dilute suspension. The Einstein model is represented by the following equation:

$$\mu_{nf} = \mu_{bf}(1 + 2.5\phi), \tag{5.5}$$

where μ_{nf} is the viscosity of nanofluid, μ_{bf} is the viscosity of base fluid (1 cP) and ϕ is the volume fraction. Equation (5.5) shows that there has an increase in viscosity of nanofluid but this increase of viscosity is not very high. With such low-volume fraction, the viscosity of the nanofluid is near to the viscosity of base fluid and as shown in Figure 5.14. Further, Said *et al.* [20] have concluded that there is no significant effect on pumping power with low-volume fraction of nanoparticles.

5.5 Summary

In summary, this chapter discusses the various options available for harnessing STE, mainly for water heating, at a community level. First, the conventional solar collectors are discussed – FPCs (for low-temperature applications) and PTC (for medium-to-high-temperature applications). Subsequently, the basic setup of the novel volumetric absorption-based solar collector (using nanofluids) is discussed. Several of the operational parameters which influence its performance are then discussed in detail.

It is found that nanoparticle volume fraction (f_v) is one of the most important parameters which influence the performance of the collector. As these nanoparticles interact directly with the sunlight (absorption and scattering), a sufficient amount of these particles must be present in the fluid to ensure adequate temperature rise takes place. Some practical problems are also associated with the volume fraction of the nanoparticles like as the nanoparticles volume fraction has been increased, the volumetric absorption system will closely resemble the solar selective plate system because the penetration depth of the incident light decreases. Due to shorter penetration depth, the heat release will be localized to the top of the receiver, where most of the losses occur, which will result to low thermal efficiency of the system. Another challenge with high-volume fraction is agglomeration of the nanoparticles which will lead to clogging of the pump.

They also affect the overall efficiency of the collector. Other design parameters studied are the geometry of the collector (height, H and length, L), mass flow rate of the fluid, and the outlet temperature.

It is concluded that such collectors are quite suitable for meeting the requirements of the community, and can be suitably sized depending on the requirements of the community as well as the availability of solar resources in that particular location.

References

[1] International Energy Agency (IEA). *Total primary energy supply by fossil* [online]. 2017. Available from https://www.iea.org/publications/free-publications/publication/KeyWorld2016.pdf [Accessed 5 July 2017].

[2] US Energy Information Administration (EIA). *Transportation sector energy consumption* [online]. 2017. Available from https://www.eia.gov/outlooks/ieo/transportation.php [Accessed 5 July 2017].

[3] BP statistical review of world energy. *Carbon dioxide emission* [online]. Available from http://www.bp.com/en/global/corporate/energy-economics/statistical-review-of-world-energy.html [Accessed 20 Feb 2017].
[4] National Oceanic and Atmospheric Administration (US). *Earth surface temperature* [online]. 2017. Available from https://www.climate.gov/news-features/understanding-climate/2014-state-climate-earth%E2%80%99s-surface-temperature [Accessed 5 July 2017].
[5] Kreith F., and Krumdieck S. *Principles of Sustainable Energy Systems*, 2nd ed. Boca Raton, FL: CRC Press; 2013.
[6] Duffie J. A., Beckman W. A., and Worek, W. M. *Solar Engineering of Thermal Processes*, 4th ed. Journal of Solar Energy Engineering; 2003.
[7] Boyle G. *Renewable Energy: Power for a Sustainable Future.* OUP, Oxford; 2012.
[8] Kalogirou S. A. 'Solar thermal collectors and applications'. *Progress in Energy and Combustion Science*. 2004, vol. 30(3), pp. 231–295.
[9] Whillier A., and Saluja G. 'Effects of materials and of construction details on the thermal performance of solar water heaters'. *Solar Energy*. 1965, vol. 9(21), pp. 21–26.
[10] Khullar V., Bhalla V., and Tyagi H. 'Potential heat transfer fluids (nanofluids) for direct volumetric absorption-based solar thermal systems'. *ASME Journal of Thermal Science and Engineering Applications*. 2018, vol. 10(1), p. 011009.
[11] Phelan P., Otanicar T., Taylor R., and Tyagi H. 'Trends and opportunities in direct-absorption solar thermal collectors'. *Journal of Thermal Science and Engineering Applications*. 2013, vol. 5(2), p. 21003.
[12] Khullar V., Tyagi H., Otanicar T., Hewakuruppu Y., and Taylor R. 'Solar selective volumetric receivers for harnessing solar thermal energy'. *Proceeding of ASME International Mechanical Engineering Congress and Exposition*, Phoenix, Arizona, USA, Nov. 2016. Paper No. IMECE2016-66599.
[13] Gorji T. B., and Ranjbar A. A. 'A numerical and experimental investigation on the performance of a low-flux direct absorption solar collector (DASC) using graphite, magnetite and silver nanofluids'. *Solar Energy*. 2016, vol. 135, pp. 493–505.
[14] Khullar V., Tyagi H., Hordy N., *et al.* 'Harvesting solar thermal energy through nanofluid-based volumetric absorption systems'. *International Journal of Heat and Mass Transfer*. 2014, vol. 77, pp. 377–84.
[15] Tyagi H., Phelan P., and Prasher R. 'Predicted efficiency of a low-temperature nanofluid-based direct absorption solar collector'. *Journal of Solar Energy Engineering*. 2009, vol. 131(4), pp. 41004.
[16] Khullar V., Tyagi H., Phelan P. E., Otanicar T. P., Singh H., and Taylor R. A. 'Solar energy harvesting using nanofluids-based concentrating solar collector'. *Journal of Nanotechnology in Engineering and Medicine*. 2013 vol. 3(3), pp. 31003.

[17] Taylor R., Coulombe S., Otanicar T., *et al.* 'Small particles, big impacts: A review of the diverse applications of nanofluids'. *Journal of Applied Physics*. 2013, vol. 113(1), pp. 0–19.
[18] The Energy and Resources Institute (TERI India). *Solar water heating systems, building regulations and guidelines to achieve energy efficiency in Bangalore city* [online]. 2010. Available from http://www.teriin.org/ResUpdate/reep/ch_9.pdf [Accessed 1 July 2017].
[19] Bhalla V., and Tyagi H. 'Parameters influencing the performance of nanoparticles-laden fluid-based solar thermal collectors: A review on optical properties'. *Renewable & Sustainable Energy Reviews*. 2018, vol. 84, pp. 12–42.
[20] Said Z., Sajid M. H., Alim M. A., Saidur R., and Rahim N. A. 'Experimental investigation of the thermophysical properties of Al_2O_3-nanofluid and its effect on a flat plate solar collector'. *International Communications in Heat and Mass Transfer*. 2013, vol. 48, pp. 99–107.

Chapter 6

Solar–water desalination for small communities

Fahad Ameen[1], Jacqueline A. Stagner[1] and David S-K. Ting[1]

Abstract

The use of solar–water desalination is one of the most promising and environment friendly means to meet the ever-increasing demand of fresh water in freshwater-scarce communities. Unfortunately, desalination of water is a highly energy-intensive process. Nonetheless, direct solar–water desalination is simple in design, low in cost, and can easily support the daily use of communities. This chapter describes various types of solar–water desalination with a focus on technologies that are suitable for small communities and remote areas. Mathematical modeling of direct solar–water desalination based on convective, radiative, and evaporative heat transfer has been provided. A case study from Pasni, District Gwadar, Balochistan of Pakistan involving design, fabrication, and cost is included to give an overview of the practical operation of technology.

6.1 Introduction

Despite advancements toward the Millennium Development Goals (MDGs), in 2015, the United Nations reported that 10 percent of the worldwide populace (approximately 663 million people) still use polluted drinking water sources, including unprotected wells, springs, and surface water [1]. It is likely that many hundreds of millions more will need feasible access to safe drinking water. Notwithstanding when municipally treated water is accessible in urban zones, access is regularly confined via proprietors who place locks on standpipes and charge occupants an expense for access to it [2].

Figure 6.1 shows the total renewable water resources (cubic meters per capita per year) in countries around the world. There are many ways to measure the levels of water scarcity and water stress in a country. Per capita renewable water per year is the best-known indicator to measure the water stress in a country [3]. If the

[1]Turbulence and Energy Laboratory, Department of Mechanical, Automotive & Materials Engineering, University of Windsor, Canada

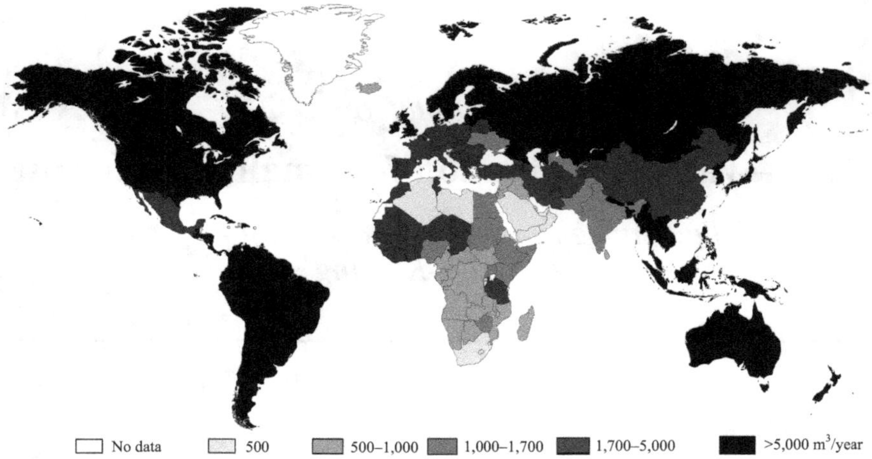

Figure 6.1 Total renewable water resources (cubic meters per capita per year), 2014 [3]

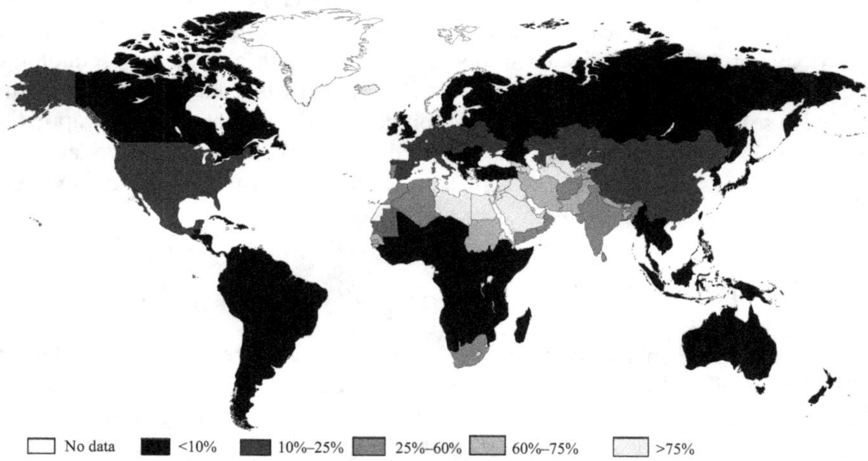

Figure 6.2 Percentage of renewable water resources withdrawn worldwide [3]

renewable water supplies drop below 1,700 m^3 per capita per year, the nation is under continuous water stress. People face high water scarcity when the per capita per year renewable water supply drops below 1,000 m^3. Absolute scarcity occurs below 500 m^3 per capita per year renewable water supply in a country [3].

Figure 6.2 shows the percentage of renewable water drawn in each country of the world. In order to show the relationship between demand and supply, MDG considers the level of human demand on the water [3]. The demand involves the usage of water through agriculture, industry, and municipalities. The stress on the supply system increases as the usage of water increases, making it difficult to meet the increasing demand for water.

The comparison of Figures 6.1 and 6.2 shows that South America, North America, Russia, and Australia have the most water resources ranging above 5,000 m^3/year, but the water withdrawn is also the least in these areas with the exception of few parts of North America. It also shows that even though the Middle East and North Africa has least amount of water resources, they are withdrawing the most water through renewable resources in the world.

The weight of waterborne infection falls excessively on groups in the developing world. Appallingly, more than 90 percent of diarrheal cases are preventable through modifications to the environment, including intercessions to increase the accessibility of clean water [4].

Water purification techniques are used to remove undesired suspended solids, gases, chemicals, and biological contaminants from water [5]. Often, water is purified for human consumption and drinking purposes; however, water purification is also used for various other purposes such as achieving the requirements of pharmacological, medical, industrial, and chemical applications [5].

Solar energy is an application of renewable energy. This "free" energy can be used for the water purification process. Solar desalination is a process that desalinates water via the use of solar energy in an environment-friendly way. Through this process, pure water is separated from pollutants.

The composition of feed water has the most important impact on the efficiency of solar–water desalination systems [6]. The type of desalination process usually depends on the amount of salinity in the water. The energy per unit mass for desalination varies with variations in salinity levels. Saline water may also contain different amounts of dissolved ions. The number of dissolved ions also varies from sea water to municipal brackish water [6]. In underdeveloped countries and regions, the municipal water more often pollutes the drinking water resources. Municipal water involves the brackish water after washing of laundry, kitchen utilities, etc.

Table 6.1 shows the concentration of ions, in ppm, in sea water and brackish ground water. These ions are responsible for the quality of water and chemical properties such as salinity. The lesser the number of undesired ions in the water, the better will be the quality of water. In Table 6.1, the Massachusetts Institute of Technology (MIT) values are the reported values of water that is safe for drinking purposes. Salinity is described as the electrical conductivity of water, and it gives the measurement of total dissolved solids in the water, typically in ppm or mg kg^{-1} [5]. Advanced standards measure the salinity of water through electrical measurements, and these measurements vary over the various oceans of the earth [5]. For water from other sources, a chemical analysis is usually done to see which ions are present in the water. For example, the rock formation from which the water is drawn will describe the ions in the ground water.

Sometimes, the concentration of ions is given by reporting the conductivity of water directly in μS cm^{-1} [5,7]. Conductivity will be roughly 0.5–3 μS cm^{-1} for distilled water and below 100 μS cm^{-1} for typical drinking water, while sea water has a conductivity of approximately 54,000 μS cm^{-1} [5,7].

Table 6.1 Ion concentrations of sea water, high and low salinity brackish water, and municipal water supply, n.r. = not reported [5]

Substance amount ($mg\ kg^{-1}$)	Standard sea water	High brackish water	Low brackish water	Massachusetts water resources authority
Sodium, Na^+	10,556	1,837	90	30
Magnesium, Mg^{2+}	1,262	130	11.7	0.8
Calcium, Ca^{2+}	400	105	96	4.5
Potassium, K^+	380	85	6.5	0.9
Chloride, Cl^-	18,980	2,970	191	21
Sulfate, $SO_4{}^{2-}$	2,649	479	159	8
Bicarbonate, HCO^{3-}	140	250	72.6	n.r.
Fluoride, F^-	1	1.4	0.2	1
SiO_2	1	17	24	3.3
Total dissolved solids	**34,369**	**5,874.4**	**651**	**69.5**

6.2 Types of solar–water desalination

Solar desalination techniques have two basic methods, direct and indirect [8]. In the direct method, the solar collector is coupled with a distillation mechanism and the desalination process is performed in one simple cycle. The indirect method uses two different systems, a conventional desalination plant and a solar collection array which consists of photovoltaic or thermal collectors [8]. Since we are focusing on small communities, only direct solar–water desalination will be discussed as they are cost effective and require less maintenance [9].

Figure 6.3 shows the types of solar–water desalination. The direct solar–water desalination process is categorized into two classes—passive systems and active systems. Passive systems include single-effect solar stills, greenhouse coupled solar stills, and multieffect basin stills [9]. Active systems involve flat plate collectors (FPCs), parallel FPCs, and parabolic solar concentrators.

6.2.1 Direct solar–water desalination systems

Direct solar–water desalination is mostly used for small production systems in areas where the demand for fresh water is as low as 200 $m^3\ day^{-1}$ [8]. The solar rays are focused through the glass cover. The absorber plate absorbs the Sun's rays and heats the brackish water present in the basin. The absorber consists of a thin absorber sheet (of thermally stable polymers, aluminum, and steel) to which a matte black or selective coating is applied [8]. The water evaporates from the basin and condenses on the inner surface of the glass. The condensed water flows down the inclined surface of the glass cover and is collected by the condensation channel. This is the oldest method to produce fresh water; however, research is on-going to improve its efficiency of water production [8].

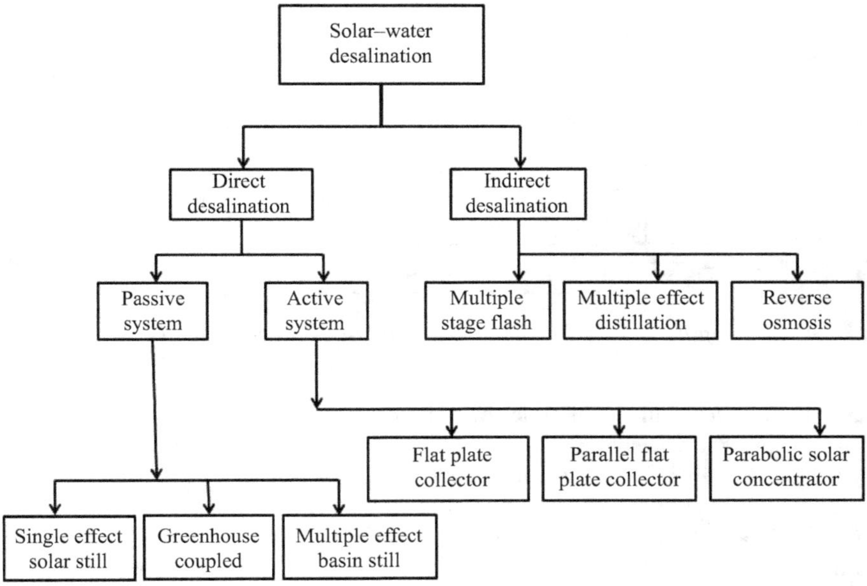

Figure 6.3 Flowchart of solar–water desalination systems [8]

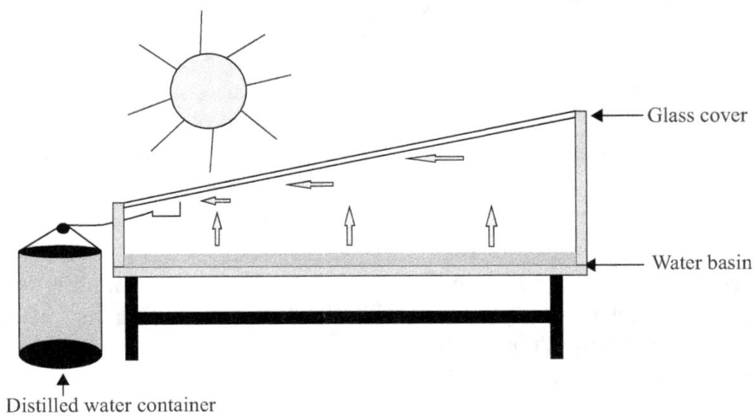

Figure 6.4 Schematic of single-effect solar still for desalination [8]

6.2.1.1 Single-effect solar stills

A solar still is a very basic device used to convert impure water into pure water fit for drinking purposes. A solar still replicates the process used by nature to generate rain, namely, evaporation and then condensation of water.

The most basic type of direct solar–water desalination system is a single-effect solar still. Figure 6.4 shows the operation of a single-effect solar still in which a

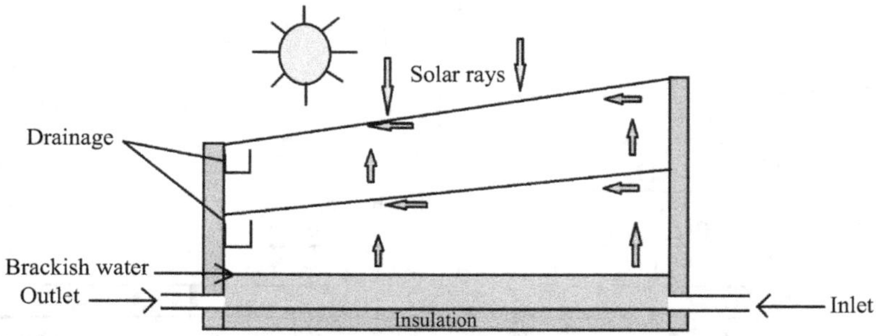

Figure 6.5 Schematic of a double-effect solar still [9]

transparent cover, which can either be glass or sheet, covers the brackish or saline water basin. The transparent cover focuses the sunlight onto the basin, thereby increasing the temperature of saline water which results in evaporation. The evaporated water then condenses on the inner surface of the transparent cover and is collected in a water container. The distilled water is of very high quality, as nearly all the impurities are left behind.

The energy required to evaporate 1 kg of water at 30 °C is 2.4×10^6 J. In the ideal case, assuming the insulation to be 250 W m^{-2} for 24-h cycle, this amount of energy can produce a maximum of 9 L m^2 day^{-1} [9]. In actual cases, however, there are conduction and convective losses, and the daily water yield drops to 4–5 L m^2 day^{-1}.

6.2.1.2 Multieffect basin stills

Multieffect basin stills have two or more basin compartments. Figure 6.5 depicts a double-effect solar still in which the basin of the upper compartment is the condensing surface of the lower basin with every compartment having its own inlet and outlet. Due to reuse of the latent heat of condensation, multieffect basin stills are more productive and efficient than single-effect solar stills. The efficiency of multieffect basin stills is typically around 36 percent, making multieffect basin stills more efficient than single-basin stills [9]. The main factor that affects the productivity of the still is the condensing surface temperature. Reducing the temperature of the condensing surface leads to increase in the productivity. The process of flowing water above the glass cover shows a good effect on the upper basin distillate output and efficiency [9]. The lower glass surface temperature increases the circulation of air inside the still and, thus, enhances the convective and evaporative heat transfer between basin water and glass [9]. However, this increase in efficiency comes with an increase in the capital cost.

An analysis was conducted to predict the effect of several basins on the productivity of the still as shown in Figure 6.6. The results show that the double-basin still gives the highest productivity peak [9]. However, the triple and quadruple basin stills produce appreciable amounts of distilled water during the night, thereby leading to higher daily productivities. But, providing a basin beyond this does not

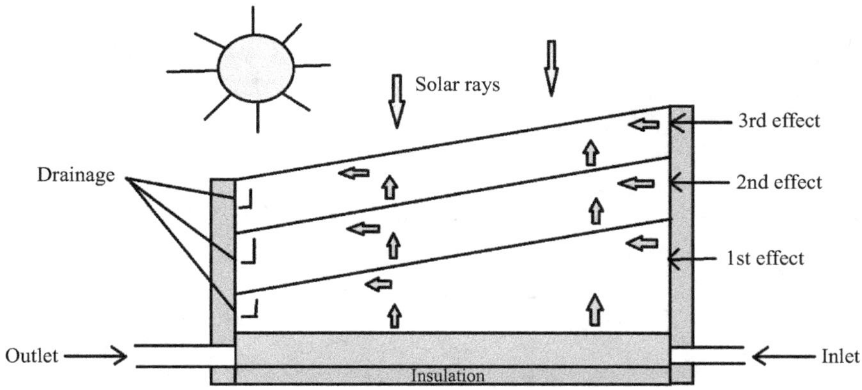

Figure 6.6 Schematic of multibasin, multieffect solar still [9]

have any significant effect on productivity. Double-basin still can yield 36 percent more distillate than a single-basin still. In addition, it has been shown that productivity increases rapidly with increasing insulation thickness [9].

6.2.1.3 Flat plate collector solar stills

An FPC solar still is used where high temperatures are required for boiling of water. An FPC either works in natural mode or forced circulation mode [8]. The difference being that in the former, water flows based on density difference, while in the latter, a pump is used to circulate the water.

6.2.1.4 Flat plate collector in forced circulation mode

An FPC provides more heat energy to the basin than single or double-basin solar collectors can collect independently, thereby increasing the temperature of the feed water. In a forced circulation system, a pump is used to circulate the water from the basin to the plate collector and then back to the basin.

A schematic diagram of an FPC equipped with a pump is shown in Figure 6.7. The still is attached to an FPC, a pump, and an insulated pipe. The pump is used to circulate the water while the FPC is used to increase the feed water temperature.

6.2.1.5 Flat plate collector in natural circulation mode

In FPC under natural circulation mode, there is an FPC and an insulated pipe connected to the solar still. The FPC is used to increase the temperature of the feed water. FPC under natural circulation mode is similar to Figure 6.7, but there is no need of a pump to circulate water. Water circulates due to the difference in densities between the water that has been heated by the FPC and the water that has been cooled by the water in the basin. Advantages of an FPC system in natural mode over an FPC system in forced circulation mode include the simplicity of its design, its reliability, reduced maintenance, and, hence, lower cost.

Single-basin solar still equipped with an FPC shows that the efficiency of a natural circulation system is greater than for a forced circulation system because

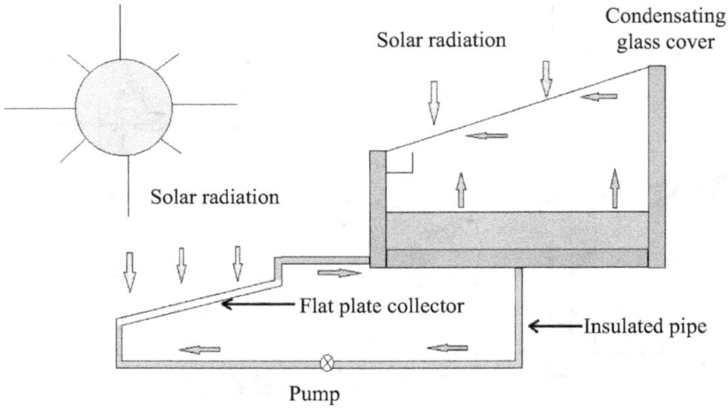

Figure 6.7 Schematic of an active solar still with flat plate collector [10]

extra energy is required to operate the pump in forced circulation mode [11]. It was also shown that the daily productivity decreases with increasing water depth for a natural circulation system. The reverse behavior was shown for a forced circulation system.

Badran and Al-Tahaineh [12] studied the results of coupling a flat pate collector with a solar still in natural circulation mode. The results showed that at a depth of 2 cm, the maximum yield is achieved. If the water depth increases in the basin, the water needs more heat to evaporate and it takes more time to distill. So, with the increase in the water depth of the basin, productivity decreases.

6.3 Mathematical modeling of an inclined solar still

Figure 6.8 shows a schematic of an inclined solar–water desalination still. It consists of an absorber plate and a glass cover which create a cavity. The cavity height, width, and length are 0.2, 1, and 1 m, respectively. The material of the absorber plate is galvanized steel and is painted with a black matte [13]. Black wick is the black cotton or the threads that are often used for the water basin to provide it with an emissivity as close as 1. Black paint is used as the coating because it can absorb more heat rather than reflecting or refracting it. Thus, water is heated rapidly. The absorber plate is insulated to decrease the heat loss from the bottom. The cavity is covered with a 3-mm glass cover [13].

The glass cover is at an angle of 30° so that the dripping water can run down the absorber plate [13]. There are studies on the relation of cover angle and the latitude that shows that optimum cover angle is somewhat equal to the latitude of the particular region [13]. This angle also ensures that the solar rays are normal to the surface at most times of the day. The feed water goes into a distribution pipe which has a 2-mm slot. The feed water then flows into the water basin which is covered with a black absorber plate. Solar energy heats up the water in the basin.

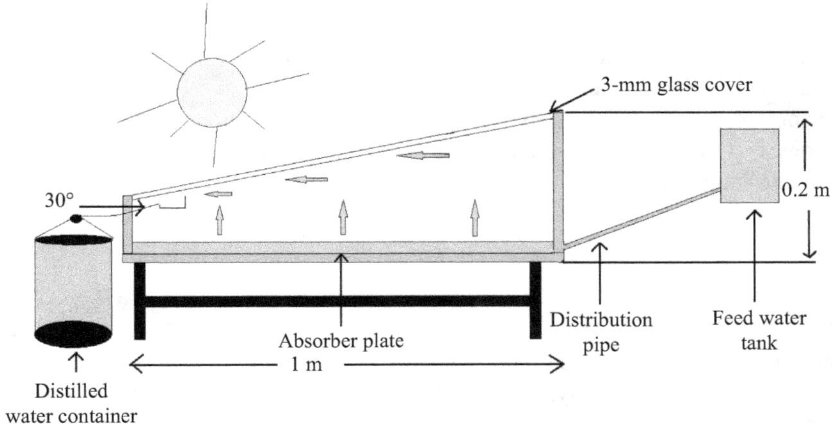

Figure 6.8 Schematic of inclined solar–water desalination unit [13]

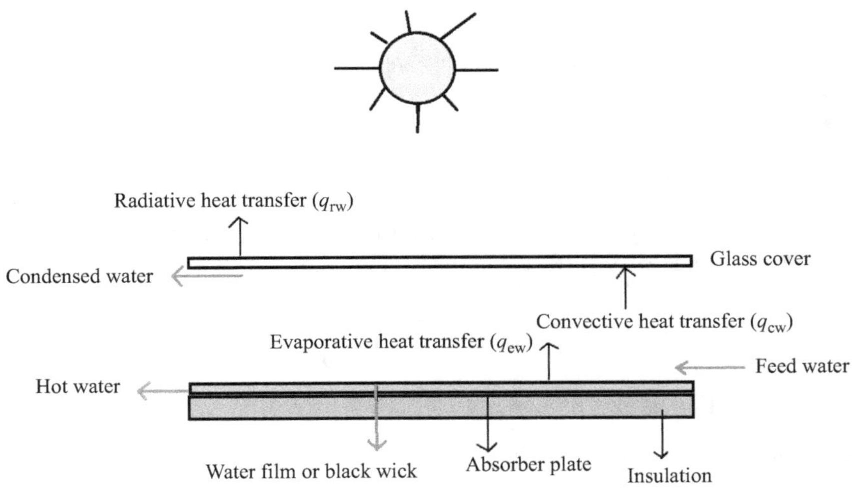

Figure 6.9 Thermal process of the inclined solar still [13]

The water evaporates and condenses on the inner surface of the glass cover. The condensate then runs through the condensation channel and is collected in the water container [13], as shown in Figure 6.9.

The performance of a solar still can be calculated based on efficiency, productivity, and heat-transfer coefficients. Heat-transfer coefficients and distillate output are directly responsible for the solar still performance. Heat transfer occurs due to evaporation, convection, and radiation [14].

6.3.1 Convective heat transfer

Convective heat transfer occurs within the solar still due to the temperature difference of humid air. It creates a density difference and is influenced by the buoyancy force. Convective heat transfer is directly proportional to the temperature difference of humid air. The convective heat transfer (q_{cw}) between the water and the condensing glass cover is given by [14,15]

$$q_{cw} = h_{cw}(T_w - T_g) \tag{6.1}$$

where the temperature of the water (T_w) is 28 °C and the temperature of glass (T_g) is taken as 32 °C for the calculations. Water is not boiling at 32 °C. The temperature usually goes as high as in the mid-1960s. The calculation is done for 8.00 AM, and data is entered accordingly. It provides with the simplest calculation of heat transfer, and as the time changes so do the values of temperature, resulting in more heat transfer.

$$\begin{aligned} q_{cw} &= h_{cw}(28 - 32) \\ q_{cw} &= -4h_{cw} \end{aligned} \tag{6.2}$$

where convective heat transfer coefficient (h_{cw}) can be calculated as

$$h_{cw} = 0.8\left[(T_w - T_g) + \frac{(\rho_w - \rho_g)(T_g + 273)}{268.9 \times 10^3 - \rho_w}\right]^{1/3} \tag{6.3}$$

where the density of water (ρ_w) is 1,000 kg m^{-3} and the density of glass (ρ_g) is 2,500 kg m^{-3}, and 268.9 $\times 10^3$ is a dimensionless quantity in the above equation.

$$h_{cw} = 0.8\left[(28 - 32) + \frac{(1{,}000 - 2{,}500)(32 + 273)}{268.9 \times 10^3 - 1{,}000}\right]^{1/3}$$

$$h_{cw} = -1.40 \text{ W m}^{-2} \text{ K}^{-1}$$

So, putting the value of h_{cw} in (6.2), we get

$$q_{cw} = 5.70 \text{ W m}^{-2}$$

6.3.2 Radiative heat transfer

Radiative heat transfer is another factor that influences the formation of pure water. Radiative heat transfer (q_{rw}) between the water surface and the condensing glass cover is given by [15,16]

$$q_{rw} = h_{rw}(T_w - T_g) \tag{6.4}$$

$$q_{rw} = -4h_{rw} \tag{6.5}$$

where radiative heat transfer coefficient (h_{rw}) can be calculated as

$$h_{rw} = \varepsilon_{effect}\,\sigma\left[(T_w + 273)^2 + \left(T_g + 273\right)^2\right] \tag{6.6}$$

where emissivity of glass (ε_g) and emissivity of water (ε_w), and σ is Stefan–Boltzmann constant.

$$\varepsilon_{effect} = \left[\frac{1}{\varepsilon_g} + \frac{1}{\varepsilon_w} - 1\right] \tag{6.7}$$

where $\varepsilon_g = 0.93$, $\varepsilon_w = 0.96$

$$\varepsilon_{effect} = 1.2$$

$$\sigma = 23.45 \times \sin\left[2 \times 3.14\left(\frac{284 + n}{365}\right)\right] \tag{6.8}$$

where n is the number of condensing surfaces.

$$\sigma = 2.10$$

Putting the values in (6.6), we get

$$h_{rw} = 4.50 \times 10^{-6}\ \mathrm{W\,m^{-2}\,K^{-1}}$$

So, putting the value of (h_{rw}) in (6.5), we get

$$q_{rw} = -1.80 \times 10^{-5}\ \mathrm{W\,m^{-2}}$$

6.3.3 Evaporative heat transfer

When solar radiation falls on the solar still, the temperature of the water increases; thus, the water evaporates and goes to the vapor phase. The evaporative heat transfer (q_{ew}) is given by [15,16]

$$q_{ew} = h_{ew}\left(T_w - T_g\right) \tag{6.9}$$

$$q_{ew} = -4h_{ew} \tag{6.10}$$

where evaporative heat transfer coefficient (h_{ew}) is calculated as

$$h_{ew} = 16.3 \times 10^{-3} \times h_{cw} \times \frac{\left(\rho_w - \rho_g\right)}{\left(T_w - T_g\right)} \tag{6.11}$$

$$h_{ew} = 16.3 \times 10^{-3} \times (-1.4) \times \frac{(1{,}000 - 2{,}500)}{(28 - 32)}$$

$$h_{ew} = -8.70\ \mathrm{W\,m^{-2}\,K^{-1}}$$

So, putting the value of (h_{ew}) in (6.10), we get

$$q_{ew} = 34.90\ \mathrm{W\,m^{-2}}$$

The overall heat transfer coefficient (h_{tw}) from the surface of the water to the condensing glass cover is calculated as

$$\begin{aligned} h_{tw} &= h_{cw} + h_{rw} + h_{ew} \\ h_{tw} &= -10.1\ \text{W m}^{-2}\ \text{K}^{-1} \end{aligned} \tag{6.12}$$

The heat transfer coefficients are responsible for the transportation of pure water in the vapor form leaving behind impurities in the water basin. They are also responsible for the condensation of pure vapor as distillate [15]. Convection, evaporation, and radiation are the heat transfer modes from water surface to condensing cover. The convective and evaporative heat transfers take place simultaneously. The convective heat transfer coefficients depend on the operating temperature range ($T_w - T_g$), geometry of the condensing cover, physical properties of the fluids, and flow characteristics of the fluids [17]. The evaporative heat transfer coefficient depends strongly on the temperature, and it is advisable to use the waste hot water either in higher temperature or during off-sunshine hours. The water at high temperature in the solar basin helps it to evaporate quickly. The radiative heat transfer coefficient is dependent on the material of the solar still. For this particular reason, the water basin is usually painted black as black color is the best absorbent of radiation. The main objective of any solar still is to maximize the evaporative heat transfer coefficient as it is directly proportional to the productivity and minimize the other energy transfers in terms of losses from basin, walls, and vapor leakages. The yield is the product of the evaporative heat transfer coefficient with the temperature difference between the condensing glass surface and the water basin ($T_w - T_g$). The increase in radiative heat transfer increases the temperature of the water basin, causing the temperature gradient to increase and hence evaporative heat transfer coefficient increases. In the above-mentioned case, the convective heat transfer (q_{cw}) is 5.70 W m^{-2}, evaporative heat transfer value (q_{ew}) is 34.90 W m^{-2}, and radiative heat transfer (q_{rw}) is -1.80×10^{-5} W m^{-2}. The evaporative heat transfer is far greater than the other two and has the most significant effect on the yield. The increase in these values results in the increase of productivity of solar still. Hence, it is essential to know the variation of these coefficients accurately [17].

The above calculations were performed based on the following assumptions [17]:

- There is no heat loss. The solar still is completely insulated.
- There is no leakage of the vapor. The chamber is leak proof.
- There is no gradient of temperature across the basin water and the glass cover of the solar still.
- The depth of the water inside the basin is constant.
- The heat capacities are negligible for the glass cover and the absorbing materials.

6.3.4 *Annual cost of water production*

The production cost of solar desalinated water per liter has been shown to be economical in Egyptian and Saudi markets for solar desalination plants. This study is applicable with different solar energy and cost characteristics. It is not confined to one particular region or a country. The payback on the capital costs depends on the production cost of fresh water. Where SFF is the sinking fund factor, i is the interest applicable every year (%) which is assumed to be 12.5 percent per year, n is the number of life years of the plant which is assumed to be 20 years, CRF is the capital recovery factor, FAC is the fixed yearly cost, S is the salvage value, ASV is the annual salvage value, P is the present capital cost, AMC is the annual maintenance cost (11% of the net present cost), AC is the annual cost, CPL is the cost of distilled water per liter, and M is the annual yield, then the cost of distilling each liter of water can be calculated as follows [18]:

$$\text{SFF} = \frac{i}{(1+i)^n - 1} \tag{6.13}$$

$$\text{CRF} = (\text{SFF}) \times (1+i)^n \tag{6.14}$$

$$\text{FAC} = P \times (\text{CRF}) \tag{6.15}$$

$$S = 0.2 \times P \tag{6.16}$$

$$\text{ASV} = (\text{SFF}) \times S \tag{6.17}$$

$$\text{AMC} = 0.15 \times \text{FAC} \tag{6.18}$$

$$\text{AC} = \text{FAC} + \text{AMC} - \text{ASV} \tag{6.19}$$

$$\text{CPL} = \frac{AC}{M} \tag{6.20}$$

6.4 Community to study

Pakistan is a developing country situated in South Asia. Pakistan has a total area of approximately 800,000 km^2 and lies in the sunny belt between latitude 24° N and 37° N and longitude 61° E and 75° E [19]. It has a contrasting topography, varying from the rugged mountainous regions in the north to the plains and fertile lands in the Indus valley, and arid and barren lands in the south and the west starting from the Cholistan through the Tharparkar desert in the Sindh. The climate also has extremes in temperatures, reaching 50 °C in the plains in the summer and below freezing in the mountains in the winter [19].

Pakistan is blessed with the abundance of sunlight for the most part of the year. A majority of the country receives 2,600 h of sunshine annually. The maximum intensity of solar radiation at noon on a clear day, in summer, may reach a value of nearly 1,000 W m^{-2}. However, during rainy and dusty days when the sky is cloudy, hazy, and dusty, the amount of solar energy input is low. The total solar radiation input on a horizontal surface may vary from 1 to 8 kW h m^{-2} day^{-1}, giving an annual daily average of 4–5 kW h m^{-2} day^{-1}.

Gwadar, at latitude 25.09° N and longitude 62.21° E, is a sea port situated in the south of Pakistan, along the Arabian Sea. It is approximately 90 km to the west

Figure 6.10 Map of Gwadar, Pakistan [19]

of the well-known seaport, Pasni. Pasni is a medium-sized town and a seaport in Gwadar district, Balochistan, Pakistan. Its population is around 33,000 [19]. It is located on the Makran coast of the Arabian Sea about 450 km from Karachi, shown in Figure 6.10. This study is for a small village in Gwadar comprising of almost 250 people.

Gwadar being a coastal city area has more than 70 percent humidity throughout the year, reaching a maximum of 80 percent in the months of July and August. The presence of humidity in the atmosphere results in a marked decrease in the intensity of total solar radiation falling over Gwadar. Figure 6.11(a) shows Gwadar's mean monthly relative humidity and wind speed, recorded at 03.00 GMT. The values are taken at 03.00 GMT because this is the starting time of the experimentation. The average monthly wind speed ranges from 2.1 to 0.6 ms^{-1} which is the average of wind speed taken over each month of the year in Gwadar. The minimum and maximum temperatures range from 11 to 35 °C [19]. Figure 6.11(b) shows the approximate daily mean of total solar radiation that falls on Gwadar, ranging from 350 to 600 cal cm^{-2} day^{-1}.

The estimated value of solar radiation is encouraging from an application point of view. The prospect of solar energy utilization is bright for Gwadar as the solar radiation received at this station is well above 300 cal cm^{-2} day^{-1}. In winter months, the sky is mostly clear and the amount of direct sunshine is high. The percentage of direct radiation can reach a maximum of 80–90 percent. Hence, the utility factor, both in summer and winter, is very optimistic [19].

A desalination unit with a capacity of 23 m^3 day^{-1} covering an area of 8,026 m^2 was installed at Gwadar, Pakistan and its performance was studied. The solar desalination plant consisted of three sections:

1. Pumping station
2. Solar still
3. Reservoir for the collection of desalinated water

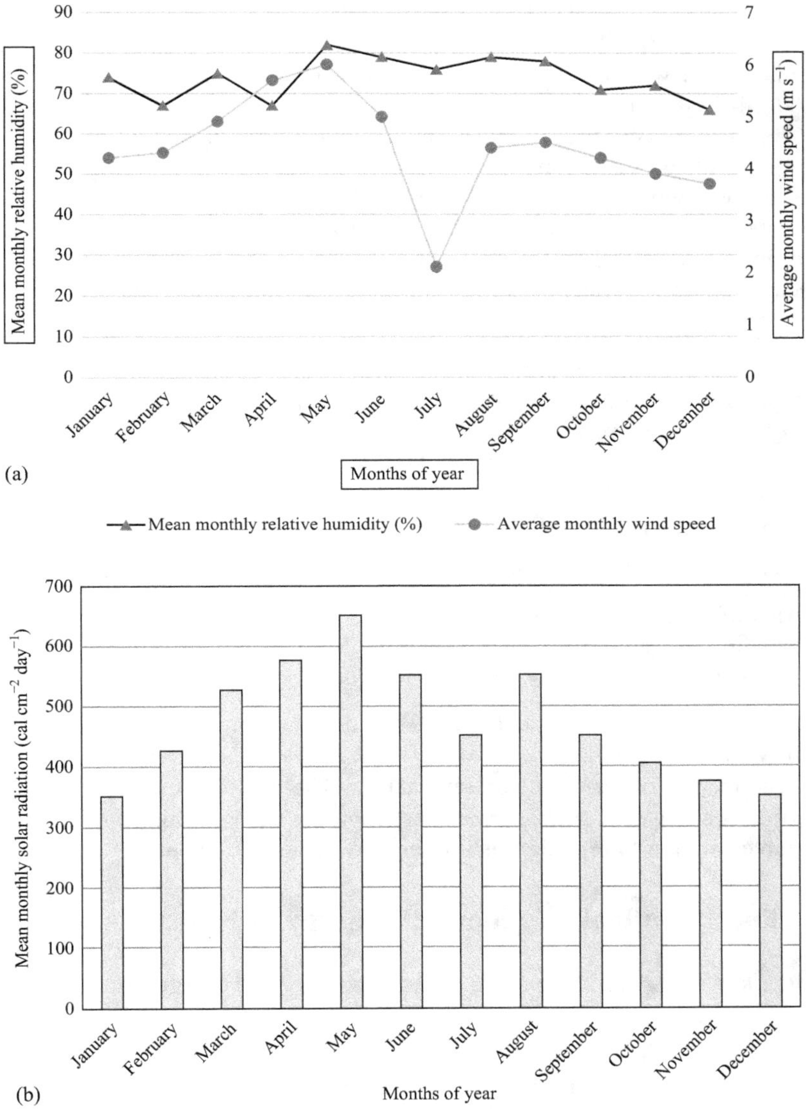

Figure 6.11 (a) Mean monthly relative humidity and wind speed at Pasni station, 90 km east of Gwadar at 03.00 GMT and (b) mean monthly solar radiation (direct and diffuse) falling at Gwadar area at 03.00 GMT [19]

The solar desalination plant consisted of 240 solar stills—arranged in 4 rows each containing 60 stills of size 24 × 1.3 m. These stills are single-effect style. The glass used was 3 mm thick. The amount of water collected was recorded at 7.00 AM–7.00 PM [19].

The basin was constructed of sheets of aluminum fitted with a glass cover and sealed with silicon sealant in order to make it air tight. From the pumping station, sea water was pumped into the stills. The Sun's radiation through the glass sheets evaporated the water. The water vapors condensed on the inner surface of the glass sheets and the fresh water film on the glass ran into an aluminum channel and was collected in the fresh water reservoir. The costs associated with this plant were [19]

- Cost of the plant = 95,000 USD
- Life of the plant = 20 years
- Maintenance cost = 1,100 USD
- Total cost of the plant = 96,100 USD
- Cost of the plant per year = 4,805 USD
- Collection of water per year = 8,290 m^3 $year^{-1}$
- Cost of water per m^3 = 0.60 USD

The yield of the plant was found to be

- Area of the glass coverage = 8,027 m^2
- Desalinated water produced = 23 m^3 day^{-1}
- Water produced per m^2 = 0.003 m^3 m^{-2} day^{-1}

The production of potable water was 23 m^3 day^{-1} (average). The estimated unit cost of water is $0.60 USD m^{-3}, which is the least expensive cost in the Gwadar area. The quality of water is also satisfactory. The desalination plant reported in this paper is a low cost, simple option for providing water in this region [19]. This is a viable solution to provide fresh water to water-scarce regions.

6.5 Future outlook of renewable energy in Pakistan

The four key drivers for developing Pakistan's renewable energy resources include

6.5.1 Energy security

Pakistan's energy mix can be diversified by mainstreaming of renewable energy and more use of indigenous resources. It reduces the country dependence on any single source, particularly imported fossil fuels, thereby mitigating against supply disruptions and price-fluctuation risk [20]. Renewable energy systems are not dependent on additional costs and risks relating to fuel stocking, transportation, and temporary substitute arrangements, except for backup purposes [21,22].

6.5.2 Economic benefits

Renewable energy options can become economically competitive with conventional supplies on a least cost basis. This is particularly true for the more remote and underdeveloped areas, where renewable energy can also have the greatest impact and the avoided costs of conventional energy supplies can be significant [20,21]. Renewable energy can thus supplement the pool of national energy supply options in Pakistan. Decentralized renewable energy systems can also help reduce energy distribution losses and result in system-wide and national efficiency gains (e.g. as measured by "energy intensity" or energy use per unit of gross domestic product). As well, growing renewable energy industry can afford new prospects for employment and business opportunities amongst local manufacturers and service providers [20,22,23].

6.5.3 Social equity

Pakistan's current low per-capita consumption of energy can be increased through greater renewable energy use. Issues relating to social equity, such as equal rights and access for citizens to modern energy supplies, improved human development indicators, poverty alleviation amongst deprived sections of society, and reduced burden on rural women for biomass fuel collection and use can also be addressed significantly through wide-spreading renewable energy [20]. Renewable energy can thus help improve the well-being of the country's poorest, which presently have little or no access to modern energy services [20,21].

6.5.4 Environmental protection

Clean renewable energy alternatives can circumvent local environment and health impacts of unsustainable and inefficient traditional biomass fuels and fossil-fuel-powered electricity generation [21]. Similarly, displaced greenhouse gas emissions carry significant global climate change benefits, toward which Pakistan has pledged action under the United Nations framework convention on climate change [20,21].

6.5.5 Future development of renewable energy

The following strategies can help to develop renewable energy sources in the country.

- Introduce investment-friendly incentives and facilitate the renewable energy market to attract private-sector interest in renewable energy projects, help nurture the nascent industry, and gradually lower renewable energy costs and prices through competition [20].
- Device measures to support the private sector in mobilizing finance and enabling public-sector investment in promotional, demonstrative, and trend setting renewable energy projects [20,21].
- Optimize the impact of renewable energy deployment in underdeveloped areas by integrating energy solutions with the provision of other social infrastructure, e.g. educational and medical facilities, clean water supply, sanitation, and telecommunication, etc. [20,21].
- Help in broad instructional, technical, and operational capacity building relevant to the renewable energy sector [20–23].

- Facilitate the establishment of a domestic renewable energy manufacturing base in the country that can help lower costs, improve service, create employment, and enhance technical skills [21–24].

Renewable energy sources not only provide stability in the manufacturing price of technology but are also easily reachable to the remote areas of Pakistan. It reduces the dependence of Pakistan on fossil fuels and biomass. Renewable sources also help Pakistan to maintain its pledge with UNO for the limited emission of greenhouse gases and play its part in resisting climate change.

6.6 Conclusion

Energy is a necessity for human beings living in the modern era. Solar-energy technology and its usage are very important for the developing and underdeveloped countries to sustain their energy demand. The use of solar energy for water desalination is one example of an application of renewable energy. An overview of solar desalination technologies, along with the theoretical and cost analysis, has been presented, focusing on those technologies appropriate for use in remote villages. Solar energy coupled to desalination offers a promising prospect for covering the fundamental needs of water in remote regions, where water scarcity is severe. Direct solar stills have become more popular in rural areas. The solar stills are friendly to the ecosystem and are easy to operate. The seashore of Gwadar, Pakistan has been studied as a small community with a working direct solar–water desalination plant of 8,290 m^3 $year^{-1}$. Pakistan is a developing country with abundant solar energy throughout the year. The study includes the cost analysis for 20 years' operation of this plant, 0.60 USD m^{-3}. The four key drivers for developing Pakistan's renewable energy resources, such as solar–water desalination, include increasing energy security, economic benefits, social equity, and environmental protection.

In brief, inclined solar stills put together some key features to the solar still which can be listed as better projection toward the sun, large effective area, simple design, low cost, longer flow of water, high rate of evaporation, flexibility to change inclination angle, etc. All these features make an inclined solar still more effective for small communities, as of now.

Even though solar energy is a clean form of energy, it is still not harnessed to the full scale of its potential due to the limitation of the efficiency and the high cost of equipment used for desalination systems. Therefore, there is further need for research on cost-cutting technologies and the equipment should be economical and consume less energy with high performance. More research should be done on technologies like multi-stage flash distillation (MSF), multiple-effect distillation (MED), and reverse osmosis (RO) so they can be more compact, automated, portable, and easily mobilized for transportation. Thus, they can be harnessed by small communities like schools, colleges, and hotels by installing them on the rooftop and playgrounds.

The major desalination technologies, along with their barriers, have been discussed in depth. The barriers can be overcome by innovative ideas and research work. The barriers in any system provide the drivers to continue experimental, modeling, and simulation research. The collaboration of academic institutions with industry is essential for transforming innovative ideas into reality and bringing the finalized product to the market.

Acknowledgment

The authors are thankful to Pakistan Council of Scientific and Industrial Research and Computerized Data Processing Centre, Pakistan Meteorological Department, University Road, Karachi 75720, for providing valuable data and information for the case study section.

References

[1] UN, "The Millennium Development Goals Report 2015," ISBN 978-1-4648-0670-4, 2015.

[2] T.A. Lawand, and J. Ayoub, "Materials for construction of solar stills," *Renewable Energy Systems and Desalination*, vol. 3, pp. 1–7, 2010.

[3] UNESCO, "The United Nations World Water Development Report 2016: Water and Jobs," UNESCO, Paris, 2016.

[4] WHO/UNICEF, "WATER FOR LIFE – Making it Happen, Joint Monitoring Programme for Water Supply and Sanitation," ISBN 92-4-1562-935, 2005.

[5] K.G. Mcguigan, R.M. Conroy, H.J. Mosler, M.D. Preez, E.U. Jaswa, and P.F. Ibanez, "Solar water disinfection (SODIS): A review from bench-top to roof-top," *Journal of Hazardous Materials*, vol. 235, no. 2, pp. 29–46, 2012.

[6] J.H. Lienhard, M.A. Antar, A. Bolton, J. Blanco, and G. Zaragoza, "Solar desalination," *Annual Review of Heat Transfer*, vol. 15, no. 15, pp. 327–47, 2012.

[7] R. Stewart, "Temperature, salinity and density," In *Introduction to Physical Oceanography*. Florida, TX: Texas A&M University, Orange Grove Text Plus, pp. 73–101, 2008.

[8] D.W. Dsilva, S. Iniyan, L. Suganthi, and P.A. Davies, "Solar stills: A comprehensive review of designs, performance, and material advances," *Renewable and Sustainable Energy Reviews*, vol. 63, pp. 464–96, 2016.

[9] P. Patelv, A.R. Patel, A.S. Solanki, and U.R. Soni, "A review to increase the performance of solar still: Make it multi-layer absorber," *International Journal on Recent and Innovation Trends in Computing and Communication*, vol. 2, no. 2, pp. 173–77, 2014.

[10] A.K. Kaviti, A. Yadav, and A. Shukia, "Inclined solar still designs: A review," *Renewable and Sustainable Energy Reviews*, vol. 54, pp. 429–51, 2016.

[11] T. Rajaseenivasan, K.K. Murugavel, T. Elango, and R.S. Hansen, "A review of different methods to enhance the productivity," *Renewable and Sustainable Energy Reviews*, vol. 17, pp. 248–59, 2013.

[12] O.O. Badran, and H.A. Al-Tahaineh, "The effect of coupling a flat-plate collector on the solar still productivity," *Desalination*, vol. 183, no. 1–3, pp. 137–42, 2005.

[13] H.S. Aybar, "Mathematical modeling of an inclined solar water distillation system," *Desalination*, vol. 193, no. 1–3, pp. 63–70, 2006.

[14] H.N. Panchal, and P.K. Shah, "Modeling and verification of single slope solar still using ANSYS-CFX," *International Journal of Energy and Environmental Engineering*, vol. 2, pp. 985–98, 2011.

[15] H.N. Panchal, M. Doshi, C. Prakash, and G. Ranvirgiri, "Effects of cow dung cakes inside basin on heat transfer coefficient sand productivity of single basin solar still," *International Journal of Applied Engineering Research*, vol. 2, pp. 985–98, 2011.

[16] M. Chandrashekara, and Y. Avadhesh, "Water desalination using solar heat: A review," *Renewable and Sustainable Energy Reviews*, vol. 67, pp. 1308–30, 2017.

[17] A. Al-Karaghouli, and L.L. Kazmerski, "Energy consumption and water production cost of the conventional and renewable energy powered desalination processes," *Renewable and Sustainable Energy Reviews*, vol. 24, pp. 343–56, 2013.

[18] E. El-Bialy, S. Shalaby, A. Kabeel, and A. Fathy, "Cost analysis for several solar desalination systems," *Desalination*, vol. 384, pp. 12–30, 2016.

[19] Pakistan Council of Scientific and Industrial Research, "Design and Fabrication of Single Slope Solar Still," 2014.

[20] United Nations Economic and Social Council, "65th session of the commission," In *Annual Report of the Economic and Social Commission for Asia and the Pacific Official Records*, United Nations Economic and Social Council, pp. 3–34, 2013.

[21] Japan International Cooperation Agency, "Data Collection Survey on Renewable Energy Development in Pakistan," Islamabad, pp. 1–141, 2013.

[22] M. Ahmed, "Fueling the future: Meeting Pakistan Energy Needs in 21st Century," Woodrow Wilson International Center for Scholars, Washington, DC, 2007.

[23] M. Asif, "Sustainable energy options for Pakistan," *Renewable and Sustainable Energy Reviews*, vol. 13, pp. 903–9, 2009.

[24] U. Mirza, N. Ahmad, K. Harijan, and T. Majeed, "Identifying and addressing barriers to renewable energy development in Pakistan," *Renewable and Sustainable Energy Reviews*, vol. 13, pp. 927–31, 2009.

Chapter 7

Community solar PV projects

Avinash Singh[1], *Paul Henshaw*[1] *and David S-K. Ting*[1]

Abstract

In today's world, global warming (GW) and the resulting climate change are a significant threat to humans, plants and animals. The main contributor to GW is greenhouse gases (GHGs) that are created from the burning of fossil fuels, mainly for electrical power. Hence, the way forward in safeguarding the future of life on planet earth is to reduce on our GHG emissions and move rapidly to the utilization of renewable energy resources that are abundantly available.

There are numerous advantages in implementing renewable energy projects versus the use of fossil fuels in meeting individual or community energy demands. With renewable energy, an individual or community will be able to have a more diverse and stable long-term energy supply, considering fossil fuel resources are slowly being depleted. Small-scale renewable energy projects, especially in remote communities which are currently served by diesel-generated electricity, offset the community's use of diesel fuel. Although there will be times when renewable energy is not available and a back-up source of power is required, the long-term cost of energy may be reduced. And the use of indigenous energy can contribute to a nation (or region's) energy security by significantly reducing its dependence on imported oil (assuming it is not an oil exporter). There are numerous renewable energy resources available globally (wind, solar, biomass, falling water, geothermal) that can be used for individual or community energy projects. *Community* energy projects are distinguished from other renewable energy projects in which members of the community are subscribers who use the electricity produced by the project, even though each individual does not solely own the equipment.

This chapter presents an overview of how energy captured from the sun can be utilized at the community level by installing solar photovoltaic systems in the form of a solar garden or solar farm or solar power plant to generate electrical energy in meeting some, if not all, of the community total energy demand. In addition, readers will be exposed to three common ownership models and their benefits, barriers affecting the adoption of such projects and selected examples of such

[1]Turbulence and Energy Laboratory, Department of Engineering, University of Windsor, Canada

projects that have been completed or are in the conceptualization or construction phase within North America, Europe, South America and East Africa.

7.1 Introduction

7.1.1 What is a community solar PV project?

A community solar photovoltaic (PV) project is when a set of solar PV panels are assembled in an identified area that is termed a solar garden or solar farm or solar power plant and from which homes within the community can receive electricity. A key feature is that the users of the electricity benefit financially or environmentally from the choice of renewable electricity generation. Hence, either the community or a third party owns the PV system and the homes in the community benefit by tapping into the electricity produced. Thus, the communities can easily be protected from major electricity outages. For example, if the grid needs to shut power off due circumstances beyond communities' control, those communities can still have electricity, at least partially [1].

With the expansion of this form of renewable energy harvesting, countries around the world will be able to meet a significant percentage of their electricity needs and experience possible savings by avoiding the purchase of fossil fuels. In addition, community solar PV projects contribute to the reduction in greenhouse gases (GHGs) emission; hence, the project owner can take advantage of the available carbon financing. Carbon financing is basically financial disbursements to community solar PV projects in the form of "entitlement certificates" issued by the United Nations Framework Convention on Climate Change (UNFCCC): one carbon credit received is equivalent to 1 t of GHGs emission. Under the UNFCCC charter, any corporation from a developed country can create an agreement with a corporation in a developing country that is a signatory to the Kyoto Protocol, in which climate change mitigation measures are documented (e.g. prevent deforestation) [2].

7.1.2 Rationale of community solar PV projects

The burning of fossil fuels around the world is the largest anthropogenic contributor to GW and climate change. In addition, the burning of fossil fuels also results in local air pollution, for example, NO_X, CO and ground-level ozone. Hence, renewable energy projects, such as community solar PV, will yield the following.

7.1.2.1 Performance guarantee

The investors in a properly engineered community PV project will have a predictable energy output and costs over a long project life. This is due to solar PV units that are presently manufactured to perform at 80% of their initial efficiency for a period of 20–25 years. The investors further benefit from guaranteed price contracts that extend over the life of the PV system. This contract may be part of a government mandated feed-in-tariff (FIT) or private power purchase agreement

(PPA). Investors benefit when the cost of the fossil fuels used to produce the electricity increases due to changes in world market prices [3].

7.1.2.2 Reduction of greenhouse gas emissions

This would mitigate the temperature rise on planet earth in the future. By mitigating the temperature increase, issues such as polar ice and glaciers melting, heat-related illness and the expanding range of pests will stabilize. In addition, through mitigation, a reduction in weather-related natural disasters (droughts, floods) is anticipated, making the habitat for humans, plants and animals safe and more comfortable [3].

7.1.2.3 Reduction in global air pollution

If solar electricity displaces producing electricity by burning fossil fuels, it will significantly reduce global air pollution-related diseases, thereby making the environment healthier for the global population [3].

7.1.2.4 Reduction in foreign exchange deficit

This applies to any country (or region) that currently imports energy carriers to generate electricity. Sunlight cannot be controlled or monopolized by a single individual or country as with fossil fuels [3].

7.1.2.5 Increase in electricity security and reliability

Homes and businesses will be able to have an available and reliable source of electricity to meet part of their demand. On most days there is sunshine and also the energy from the sun can be stored for future usage. Furthermore, sunlight will continue to be available for many generations to come, as opposed to fossil fuels, which are nonrenewable and therefore will be depleted at some point [3].

7.1.2.6 Local control

The local community is able to control and manage community solar PV projects; hence, the community will be empowered to make decisions on matters such as the size of the project and siting. In addition, if the project is to be developed by a sector outside of the community, then forming a community initiative may be considered as a tool for local control [4].

7.1.2.7 Job creation

The installation of solar PV will provide support to the local solar industry and create green jobs locally [5].

7.1.2.8 Utilization of prospective sites

Many sites (parking structures, warehouses, landfills, brownfields, etc.) that are not developed for various reasons, such as regulatory barriers or lack of financial inputs, are available for community solar PV projects [5].

The main purpose of installing a solar garden/solar farm/solar power plant within a community is to provide the individuals within that community the opportunity to share the benefits listed above, even though they might not be able to install a solar PV system on their individual property due to constraints, such as

1. Their house roof is too small to install an economically feasible system;
2. The roof is at a wrong orientation or is structurally unable to support a PV system;
3. The individual is unable to afford a full-roof PV system, making it uneconomical;
4. The individual does not own his/her own residence;
5. Homes are unable to sell the power to the local utility grid;
6. The individual has already maximized his/her rooftop PV and wants more benefits.

In short, community solar PV projects are a means of giving individual investors, homes and businesses, the opportunity to get involved in the vastly expanding clean energy economy. This opportunity has become available recently as the cost of PV systems has dropped, utilities have become more accepting of distributed generation and citizens have become more aware of the environmental benefits of renewable energy.

7.1.3 Variations in community solar PV projects

The basic concept is that numerous customers share the benefits and costs of a solar system; however, community solar PV can take various forms. Outlined below are some of the key variables for community solar PV projects [5].

7.1.3.1 Ownership

Who is responsible for the development and operation of the community solar PV project? Some of the projects have been developed and owned by private investors (either for profit or nonprofit) and others by local utility companies or electricity customers themselves. In addition, such a project could be installed either on the property of the project owner or on the property of a third party.

7.1.3.2 Participation

What role does the customer play? In some of the projects, the customers are given the opportunity to have a stake in the project, whilst in other situations, the customers can only subscribe rather than own. In addition, there are some projects allowing the customer to purchase a share of the capacity, for example 1 kW, or share of the output, for example 1,000 kW h.

7.1.3.3 Valuing the solar energy

How is the cost of the energy credited to the participants? In a net metered system, the energy produced (kW h) is credited to the customers' energy utility bill: customers utilize their own generation from the on-site solar PV system to offset their energy consumption over a period. This requires meters that normally measure the flow of electricity to the customer to function in a reverse direction when electricity is generated in excess of the customer's demand. In a utility-owned system, subscribers pay the utility for the energy produced from the community solar PV system, and this fee pays for the investment made by the utility. In FIT systems, the system owner sells electricity to the utility at a fixed rate.

7.1.3.4 Geographic location

The participants in a community solar PV project may be required to be [5]

1. In the same territory of the utility and solar PV installation; or
2. In the same county of the utility and solar PV installation; or even
3. In the same neighborhood of the utility and solar PV installation.

7.1.3.5 Utility compensation

The utilities may receive compensation for the program administration and distribution of the electricity generated by the community solar PV. Compensation can either be through a fee that is embedded into the agreement with the developer of the solar PV project or controlled by a state public utility commission, by either retaining the charges for the distribution of the electricity generated or, as is often done in the United States of America (USA), by allowing the utility to receive the renewable energy credits (RECs) from the project [5].

7.2 Community solar PV models

Physically, all community solar PV projects have some commonalities. Solar PV panels are mounted where they will be exposed to the daylight, for the greatest duration possible. Ground-mounting and rooftop mounting are the two main options. PV panels are connected in series and parallel to increase the voltage and current from the PV array, respectively. The direct current output from the panels is converted to alternating current power by an inverter. Inverter sizes range from microinverters (on each PV panel) to a single inverter for the array (typical for household rooftop systems) to larger systems with multiple inverters to provide reliability. The power must be further conditioned to match the voltage and phase at its tie-in to the grid. In addition, in the installation of community scale solar PV projects, it may be advantageous but costly to mount the PV panels on a solar tracker system so as to maximize the harvesting of solar irradiance especially when the days are shorter. A solar tracker system enables the array to harvest approximately 40% more of solar irradiance than when fitted in a fix position [6].

Community solar PV projects may be classified as grid/utility sponsored, special purpose entity sponsored and nonprofit sponsored. However, there are hybrids of these main models in existence. Below are more details on the main models.

7.2.1 Grid/utility sponsored community solar PV projects

7.2.1.1 What is grid/utility sponsored community solar PV project?

A grid/utility sponsored community solar PV project is when the grid/utility develops, invests in the solar garden or solar farm or solar power plant and operates it and then opens it to utility customers (rate payers) to voluntarily join and purchase electricity. In addition, the utility customers participate in the project by contributing an advance or on-going payment to give their support and in exchange, the customers are given a payment or credit that is proportional to their contribution and the amount of electricity produced from the community solar PV project. However, the participating customers do not have any ownership claim to the PV system in this model: the project is owned by a grid/utility company or a third party [7].

7.2.1.2 Grid/utility sponsored structure

As can be seen in Figure 7.1, electricity flows from the solar PV system, through the grid and to the customer (rate payer). The rate payer voluntarily purchases electricity from the solar PV project and may receive a credit for the amount of prepaid electricity. In the USA, RECs may be sold to rate payers but more often are retained by the utility to document their efforts to meet renewable energy generation quotas.

7.2.1.3 Benefits of grid/utility sponsored model

There are several benefits to executing a solar PV project through a grid or utility. First, the utility company is considered to have the necessary revenue, legal status and infrastructure management that are required to successfully develop a community solar PV project [7]. In some cases, grid/utilities are controlled by the

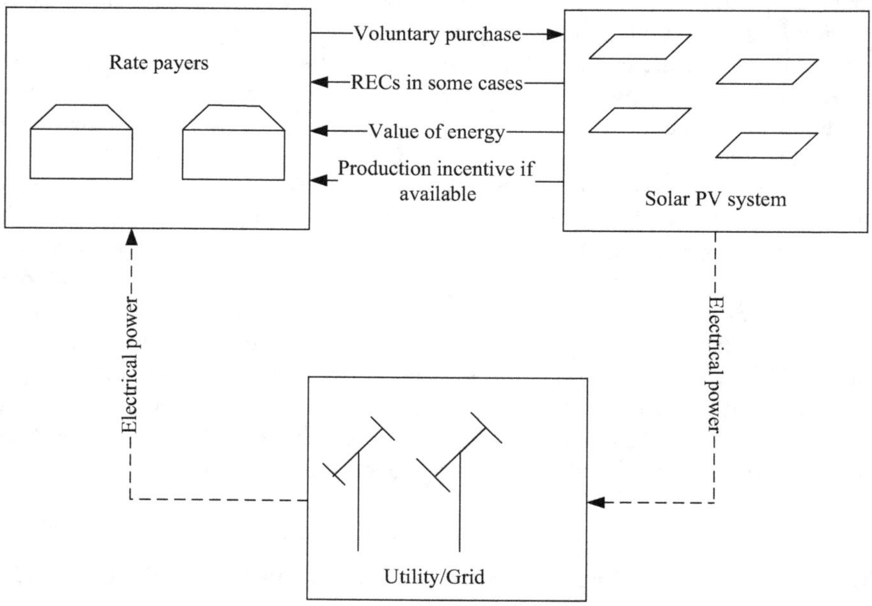

Figure 7.1 Generic grid/utility sponsored community solar PV model [7]

member customers and, hence, can be directed to develop such projects on behalf of the member customer. It is said that a quarter of the USA population own their own electricity company either through co-ops, or city or county-owned utilities. In this case, the customers are arm's length owners [7]. It has been shown that publicly owned electricity grids/utilities have taken the leadership role in driving community solar PV. Even if the grid/utility is privately or investor owned, it may aim to expand its customers' choices with an option for a renewable source of energy in the form of solar PV [7]. It should be noted that in some jurisdictions, only private utilities can make use of any tax credits flowing from the installation of the solar PV system. Even then, such utilities must normalize the benefits over the life of the solar PV system.

7.2.2 Special purpose entity (SPE) sponsored community solar PV

7.2.2.1 What is a special purpose entity sponsored community solar PV?

A special purpose entity sponsored community solar project occurs when an investor or a group of investors comes together in a business enterprise to develop the project and operate it after the construction. In addition, the investor or group of investors must be able to navigate the legal and financial hurdles and raise the required capital and at the same time to comply with regulations. Also, it must be able to negotiate contracts among the participants/owners, site host and utility (through a PPA), put legal and financial processes in place for the sharing of the electricity produced and administer the business operations [7]. Baywind was responsible for setting up the very first special purpose entity sponsored community model, which was established in the United Kingdom in the later part of the 1990s [4].

7.2.2.2 How does a special purpose entity sponsored project operate?

The solar garden or solar farm or solar power plant is developed by an investor or a group of investors, sometimes a cooperative, who supply and install all the materials required for making the community solar PV project a reality. Rates are then computed by the investor or investors and offered to residents in the community to join in and purchase their daily power demand from the project investors. The electricity generated from the project can either be sold directly by the investor or group of investors to the residents or fed into the grid/utility. It may be less expensive to feed the electricity produced into the grid/utility than to sell direct to the residents because the grid/utility will already have the infrastructure in place for transmission and distribution instead of having to install new infrastructure [7].

7.2.2.3 Special purpose entity sponsored community solar PV structure

In this model (Figure 7.2), electricity flows from the solar PV system to the utility and ultimately to the rate payers (not shown). Electricity can also be used by the

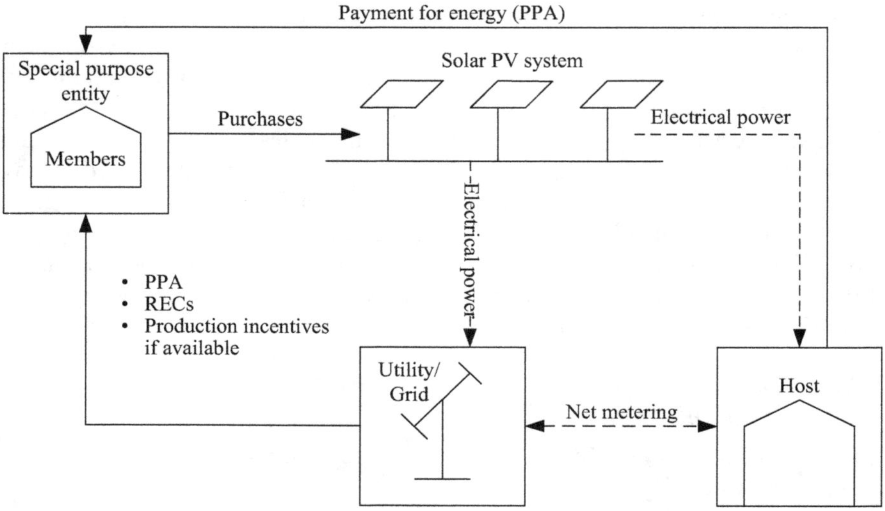

Figure 7.2 Generic special purpose entity sponsored community solar PV model [7]

host to reduce the energy it purchases from the grid (net metering). Part of the revenue collected by the utility goes to the SPE through a PPA. If the host employs net metering, then the SPE must have a PPA with the host for the host's purchase of power from the solar PV system.

7.2.2.4 Benefits of SPE sponsored community solar PV projects

There are a few additional benefits to a SPE sponsored project. One is that, although the solar garden or solar farm or solar power plant is owned by the SPE, in the long term, the system can be sold to the community or host [7]. The other advantage is that investors who are members of the SPE are more able to capitalize on tax incentives for solar PV projects, than are utilities [7].

7.2.3 Nonprofit sponsored community solar PV

7.2.3.1 What is a nonprofit sponsored community solar PV project?

A nonprofit sponsored community solar PV project is when financing for the development and installation of the project comes from a donor agency, and the PV system is operated and managed by a nonprofit corporation or body [7].

7.2.3.2 How does a nonprofit sponsored community solar PV operate?

In a nonprofit sponsored community solar PV project, a nonprofit corporation or body will provide their ideas and project drawings to potential donors of a

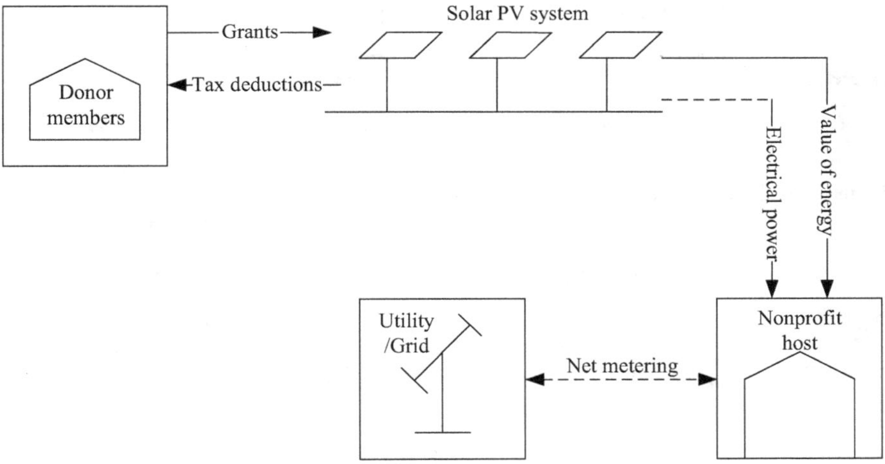

Figure 7.3 Generic nonprofit sponsored community solar PV model [7]

renewable energy resource and in doing so try to solicit funding for the project. After installation, the PV system is operated by the charitable nonprofit corporation or body to reduce electricity purchased from the utility or to sell excess electricity to the utility (Figure 7.3) [7].

7.2.3.3 Nonprofit sponsored community solar PV structure

Electricity from the solar PV system is used by the host to offset its utility costs, ideally through net metering. Finances come from donors to build the PV system. The nonprofit host of the system then has a legacy which will reduce or eliminate its utility costs over the life of the project.

7.2.3.4 Benefits of a nonprofit sponsored community solar PV model

In the nonprofit sponsored community solar PV model, donors may be able to receive from their government a tax benefit for contributing to a charitable organization, in the form of a tax deduction. In addition, the nonprofit host is largely sheltered from electricity price fluctuations in the future, because it has its own source of power. In effect, donors pay for an asset, which in the long term may provide a benefit to the nonprofit body [7].

7.2.4 Comparison of the community solar PV project models

A summary of the features of the three models for community solar PV projects is presented in Table 7.1. None of the models is superior to the others in all situations, and the choice of model depends on the intentions, willingness to take risks, availability of capital and tax appetite of the investors.

Table 7.1 Comparison of the three community solar PV models [7]

Description	Community solar PV project models		
	Utility sponsored	***Special purpose entity sponsored***	***Nonprofit sponsored***
System owner	Utility company or 3rd party	Members of SPE	Nonprofit organization
System financer	Consumers, government grants, utility	Members of SPE	Nonprofit organization, using funds from donors
Host	Utility company or 3rd party	3rd party	Nonprofit organization
Subscriber	Utility customers (rate payers)	Members of SPE	Donors, usually members of nonprofit organization
Motive of subscriber	Offset own electricity use	Investment gains and offset own electricity use	Charitable
Motive of owner	Give customers renewable options, meet renewable quotas	Continue electricity generation or sell to the host	Continue electricity generation for the project life
Example	Sacramento Municipal Utility District, California, USA	Agahozo-Shalom Youth Village, Rwanda, East Africa	Kingston, Timiskaming and Earlton, Ontario, Canada

7.3 Community solar PV projects implementation barriers

Community solar PV projects face implementation barriers. Below are some examples:

7.3.1 High acquisition and installation cost

The cost of acquiring of solar panels and installation is considered to be high at present, but prices are dropping. Early 2016 estimates by NREL put the installed cost of a 5,600-W PV system at US$2.93 per Watt [8], which is down from US$7.6/W for a 1,500 W system in 2009 [9].

7.3.2 Space

In setting up a solar garden or solar farm or solar power plant within a community, sufficient space is needed for the required solar collectors and any future expansion and developments. An average of 2 ha (0.020 km^2) is required to construct and manage a solar garden/solar farm/solar power plant with a capacity of 1 MW [9].

7.3.3 Investors

In setting up a solar garden or solar farm or solar power plant within a community, there may not be sufficient investors to contribute to the project, due to security laws in various countries that are not favorable to attract investors. Hence, individuals may not have the relevant knowledge or cannot afford to employ the skills to design, install and operate a project of this magnitude [9].

7.3.4 No grid connection

In setting up a solar garden or solar farm or solar power plant within a community, grid connection is very important since the grid will already have the infrastructure in place to distribute the electric energy for utilization by the prospective beneficiaries of the project. In various countries around the world, the absence of a grid or the lack of ability to connect a PV system to the grid can be major obstacles [9].

7.3.5 Lack of government policies

In setting up a solar garden or solar farm or solar power plant within a community, having a FIT, PPA or net metering policy is a key factor in making a community solar PV project a financial success. However, there are plenty of countries that have great potential in solar energy but do not have any of the mentioned programs, making it difficult to have community solar PV projects since investors are not clear on what the future policies will be. Guyana is an example of a country potentially rich in solar energy, but with no solar programs or policies in place [9].

7.3.6 Lack of government incentives

In setting up a solar garden or solar farm or solar power plant within a community, incentives from the government is an important factor for community solar PV projects to be a success. However, many countries around the world have great potential in solar energy, but do not have any incentive programs in place, making the investment payback time longer. Guyana is an example of such a country: with no incentive programs or policies in place, there is very little investment in solar energy [10]. In addition, many renewable energy advocates believe that small-scale privately owned community renewable energy projects, such as community solar PV, are not strongly incentivized [11].

7.3.7 Complexity issues

Setting up a solar garden or solar farm or solar power plant within a community can be very complex to many individuals especially when having to make the technical choices such as sizing the inverter, the amount of PV modules required and setting the best tilt for the array. In addition, the owner must identify the right contractor for the job, especially for owners who do not have technical knowledge [12].

Table 7.2 TREC solar share project summary, Ontario, Canada [13]

Site/ Location	Quantity (no.)	Type of installation	Generated capacity (kW)	Total capacity (kW)
Kingston	2	Ground mounted	600	1,200
Timiskaming	8	Ground mounted	600	4,800
Earlton	1	Roof mounted	300	300
Total capacity (kW)				**6,300**

7.3.8 Customer inertia

In setting up a solar garden or solar farm or solar power plant within a community, customer inertia may be a factor in deterring investors. For commercial PV systems, the period from first inquiry to installation is usually over 2 years [12].

7.4 Selected examples of existing/future community solar PV projects

7.4.1 Ontario, Canada

On May 2, 2016, it was announced by UGE International, through its wholly owned subsidiary Endura Energy, that it was going to develop, design and construct a community solar PV Project in Ontario, Canada with a capacity of 6.3 MW. After the completion of the Project, it will be owned and operated by Canada's largest nonprofit sponsored entity, TREC Solar Share Cooperative (No. 1) Incorporated. Hence, this project follows the special purpose entity sponsored model. The project consists of a total of 11 sites for which 10 will be ground-mounted facilities and the other a roof-mounted facility. The project is estimated to give UGE International revenue of CAD$18 million to $20 million when completed [13].

The electrical energy produced from this project will feed the Ontario grid. The Green Energy Act was first introduced by the province of Ontario in 2006 and revised in 2009 and 2010. One component of this Act, the FIT program, was introduced to encourage and promote the usage of renewable energy resources (on shore wind, solar PV, landfill gas, biogas, water power and renewable biomass) in meeting Ontario's electrical energy demand. The Ontario FIT was North America's first wide-ranging price guarantee structure for electrical power produced from renewable energy sources, offering stable prices under a long-term contract agreement. Under the FIT program, PV system owners who enter into a 20-year contract to supply electricity to the Ontario grid are guaranteed a fixed price per kW h depending on when they first applied to the program. The guaranteed rate for the electricity produced has decreased in keeping with the decrease in installed PV system costs (Table 7.2). Eventually, the FIT should reach "grid parity," such that the tariff represents the average cost of electricity generated by all inputs to the grid [14].

Table 7.3 Feed-in-tariff rates in Ontario, Canada for 50 and 500 kW solar rooftop installations

Date	FIT rates (¢/kW h)	
	50 kW	500 kW
September 2009	71.3 [16]	53.9 [16]
April 2012	54.3 [17]	48.7 [17]
August 2013	34.5 [18]	32.9 [18]
January 2014	34.5 [19]	32.9 [19]
January 2015	34.5 [20]	31.6 [20]
June 2016	24.2 [21]	22.5 [21]
January 2017	22.3 [22]	20.7 [22]

As of April 24, 2017, the Independent Electricity System Operator (IESO) had completed evaluation stages 1 and 2 for a total of 1,120 FIT applications received by them during the fifth cycle period of applications, totaling approximately 395 MW of electrical energy from across the province of Ontario. However, after thorough evaluation by an independent body, a total of 698 applications, which represent approximately 248 MW of electrical energy, were given the go-ahead to the testing and connection stage [14]. The program is administered by the IESO, and it has been split into two streams; PV systems designed to generate 10 kW or lower are considered in the micro FIT program and projects above 10 kW are considered to fall under the FIT program [15].

As shown in Table 7.3, when the FIT started in 2009, it only catered to solar PV installations on roof tops [16]. However, in August 2013, it was modified to include nonroof top solar PV installations, and this policy is still in use as of January 2017.

7.4.2 California, United States of America

In 2008, the Sacramento Municipal Utility District (SMUD) contracted out its Community Solar PV Project to a solar developer named enXco. The developer built, owns and operates the project and the electricity that is produced is then sold to SMUD. Hence, the electricity produced is fed straight into the grid and SMUD in turn uses the electricity as the basis for its SolarShares program. In addition, SMUD purchases 100% of the electrical output energy produced by enXco under a 20-year PPA. Hence, this is a utility/grid community solar PV model [23].

Customers attached to SMUD are asked to pay a fixed charge monthly, which is calculated based on the amount of energy they subscribe for that month (ranging 0.5–4 kW) and the mean electricity utilized. In addition, SMUD offers to its customers the SolarShares program at less than a normal fixed rate if they can reduce their electricity usage. The program offered by SMUD is fully booked, and approximately 700 homes are on the SolarShares program. Now homes wanting to join the SolarShares program are added to a waiting list and can only join when current customers decide to opt out from the service [23].

In June 2015, community solar PV in California, USA was split into three sections by State Bill 43. The SMUD SolarShares program was the first community solar PV, which was introduced in California in 2008. SMUD customers were offered an opportunity to become subscribing members to the program offered at a fixed fee which was based upon the customer's energy history and share size [24]. The California Public Utilities Commission approved the state's investor-owned utilities (IOUs) for community solar PV in June 2015. The IOU program has two sections. The first of the two sections allows customers to subscribe from 50% to 100% of the electricity from the projects purchased by the utility under what is known as the SB43 green tariff shared renewable program. The second of the two sections gives customers the opportunity to enroll in a SB43 enhanced community renewable (ECR) program which requires that the customers execute an agreement (a PPA) with a community solar developer in order to purchase a specified portion of a local community solar PV project output [24]. Electrical PPAs for ECR programs are also agreed between the utility and the community solar PV project developer and are different from the agreement between the customers and a community solar PV developers. Hence, this is considered a utility/grid sponsored model [24].

In addition, SMUD has in its future plan to increase its solar generating capacity from 1 to 60 MW. This expansion clearly shows that the demand for the SolarShares service within the district is increasing rapidly. At the beginning of 2017, SMUD signed a deal with Recurrent Energy, a Canadian Solar subsidiary, to develop, construct and operate a 60-MW community solar garden, and the electrical energy produced will feed into SMUD with a 20-year PPA. The project is named "Tranquility 8 Verde Solar PV" and is located in Fresno County, California. Construction is expected to commence in the later part of 2017, and the installation is expected to provide electrical energy to approximately 15,000 homes. This project is the fifth done by Recurrent Energy for SMUD over the years [25] (Tables 7.4 and 7.5).

Table 7.4 Existing solar shares project summary, California, USA [24]

Site/ Location	Quantity (no.)	Type of installation	Generated capacity (kW)	Total capacity (kW)
Sacramento	1	Ground mounted	1,000	1,000
Total capacity (kW)				**1,000**

Table 7.5 Planned solar shares project summary, California, USA [25]

Site/ Location	Quantity (no.)	Type of installation	Generated capacity (kW)	Total capacity (kW)
Sacramento	1	Ground mounted	60,000	60,000
Total capacity (kW)				**60,000**

7.4.3 Guyana, South America

The Government of Guyana has included in its 2017 budget estimate community solar PV projects at Mabaruma, Bartica, Mahdia and Lethem, which are all remote communities within Guyana. The projects' aim is to provide more reliable and available electricity to the residents of these communities, and the objective is the reduction of GHG emissions and the importation of fossil fuel. All four locations will be ground mount PV systems, since land accessibility is in abundance. The 400-kW system in Mabaruma is expected to provide service approximately to 3,000 residents at a cost of Guy$264M or CAD$1.760M, making the cost CAD $4,400/kW [26].

However, to date, it has not been decided which of the three community solar PV models Guyana will adopt in managing the four projects. The Government of Guyana will advertise for expressions of interest for private developers to design and construct these solar PV projects. The electrical energy that is produced will be fed into the existing government-owned utility companies who already have infrastructure in place and then it will be sold to the utility customers. By the time construction starts, the Government of Guyana will have a program in place to have the residents involved in contributing to the projects (Table 7.6).

Table 7.6 Project summary, Guyana, South America [26]

Site/ Location	Quantity (no.)	Type of installation	Generated capacity (kW)	Total capacity (kW)
Mabaruma	1	Ground mounted	400	400
Lethem	1	Ground mounted	800	800
Mahdia	1	Ground mounted	400	400
Bartica	1	Ground mounted	1,500	1,500
Total capacity (kW)				**3,100**

7.4.4 Germany, Europe

In 2016, approximately 7.0% of Germany's net electric power was derived from solar PV systems. In addition, the country has been amongst the world's top installers of solar PV systems for many years, having a total of 40,782 MW installed by the end of November 2016. The only country that is presently leading Germany in solar PV system installation is China [27].

In 2012, Germany had a total of 7.6 GW of solar PV systems installed, which in 2011 provided 18 TW h of electricity, which was about 3% of the country's net electricity use in that year. At noon on May 26, 2012, solar PV systems supplied 40% of the country's total electricity consumption and 20% for the entire day (24 h). The federal government of Germany has set a target of having 66 GW of solar PV systems by the year 2030 [28].

The main contributor in the growth of solar PV systems in Germany is the introduction of the FIT policy which was introduced by the German Renewable

Table 7.7 Recklinghausen project summary, Germany, Europe [30]

Site/ Location	Quantity (no.)	Type of installation	Generated capacity (kW)	Total capacity (kW)
Recklinghausen	1	Roof mounted	400	400
Total capacity (kW)				**400**

Energy Source Act in 2000 and was modified in 2008, which gives priority for the utilization of renewable energy sources versus the use of fossil fuels in meeting the country's net electric energy demand. The introduction of the German Renewable Energy Source Act led to a decrease in the cost of solar PV systems by more than 50% since 2006 [29].

According to the solar power industry, a FIT policy is the best practice for the development of solar PV system projects. It is considered to be the same as a power procurement agreement; however, it is much more beneficial for investors. In addition, the FIT policy gives investors price certainty so that the return on their investment is predictable [29].

An example of community solar PV project in Germany is in the city of Recklinghausen. In this city, the residents in 2011 decided to exploit the idea of having solar PV systems installed onto their house roofs for electric energy generation. The electric energy that is produced from the solar PV systems on the roofs is fed into the utility for consumption by the grid connected customers. The amount of electric energy generated per year is approximately 195,000 kW h, which is sufficient to power approximately 60 homes through the year in Recklinghausen. In addition, the city of Recklinghausen decided to give its support to the residents by leasing its building roofs for the installation of solar PV systems which will be funded by the residents. Hence, citizens of the city were given the opportunity to make an investment of no less than €500 in this project. The PV systems on the residents and city rooftops fall under the special purpose entity sponsored model. The investment cost of the project was €220,000–€260,000 resulting in an initial cost of €550–€650/kW [30] (Table 7.7).

7.4.5 Rwanda, East Africa

In Rwanda during 2015 and 2016, an 8.5 MW solar garden was constructed at the Agahozo-Shalom Youth Village which is east of Kigali. This project is considered to be one of the largest community solar PV systems in East Africa. Upon completion of the project, it will increase the country's energy supply by 6% and the electric energy produced from this solar garden will be fed into a grid/utility and then distributed to the nation for utilization. The cost of this system works out to US$2,800/kW [31].

In Rwanda, less than 15% of the population have access to electricity, and out of that 15%, only 1%–2% are in rural areas. Since 90% of the country's total population lives in rural areas, citizens with access to electricity are rare in

Table 7.8 Agahozo-Shalom Youth Village project summary, Rwanda, East Africa [31]

Site/Location	Quantity (no.)	Type of installation	Generated capacity (kW)	Total capacity (kW)
Agahozo-Shalom Youth Village	1	Ground mounted	8,500	8,500
Total capacity (kW)				**8,500**

Rwanda. In addition, due to low accessibility to electricity, the economy is unable to develop, resulting in major suffering for its citizens [31].

> *"Funding for the project comes from various agencies such as FMO (Netherlands Development Finance Company), the London-based EAIF (Emerging Africa Infrastructure Fund), Norfund (The Norwegian Investment Fund for Developing Countries), Scatec Solar ASA (equipment provider and operator), KLP (the largest pension fund in Norway), the United States Government via a ACEF (Africa Clean Energy Finance) grant, and from Finland's EEP (Energy and Environment Partnership)"* [31].

The project is owned by third-party investors and donors, and the total electric energy generation output is fed into the grid, through a PPA between the grid and owners. Electric energy is then supplied to customers connected to the grid/utility company. So this could be considered a special purpose entity sponsored community solar PV project, in that the "community" is a consortium of international donors [31] (Table 7.8).

7.5 Summary

The type of ownership structure for a community solar PV system depends on various factors such as the capital available to members of the community who want to contribute incentives, policies and the regulatory environment in which the project is planned. The primary difference that exists between a FIT policy and a tax credit is that a tax credit is borne in the year of installation, while with a FIT, it is spread out over a period of years, meaning that a tax credit offers to the owner a reduction on federal or government taxes during the year of installation, whilst the FIT incentivizes the renewable electricity generation by offering to the investor or owner of community solar PV projects an agreeable rate for a specified duration for the electricity sold to the grid (the FIT in Ontario, Canada has a 20-year duration). In countries such as Denmark and Germany, that the FIT programs are an excellent method of generating investment in renewable energy generation and it is highly believed by advocates for renewable energy that this policy can greatly revolutionize community solar PV projects. However, it is observed that costs per kW vary widely but are linked to factors such as economy of scale, the availability of skilled labor, project interest rate, type of technology utilize, incentives and cost of equipment (including shipping) [3,32].

The Rocky Mountain Institute of Colorado, USA has identified three scales of solar PV installations and list community solar PV projects in the middle category, which has a range from 0.5 to 5 MW, where all the electrical energy generated is fed into a utility/grid which is then distributed to paying customers. These scales are based on "behind-the-meter" capacities. Where solar PV panels are installed and the electric energy generated is for on-site utilization, that is, the home, business building or other commercial facility where the installation is hosted, net metering allows for the option of connection to the utility/grid [33].

7.6 Recommendations

For community solar PV projects to be successful, the following policies, regulations and funding options are recommended below:

7.6.1 Policies and regulations

1. Develop and implement consistency policies for renewable energy at the community level. These policies will be used to protect investors from any major losses on investment, should there be any changes for reasons such as, and not limited to, the change of government. In addition, more investors will come onboard, resulting in the growth of community solar PV projects [34].
2. The policies developed by governments for renewable energy, such as community solar PV, must have consideration for public support. Without such consideration, it is doubtful that the public will support development of community renewable energy projects [34].
3. A policy should be developed that will create a supportive and stable regulatory environment which will take into consideration the social, total economic and environmental values of community solar PV projects, either by the government or energy stakeholders in the country [35]. The intention is to create a level playing field for all means of electricity generation.
4. Establishing relevant laws that will make way for equitable utility/grid access and pricing for electrical energy generated from the community solar PV [35]. Access to distribution is obviously a key enabler for these projects.
5. Establishing relevant laws that will give clear guidance on community ownership mandates for commercial community solar PV projects. This would provide clear guidance on what becomes of the project when the PPA or FIT period expires [35].
6. Establish a policy that will give the option of virtual net metering to the citizens. This should include the case where the citizens own a system that is not located or installed on their property, such that the energy credits will be reflected in their utility bill [35].

7.6.2 Start-up capacity

1. Establish a special fund for citizens who wish to have feasibility studies for community solar PV performed within their community, either in the form of a grant or a revolving loan. In some jurisdictions where cooperatives are responsible for community renewable energy projects, the mission of the co-op includes providing guidance to other start-up cooperatives [35]. In England, the community renewables initiative provided expertise and advice to communities wishing to develop local renewable energy projects [3].
2. Establish legislation that will offer to the developer(s) of community solar PV a reduction in security payments, making the payments very minimal so that more developers will consider undertaking community solar PV projects [35].
3. Establish a policy that will give direct support in the form of grants for community solar PV for the aboriginal or first nation communities so that their community electrical power capacity will increase. The first people or natives are those who were the first to occupy the land before the arrival of other migrants. For example, in Guyana, the first people are called the Amerindians and their communities are located in remote locations that are very far away from the utility/grid. Hence, the most feasible method of getting electrical power to their homes is through solar PV projects [35].

7.6.3 Funding

1. Provide tax holidays for community solar PV developer(s) by putting in place flexible financial structures [35].
2. Relax the barriers affecting community solar PV by having access to low-cost debt. For example, in the Province of Nova Scotia, Canada funds are available through Community Investment Funds (CIFs) and through insured loans. Community Investment Funds or CIFs are local funds which are contributed by individual investors within a specific location or community [35]. Similarly, Development Trusts in England provided financing for community renewable energy projects [3].
3. Establishing bonds for clean community renewable energy programs such as community solar PV projects. Hence, the bonds will purchase by investors and these investors will be provided bank financing at a 0% interest rate such that the investor only needs to pay the principal borrowed from the bank. This would motivate investors, which will lead to the growth of community renewable energy program such as community solar PV [36].

7.7 Conclusion

In conclusion, there are numerous forms of renewable energy available that can be used to generate electrical energy to satisfy a great amount of our individual, community and national energy demands. Solar energy shows a great potential for the future since the sun is available for most of the time throughout the year and cannot be controlled or monopolized by a single individual or nation.

Three possible models for community solar PV projects have been presented: grid/utility sponsored, special purpose entity sponsored and nonprofit entity sponsored. However, it is very important to have policies in place that will attract investors to make community solar PV projects a success. For example, in Ontario (Canada) and Germany, there is an FIT. This encourages investment in renewable energy projects such as community solar PV projects. However, in California, USA investors are incentivized by the investment tax credit, net metering policies and PPAs (which act like an FIT but the price is not mandated by the government).

Community solar offers an alternative pathway for individuals to access electricity generated from renewable sources. No single model of community solar PV ownership works in all situations, but it is important that countries, provinces, states and cities develop policies that enable investment in community solar PV projects, in order to prepare for a future in which the overall costs of solar-generated electricity are less than those of fossil fuel generation.

Abbreviations

AC	alternating current
CC	climate change
CPUC	California Public Utilities Commission
DC	direct current
EA	East Africa
EAIF	Energy Africa Infrastructure Fund
ECR	enhanced community renewable
EEP	energy and environment partnership
FIT	feed-in-tariff
GHG	greenhouse gas
GW	global warming
GRESA	German Renewable Energy Source Act
IESO	Independent Electricity System Operator
IOUs	investor-owned utilities
PPA	power purchase agreement
PV	photovoltaic
REC	renewable energy credit
RE	recurrent energy
ROI	return on investment
SA	South America
SB43	State Bill 43 (California)
SMUD	Sacramento Municipal Utility District
SPE	special purpose entity

USA	United States of America
UNFCCC	United Nations Framework Convention on Climate Change
VOLT	voltage

References

[1] Farrell J. *Community Solar Power: Obstacles and Opportunities* [online]. 2010. Available from https://www.ilsr.org/wp-content/uploads/files/community solarpower2.pdf [Accessed on 28 May, 2017].

[2] Natarajan P and Nalini GS. Social Cost Benefit Analysis of Solar Power Projects. Prabandhan: Indian Journal of Management. 2015; 8(5). http://dx.doi.org/10.17010/pijom%2F2015%2Fv8i5%2F68772.

[3] Maehlum MA. *Top 10 Benefits of Going Solar* [online]. 2015. Available from http://energyinformative.org/benefits-of-solar-panels/ [Accessed on 21 May, 2017].

[4] Walker G. What are the Barriers and Incentives for Community-Owned Means of Energy Production and Use? Journal of Energy Policy. 2008; 36: 4401–5.

[5] U.S. Department of Energy (U.S.A). *SF Environment. Our Home. Our City. Our Planet: Community Shared Solar* [online]. 2012. Available from http://sfenvironment.org/sites/default/files/fliers/files/community_shared_solar.pdf [Accessed on 21 May, 2017].

[6] Banerjee R. Solar Tracking System. International Journal of Scientific and Research Publications. 2015; 5(3). ISSN 2250-3153.

[7] Coughlin J, Grove J, Irvine L, *et al. A Guide to Community Solar: Utility, Private, and Non-Profit Project Development* [online]. 2010. Available from http://nrel.gov/docs/fy11osti/49930.pdf [Accessed on 21 May, 2017].

[8] Fu R, Chung D, Lowder T, Feldman D, Ardani K and Margolis R. *U.S. Solar Photovoltaic System Cost Benchmark: Q1 2016, National Renewable Energy Laboratory, Technical Report, NREL/TP-6A20-66532* [online]. 2016. Available from http://www.nrel.gov/docs/fy16osti/66532.pdf?utm_source=NREL%20Report%20Shows%20U%2ES%2E%20Solar%20Photovoltaic%20Cost%20Continuing%20to%20Fall%20in%202016&utm_medium=email&utm_content=nrel&utm_campaign=NewsRelease [Accessed on 21 May, 2017].

[9] Wiser R. *Tracking the Sun: The Installed Cost of Photovoltaics in the US from 1998–2007* [online]. 2009. Available from http://escholarship.org/uc/item/9qc4b7qt#page-2 [Accessed on 21 May, 2017].

[10] Margolis R and Zuboy J. Nontechnical Barriers to Solar Energy Use: Review of Recent Literature. NREL/TP-520-40116. 2006.

[11] Hain JJ, Ault GW, Galloway SJ, Cruden A and McDonald JR. Additional Renewable Energy Growth through Small-Scale Community Oriented Energy Policies. Energy Policy. 2005; 33: 1199–212.

[12] Irvine L, Sawyer A and Grove J. The Solarize Guidebook: A Community Guide to Collective Purchasing of Residential PV Systems. NREL Report DOE/GO-102012-3578. 2012.

[13] UGE. *UGE Announces Agreement to Develop and Design-Build 6.3 Megawatts of Solar PV in Ontario* [online]. 2016. Available from http://ugei.com/prmay-2nd-2016 [Accessed on 22 May, 2017].

[14] IESO. *Feed-in Tariff Program* [online]. 2017. Available from http://ieso.ca/en/sector-participants/feed-in-tariff-program/overview [Accessed on 23 May, 2017].

[15] Ontario Ministry of Energy. *FIT and microFIT Program* [online]. 2017. Available from www.energy.gov.ca/en/fit-and-microfit-program/ [Accessed on 23 May, 2017].

[16] IESO. *Feed-in Tariff for Renewable Energy Projects in Ontario* [online]. 2009. Available from http://www.ieso.ca/en/get-involved/mircofit/-/media/files/ieso/document-library/mircofit/version-1/Price-Schedule-September-24-2009.pdf [Accessed on 27 June, 2017].

[17] IESO. *Feed-in Tariff for Renewable Energy Projects in Ontario* [online]. 2012. Available from http://www.ieso.ca/en/get-involved/mircofit/-/media/files/ieso/document-library/mircofit/version-2/FIT-mFIT-Price-Schedule-V-2-0.pdf [Accessed on 27 June, 2017].

[18] IESO. *Feed-in Tariff for Renewable Energy Projects in Ontario* [online]. 2013. Available from http://www.ieso.ca/en/get-involved/mircofit/-/media/files/ieso/document-library/mircofit/version-3/2013-FIT-Price-Schedule-Tables.pdf [Accessed on 27 June, 2017].

[19] IESO. *Feed-in Tariff for Renewable Energy Projects in Ontario* [online]. 2014. Available from http://www.ieso.ca/en/get-involved/mircofit/-/media/files/ieso/document-library/mircofit/version-3/2014-FIT-Price-Schedule-Final-2013-11-07.pdf [Accessed on 27 June, 2017].

[20] IESO. *Feed-in Tariff for Renewable Energy Projects in Ontario* [online]. 2015. Available from http://www.ieso.ca/en/get-involved/mircofit/-/media/files/ieso/document-library/mircofit/version-3/FIT-Price-Schedule-2014-09-30.pdf [Accessed on 27 June, 2017].

[21] IESO. *Feed-in Tariff for Renewable Energy Projects in Ontario* [online]. 2016. Available from http://www.ieso.ca/-/media/files/ieso/document-library/fit/fit-Price-Schedule-2016-06-21.pdf?la=en [Accessed on 27 June, 2017].

[22] IESO. *Feed-in Tariff for Renewable Energy Projects in Ontario* [online]. 2017. Available from http://www.ieso.ca/-/media/files/ieso/document-library/fit/2017-fit-price-schedule.pdf?la=en [Accessed on 27 June, 2017].

[23] Coughlin J, Grove J, Irvine L, *et al. A Guide to Community Shared Solar: Utility, Private, and Nonprofit Project Development* [online]. 2012. Available from http://nrel.gov/docs/fy12osti/54570.pdf [Accessed on May 23, 2017].

[24] Kaatz J and Andres S. *Community Solar in California, Center for Sustainable Energy* [online]. 2015. Available from http://www.sandiego.edu/law/documents/centers/epic/Community%20Solar%20Final_2.2.16.pdf [Accessed on 24 May, 2017].

[25] Recurrent Energy. *Recurrent Energy Signs 60 Megawatt Solar Power Purchase Agreement with SMUD* [online]. 2017. Available from http://recurrentenergy.com/press-release/canadian-solar-subsidiary-recurrent-energy-signs-60-megawatt-solar-power-purchase-agreement-smud/ [Accessed on 28 May, 2017].

[26] Guyana News. *Demerara Waves* [online]. 2016. Available from http://demerarawaves.com/2016/11/28/guyana-unveils-solar-energy-plan-mabaruma-to-get-first-solar-farm/ [Accessed on 28 May, 2017].

[27] Fraunhofer ISE. *Electricity Generation in Germany in 2016* [online]. Available from http://www.energy-charts.de/energy_pie.htm?year=2016 [Accessed on 28 May, 2017].

[28] NuWire Investor. *Germany Reducing Incentives for Solar Property Investment* [online]. Available from http://www.nuwireinvestor.com/germany-reducing-incentives-for-solar-property-investment/ [Accessed on 28 May, 2017].

[29] Hall M. *The US Needs a Feed-in-Tariff* [online]. 2010. Available from http://www.power-eng.com/articles/print/volume-114/issue-10/departments/View-on-Renewables/the-us-needs-a-feed-in-tariff.html [Accessed on 28 May, 2017].

[30] Mergner R and Rutz D. *Community Energy in Germany: Existing Models Public–Private Funding and Good Practices Examples* [online]. 2016. Available from http://www.communitypower.eu/images/GermanyD32.pdf [Accessed on 28 May, 2017].

[31] Gigawatt Global. *A Founding Partner of Power Africa* [online]. 2016. Available from http://www.gigawattglobal.com [Accessed on 28 May, 2017].

[32] Brehm K, Brouski P, Coleman K, *et al. Community-Scale Solar: Why Developers and Buyers Should Focus on this High-Potential Market Segment* [online]. 2016. Available from https://d231jw5ce53gcq.cloudfront.net/wp-content/uploads/2017/03/Shine-Report-CommunityScaleSolarMarketPotential-2016.pdf [Accessed on 28 May, 2017].

[33] Alberta Green Economy Network (Canada). *The Case for Small-Scale and Community owned Renewable Energy in Alberta* [online]. 2016. Available from http://www.albertagen.ca/wp-content/uploads/2016/05/20MayeditedCORE [Accessed on 28 May, 2017].

[34] White W, Lunnan A, Nybakk E and Kulisic B. The Role of Governments in Renewable Energy: The Importance of Policy Consistency. Biomass and Bioenergy. 2013; 57: 97–105.

[35] Konkle D. *A Guidebook for Community Solar Programs in Michigan Communities* [online]. 2014. Available from https://www.michigan.gov/documents/mdcd/Michigan_Community_Solar_Guidebook_437888_7.pdf [Accessed on 28 May, 2017].

[36] Bonneville Environmental Foundation. *The Northwest Community Solar Guide* [online]. 2013. Available from https://www.nwseed.org/wp-content/uploads/2013/05/NW-Community-Solar-Guide.pdf [Accessed on 31 July, 2017].

Further reading

[1] Walker G and Devine-Wright P. Community Renewable Energy: What Should it Mean? Energy Policy. 2008; 36: 497–500.

[2] Mitchell C and Connor P. Renewable Energy Policy in the UK 1990–2003. Energy Policy. 2004; 32: 1935–47.

[3] Wood G and Dow S. What Lessons have been Learned in Reforming the Renewable Obligation? An Analysis of Internal and External Failures in UK Renewable Energy Policy. Energy Policy. 2011; 39: 2228–44.

[4] Strong SJ and Scheller WG. The Solar Electric House: Energy for the Environmentally-Responsive, Energy-Independent Home. Still River, MA: Sustainability Press, 1993, pp. 29–52.

[5] Hille G, Roth W and Schmidt H. Photovoltaics Systems. Fraunhofer Institute for Solar Energy System. Freiburg, Germany, 1995, pp. 194.

[6] Bahaj AS. Means of Enhancing and Promoting the Use of Solar Energy. Renewable Energy. 2002; 27(2): 97–105.

[7] Shahan Z. *Advantages and Disadvantages of Solar Power* [online]. 2013. Available from http://cleantechnica.com/2013/10/08/advantages-disadvantages-solar-power/ [Accessed on 29 May, 2017].

[8] Nangia D, Niranjan S and Chauhan YK. *Potential of Solar Thermal Technology and its Status in India* [online]. 2014, Vol.4, special issue 1. Available from http://www.ijeta.com [Accessed on 29 May, 2017].

[9] Energy Sage. *Community Solar* [online]. 2017. Available from http://www.energysage.com/solar/community-solar/community-solar-power-explained [Accessed on 29 May, 2017].

[10] The Guardian. *How Africa's Fastest Solar Power Project is Lighting Up Rwanda* [online]. 2015. Available from https://www.theguardian.com/environment/2015/nov/23/how-africas-fastest-solar-power-project-is-lighting-up-rwanda [Accessed on 30 May, 2017].

Chapter 8

Assessing wind loads for urban photovoltaic installations

David Kazmirowicz[1], Jesse Bridges[2], Jonathan Whale[1] and David Wood[2]

Abstract

Rooftop solar photovoltaic installations are becoming common around the world but knowledge of the wind loads on them is not as advanced. The wind loads are important as they determine the cost of the structure needed to hold the photovoltaic modules in place, which can be a significant proportion of the total project cost. This chapter discusses these wind loads and reviews the few international building codes that cover them. Wind loads can be estimated from wind tunnel tests on model installations or by computational fluid dynamics simulations. We show that there are problems with both methods, due largely to scaling issues for the former and limitations in turbulence modeling for the latter. Wind tunnel tests show significant scatter in module pressures but improvements in test methodologies is likely to reduce these in the future. Future full scale rooftop measurements are also likely to improve our knowledge.

8.1 Introduction

Rooftop solar photovoltaic (PV) installations in urban communities have become increasingly prevalent in recent years. The reasons for this include:

- rapid urbanisation of the world's population,
- the advantages of decentralised, clean power competing with the retail price of conventional electricity,
- uncertainty in grid electricity delivery and
- the rise of affordable batteries.

The worldwide rooftop solar PV market is very large, with California currently having over 670,000 PV systems with a combined capacity over 4.5 GW [1]. Singapore is

[1]Electrical Engineering, Energy and Physics, School of Engineering and Information Technology, Murdoch University, Australia
[2]Department of Mechanical and Manufacturing Engineering, University of Calgary, Canada

noteworthy for its high density of rooftop PV, with a number of large capacity systems, between 1 and 10 MW [2] on the rooftops of 15–25 storey commercial and industrial buildings. Australia has the highest percentage of residential PV systems in the world (16.5%) with over 5 GW of installed PV in systems less than 10 kW in rated power [3]. Over the next 5 years, the greatest growth in rooftop solar PV systems is expected to be in the Asia-Pacific region, in countries such as China, Japan and India.

Increasing demand for this technology has resulted in efforts to reduce system complexity and cost. Since 1980, the cost of PV modules[1] has decreased at an average rate of around 10% per year, and this trend is likely to continue in the coming years [4]. Further decreases in system cost could be achieved by reducing the complexity and cost of module structural support, but this requires detailed knowledge of the urban operational environment, specifically the extreme wind loads. For the Asia-Pacific region, South East Asia and the China Sea are areas where urban environments are affected by tropical cyclones, and more than 30% of all tropical typhoons pass through Japan and South East Asia every year. In 2015, extreme winds due to two typhoons caused damage to 25 MW of PV at 30 sites across Taiwan, including NTD 30 million damage to a single solar installation in southern Taiwan [5]. At some sites, modules were observed to have become detached from their mounting racks due to loosening of clamps.

This chapter concentrates on the structural loads caused by extreme winds on PV systems. We concentrate on PV systems attached to buildings, which usually experience the highest wind loads. This chapter is also relevant to building-mounted PV installations in rural communities.

There are two main categories of urban PV installations. The first is typical of single houses where a small number of modules are flush-mounted to the roof. This arrangement often expedites development approval for the project and reduces the installation cost. The disadvantages can be poor orientation of the modules and shading, which both lead to lower power output. The second category is larger, mainly flat roof areas that occur extensively in cities: school and university buildings, shopping malls and factories are common examples around the world. These have the potential for much larger installations with the modules installed at optimum angles of tilt,[2] which are usually 5°–10° less than the local latitude for mid to high latitudes. An important consequence is that, in general, the higher the tilt, the higher the wind loads. And thus this second arrangement can lead to large wind loads dominating the design of the supporting structure. Further, it is often necessary, during installation, to avoid penetration of the roof. This can lead to the array being held in place by ballast or a complex mounting structure, which can

[1] In this chapter, 'module' means a PV device, usually solid and made up of 60 or more solar cells. The term 'panel' although widely used as a synonym for 'module' is reserved for solar thermal (heating) devices. Panels may be solid or porous and so can have very different wind loadings to modules. An array is a collection of modules.

[2] In PV installations, the tilt is usually measured from the horizontal in the north:south plane, with the azimuth relative to east or west. Both angles affect electricity production but have no separate meaning for wind loads. 'Tilt' is used here to denote the single angle of inclination relative to the locally flat, but not necessarily horizontal, roof.

limit the number of modules installed. Smaller versions of solar farms are also possible in urban areas and for community power schemes. In general, ground mounting of modules is considerably easier than roof mounting, partly because the wind loading is easier to determine and there is no need for ballast.

A third category of PV installation is becoming increasingly popular but is still small in total installed capacity. A range of PV technologies can produce power from windows and other elements of building facades, sometimes in combination with utilisation of the waste heat – all commercial PV technologies are less than 20% efficient – for building heating and ventilation, e.g. Fortuin *et al.* [6].

This chapter considers only roof-mounted PV installations, primarily those in the second category. PV mounting structures can contribute as much as 40% to the material cost of an installation [4] and, as the costs of PV modules and inverters fall, will account for an increasing fraction of the overall system cost. Therefore, it is critical to understand the nature of the wind loads on the second category of structures in order to optimise system design [2]. This review will show that understanding the wind loads is not simple. This chapter reviews the calculation of PV wind loads using appropriate national standards, wind tunnel and site measurements and computational fluid dynamics (CFD). One of the surprising results of our review and the literature searching for it is the lack of field measurements of wind loads on PV structures. We could not find any data in the open literature. The reason is probably that extreme wind loads occur rarely so that a measurement campaign must be very long, especially in comparison to the time over which PV systems have been installed in large numbers. However, dynamic mechanical loading tests have been carried out on solar modules by institutes such as the Industrial Technology Research Institute of Taiwan. These tests attached 16 computer-controlled vacuum cups to modules that deliver push–pull forces on the module, over a number of cycles, to simulate the effects of strong winds due to typhoons [7].

In the next section, we discuss the use of the Australian standard for wind loads as it is one of the few that specifically includes PV installations. The new edition of the ASCE wind loading standard, ASCE 6-17 [8], covers PV installations using design loads obtained from wind tunnel tests. It is more thorough and complex than the Australian standard but uses a similar methodology. In Section 8.3, we consider some of the more important features of the urban wind environment. Section 8.4 describes features of PV mounting practice. Wind tunnel measurements of PV wind loads are reviewed in Section 8.5 and the increasingly popular determination of these loads using CFD is covered in Section 8.6. The conclusions follow.

8.2 Wind loading of PV installations using Australian Standard 1170.2

Australian Standard *AS 1170.2:2011 Structural Design Actions; Part 2: Wind Actions* (hereafter: AS1170.2) [9] is typical of modern wind loading standards and provides specific information on loads on PV modules. We examine the calculation of wind loads on the basis of the requirements and recommendations in AS1170.2.

Characteristics of urban and open terrain wind conditions are investigated, and we compare mean wind speeds and turbulence levels given in the standard to those provided elsewhere. The background to AS1170.2 is described in Holmes *et al.* [10], and a more general treatment of wind load estimation around the world is given by Holmes [11].

The procedure described in Clause 2.4 of [9] first requires determination of a site's wind speeds for each cardinal direction, V_{sit}. This value is derived by obtaining the site regional gust wind speed, V_R, from Table 3.1, as given by Australian Building Code Board [12]. Values given for V_R represent extreme speeds expected from gales, tropical cyclones and thunderstorm downdrafts. Several multipliers are then applied to V_R to account for effects of wind direction, terrain and height, downstream shielding and the local topography.

The design wind speed, V_{des}, is defined as the maximum value of V_{sit}, acting on a structure's face within a ±45° arc from the normal direction. The pressure p acting on the structure is given by

$$p = \frac{1}{2}\rho V_{des,\theta}^2 C_{fig} C_{dyn} \tag{8.1}$$

where ρ is the air density, C_{fig} is the aerodynamic shape factor and C_{dyn} is the structure's dynamic response factor. C_{fig} is the product of the pressure coefficient, C_P, the net pressure acting on the structure divided by the dynamic pressure $1/2\rho V_{des,\theta}^2$, and factors that account for the combined effect of different building elements and other factors. Determination of p in turn allows calculation of the wind load on a structure when multiplied by its projected area. Other standards use very similar procedures, as can be seen in the very useful SEAOCC [13] guidelines for PV wind load calculations using the US code, ASCE 7-10 [14]. When extreme wind loads are defined by (8.1) or similar, there are two main ways in which the wind influences the load: through the design wind speed and the shape factor. This review concentrates on the latter but the conservativeness of many codes that is noted in Section 8.4 may arise partly from the definition of the extreme wind speeds. For example, the Canadian Building code [15] uses a 'gust effect factor' of 2.0 to convert specified hourly wind speeds to gust speeds. Analysis of many hours of wind data from Calgary by Osvaldo [16] suggests that this value of the factor is too high. In using an equation like (8.1), it is not necessary to measure or collect any wind data; all figures and information necessary for derivation of wind loads are provided within the standard. According to Holmes *et al.* [10], it is likely that this design data will yield conservative results in comparison to other methods such as wind tunnel testing. Therefore, scope may exist for further refinement of these values and the reduction in cost of PV and other structures.

AS 1170.2 accounts for terrain-related wind variations in the determination of $V_{des,\theta}$ and p. The terrain categories include Category 4: '... with numerous large, high (10 m to 30 m tall) and closely spaced constructions, such as large city centres and well-developed industrial complexes' [9]. The normalised wind speed profiles for differences in terrain roughness are given in Table 4.1 of [9] as gust wind speed terrain multipliers, $M_{z,cat}$, whose use is demonstrated in Figure 4.1 of [9]. Turbulence intensity profiles for various terrain categories are given in Table 6.1 of [9], forming inputs for the calculation of C_{dyn} in 6.1 of [9]. It is clear from these tables that wind speeds decrease and turbulence intensities increase in built-up areas relative to those experienced in open areas.

Cheng *et al.* [17] aimed to construct realistic wind speed and turbulence intensity profiles for various terrain categories. Field measurements for three different sites and wind tunnel simulations were conducted at a scale of 1:500. Comparison of field measurements and wind tunnel results showed a close correlation between actual and simulated wind profiles for each location. The results imply that the assumed distributions of mean wind speed and turbulence in *AS 1170.2* were overestimates at all heights, reinforcing the conservativeness of the standard. We return to this issue in Section 8.3.

8.2.1 PV wind loading

PV modules and solar thermal panels are considered in Appendix D6 of [9] which is the only national wind loading standard to do so. Here, C_{fig} values are provided for pitched roof PV installations, as commonly found on residential houses. $C_{dyn} = 1.0$ as these structures are not dynamically sensitive, i.e. they are very, but not perfectly, stiff. PV arrays, whose structures are not explicitly covered in the standard, can have interesting and complex structural characteristics, e.g. Schellenberg *et al.* [18]. However, the applicability of the shape factors in the standard is limited to modules mounted parallel to sloped roofs of angle a on buildings of aspect ratio (breadth:height) <0.5. Module installations must have spacing, S, within 50–300 mm from module to rooftop and a distance of at least 2S between the edge of a module and the edge of the roof, as per Figure 8.1. It is also important to note that development approval for roof-mounted systems is often predicated on this type of flush mounting.

Holmes *et al.* [10] provide the rationale and background for the data presented in Appendix D6 of [9]. Prior to publication of the standard, information on PV module wind loading was scarce, often requiring assumptions to be made in the design process. An example of this is seen in the 2004 design guide presented by the UK Building Research Establishment, Blackmore [19], which provided a method for

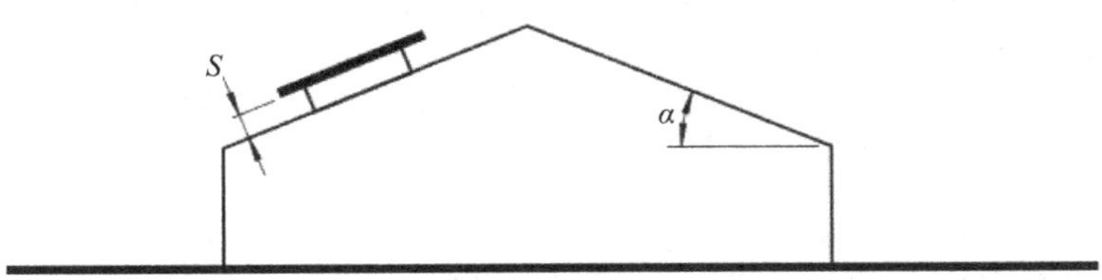

Figure 8.1 PV module wind loading limitations as per AS 1170.2, Appendix D6 [9]. The module has no tilt, and for small S can be considered to be mounted flush with the roof

determining wind forces on roof-mounted solar installations. Due to the lack of PV-specific information at the time, the pressure coefficients given in the guide were based on 'expert judgement and data from other published sources' [19].

The nature of wind forces on PV modules is briefly discussed in Holmes *et al.* [10] regarding the influence of array size, layout and dimensions of the building, and that due to differences in airflow, aerodynamic pressures on the modules cannot be derived in the same way as for the underlying roof surface. Given this level of system complexity, it is likely that design assumptions will lead to conservative predictions of the wind loads.

AS 1170.2 used the PV wind loading research that was available: the atmospheric boundary layer (ABL) wind tunnel tests by Wood *et al.* [20], Stenabaugh *et al.* [21] and Ginger *et al.* [22] were consulted, and the values provided for C_{fig} in Appendix D6 of [9] were based on their results. An ABL tunnel is designed to produce a representative mean wind profile for the ABL such that a scaled model of the structure is tested at the correct ABL scale. The wind tunnel studies referenced here investigated the effects of positioning, module height, roof angle, tilt angle and wind direction on the pressure coefficients, with the goal of providing new or updated data for codification of design guidelines.

Wood *et al.* [20] modelled an array parallel to the flat roof of an industrial building at 1:100 scale, whilst Ginger *et al.* [22] considered a tilted array on a similar building at 1:20 scale. All tests were carried out under open terrain wind profiles. The conclusions of these studies are consistent. The highest wind loads on the modules were found to be suction (uplift) loads, occurring for modules located closest to the roof edge normal to the wind direction. Wind loads were found to increase as the arrays were moved closer to the roof edges for all cases. Stenabaugh *et al.* [21] and Ginger *et al.* [22] found peak suction for wind direction parallel to the roof ridge for sloped-roof buildings, with both determining that increased module tilt or roof angle resulted in higher loads. It is apparent that pressure coefficient results from Ginger *et al.* [22] provided the majority of C_{fig} data found in Appendix D6 of [9]; shape factors given in [9] for roof angles between 5° and 30° are close to those presented in Ginger *et al.* [22].

Wind loads on tilted modules are not addressed specifically in *AS 1170.2*. It is recommended in Holmes *et al.* [10] that such loads are determined according to the procedure presented by Ruscheweyh and Windhövel [23]. Here an example wind tunnel test modelled an industrial building with a tilted, PV array at 1:50 scale for an open terrain wind profile. They measured peak C_P values on modules under various wind directions, using these figures to derive design pressures and resultant maximum forces on the modules.

Building-specific characteristics of wind flow in and around PV arrays are briefly discussed in Holmes *et al.* [10], concerning the influence of proximity of an array to the edge of the building, ratio of module size to building height and details of roof geometry. Due to the possible complexity of array installation, it is claimed that standardisation of wind loads for such arrays should not be prioritised; rather, design innovations aimed at reducing the wind loads should be encouraged.

8.3 The urban wind environment

Urban wind behaviour is complex and has been the subject of various studies. Reviews by Walker [24], Millward-Hopkins *et al.* [25] and Emerjeamara *et al.* [26] illustrate factors that complicate wind flow in urban areas, including wind shadowing and funnelling between neighbouring buildings, interacting airflows, local heat sources and the street canyon effect. The latter, according to Marciotto and Fisch [27], is further impacted by variations in aspect ratio between building height and street width. Razak *et al.* [28] argue that a significant factor influencing the urban wind environment is the ratio between building and ground surface areas for a particular building array.

While not the main focus of this chapter, it is noted that many studies of urban wind flow have been conducted for wind energy applications. There is a natural mutual benefit here: wind turbines require high average wind speeds to be cost-effective, whereas PV arrays need low speeds. Walker [24] emphasised the difficulty in predicting urban wind flows from a wind energy perspective, as buildings form a highly turbulent and possibly recirculating surface roughness layer. The famous Warwick Wind Trials on urban turbines support these findings [29]. Tabrizi *et al.* [30] analysed the power spectral density (PSD) of measured wind components at various heights on the rooftop of a warehouse in a commercial estate. Figure 8.2 shows a comparison between the average PSD, $S_1(f)$, for the longitudinal wind component with values predicted by von Karman and Kaimal spectra models, as defined by the international wind energy standard IEC61400-2 [31]. The spectral functions are part of a normal turbulence model that is based on non-intermittent flow in open terrain and Figure 8.2 shows, for a large part of the spectra, that the measured wind data from the urban environment contains more turbulent power. Toja-Silva *et al.* [32] and Emerjeamara *et al.* [26] describe the multidirectional nature of wind flow in areas with high building density, indicating the large local variations in wind speeds caused by building layout and geometry. Similarly, Millward-Hopkins *et al.* [25] state that, due to these factors, wind characteristics are unique for each individual building. A number of authors have emphasised the usefulness of multiple strategies for determining wind flow. For example, Karthikeya *et al.* [33] conducted a wind survey for a particular urban renewable energy installation using a combination of CFD modelling, anemometer measurements and wind tunnel testing.

8.4 Australian mounting system design practices

Application of the methodology and design values presented in *AS 1170.2* was investigated by a review of design practices used by Sunlock™ [34], an Australian manufacturer of PV module mounting systems. Sunlock's Rooftop Installation Manual [35] provides examples of mounting componentry, including support rails, brackets, clamps and adjustable module tilt legs if needed.

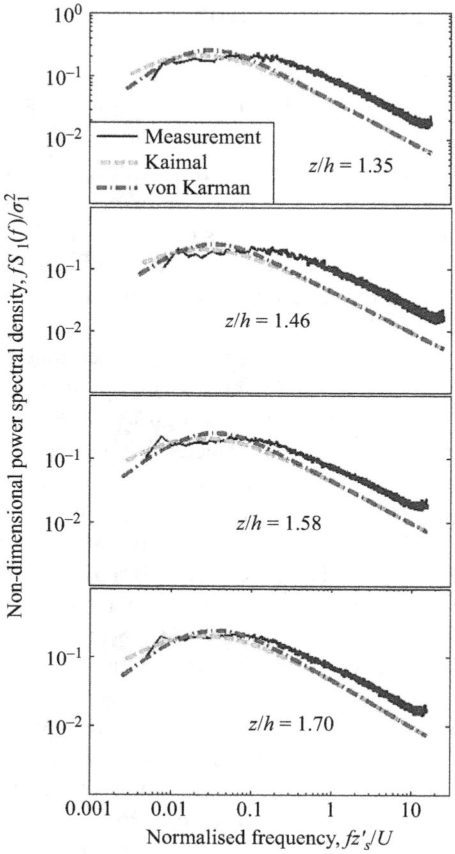

Figure 8.2 Normalised power spectral density of longitudinal wind speeds measured in the urban environment in comparison to stochastic turbulence spectral functions associated with open terrain as defined in IEC61400-2, Appendix C [31]. (U is the magnitude of the wind speed, σ_1 is the standard deviation of the longitudinal wind speeds, f is frequency and $z'_s = z - z_d$, where z is the height of the anemometer and z_d is the zero plane displacement height for the building [31].)

Sunlock provided the authors with the company's *Residential Framing Calculator*, which applies *AS 1170.2* to PV mounting systems. This calculator is a spreadsheet that uses input data in the form of building and array dimensions to calculate loads based on the regional gust wind speeds and pressure coefficients given in [9]. Limitations of the standard are apparent in this calculator; for cases in which the standard is not applicable, such as for tilted module support structures, assumptions regarding pressure coefficients are made. Discussions with Sunlock personnel indicated dissatisfaction regarding the applicability of *AS 1170.2* to PV mounting structures, particularly in regard to urban installations. It is the company's belief that *AS 1170.2* produces conservative PV module wind loading values and refinement of these should be possible based on recent wind tunnel test results, as described in the next section.

8.5 Wind tunnel test methods

The wind tunnel studies informing the design requirements of *AS 1170.2* are representative of early methods of modelling structures at large scales in ABL wind tunnels, designed to model large buildings at 1:300–1:500 scale. However, scales much larger than this are required for testing small structures such as PV modules, in order to suitably instrument them with pressure taps and provide sufficient data resolution, as per Stenabaugh *et al.* [21] and Banks [36]. It is stated in Stenabaugh *et al.* [21] that the 1:30 scale used by them is not optimal for simulating accurate wind characteristics, as the model will be exposed to an incorrectly scaled section of the ABL. This problem is greater for the large scales used in Wood *et al.* [20] and Ginger *et al.* [22]. Therefore, it is probable that the pressure coefficients presented in the studies and adopted by *AS 1170.2* are unrealistic.

Recent studies have addressed wind tunnel scaling issues with specific regard to simulation of PV modules. Stathopoulos *et al.* [37] modelled a small roof-mounted array at 1:200 scale, which is closer to the intended wind tunnel scale range. To achieve adequate data resolution, multiple modules were combined by removing gaps between rows and creating more space for pressure tapping.

Banks [36] addressed scaling issues directly, arguing that typical wind tunnels are too small to correctly simulate full-scale low frequency turbulence when large scale solar array models are tested, resulting in a test environment with insufficient turbulent energy. This is demonstrated in Figure 8.3, illustrating the discrepancy between full scale and simulated wind characteristics when a 1:40 scale model is tested in a wind tunnel designed for 1:300 scales. The square of the turbulence intensity, shown by the area under the curve, is too low for the modelled case.

To correct this, Banks [36] argued that it is sufficient to match the high frequency portions of the spectra only. This is because, in many countries, design wind speeds are given as three second averages, as is the case for the values of V_R provided in *AS 1170.2*. Therefore, a simulation need only be concerned with three second events at full scale, shown by the dashed line in Figure 8.3. In the example provided in Banks [36], this is equal to matching the spectra at a wind gust frequency of 5 Hz and is achieved by reducing the turbulence intensity in the wind tunnel. If data is required for wind gusts exceeding 3 s, Banks [36] states that this can be obtained by combining the high frequency spectrum matching process with quasi-steady analysis, as described in detail by Banks *et al.* [38]. Aly and Bitsuamlak [39] confirmed the applicability of high frequency spectrum matching for PV module wind tunnel modelling. Large scale models of a ground-mounted PV module were simulated ranging in size from 1:30 to 1:5, matching the 3-s gust

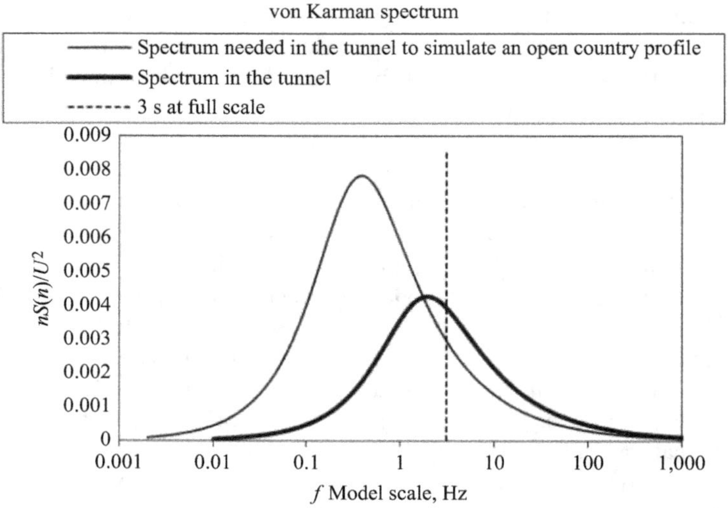

Figure 8.3 Normalised power spectral density, $nS(n)/U^2$, vs. wind gust frequency, f, for 1:40 scale model and full scale from Banks [36]

spectra with full-scale values. The resulting wind loading coefficients were consistent across the modelled range.

In recent years, many wind tunnel tests have been conducted on wind loading on PV arrays. These studies were typically carried out to provide data in the absence of specific standards. Given that substantial differences exist from one installation to another, dictated by factors such as building geometry and orientation, roof angle and system output requirements, a wide variety of wind tunnel test configurations are encountered in the literature.

8.5.1 Flat roof experiments

Most studies examined PV arrays on flat roofs of large, low-rise industrial buildings. The wind tunnel tests by Radu *et al.* [40] provided the first measurements of the module pressure coefficient, C_P. These used 1:50 scale building models with highly tilted PV modules and varying roof coverage, under open terrain wind characteristics. Module uplift occurred at the highest wind loads, for northerly winds.[3] Modules in the north corners of the buildings experienced the highest pressures, and central rows of the arrays were lightly loaded due to wind shadowing by the first windward row.

Several later studies used similar test models to produce data for codification or comparison with existing wind standards. Both Ginger *et al.* [22] and Cao *et al.* [41]

[3] Because of the hemisphere dependence of array installation angles, the convention adopted in this chapter is to consider northerly winds as blowing against the back face of a module, consistent with the literature and shown in Figure 8.5.

generated C_P data for comparison with Australian and Japanese standards, respectively, by examining both small and large, tilted arrays mounted on large, flat roofs modelled at 1:20 and 1:50 scale, respectively. Stathopoulos *et al.* [42] modelled a tilted array on a small, flat-roofed building at 1:200 scale to provide design guidance for the National Building Code of Canada. These studies agreed with the findings of Radu *et al.* [40] and Ruscheweyh and Windhövel [23] that the largest loads are experienced by the modules near the building edge facing a northerly wind, and with Stathopoulos *et al.* [42] in determining that wind directed at a building's north corner is critical. Each study found that a higher module tilt resulted in higher wind loads. Again, all tests were run under open terrain wind characteristics.

Various wind tunnel tests have examined the effects of altering array geometry on PV module wind loads on flat roofs. Wood *et al.* [20] modelled the effects of the height and inter-row spacing of an array at 1:100 scale. Warsido *et al.* [43] investigated the effects of varying the distance of a tilted array from the building edge, modelled at 1:30 scale. Multiple factors were explored by Stathopoulos *et al.* [37] using a 1:200 scale tilted array model, including building height, array position and module tilt. Results of these tests were in agreement with those above regarding peak module loading conditions, finding in particular that loading on an array decreased as it was moved further from a building's edge. It was found, however, that varying row spacing and module height above the roof did not have significant impact on wind loads.

Recent studies have aimed to understand the role of building and solar array interactions on local wind flow patterns and subsequent module loading in detail. These effects were modelled by Kopp *et al.* [44] using a 1:30 scale model of a tilted module array covering the flat roof of a large building under open wind terrain conditions. This study described how the critical loads created by northerly corner winds were caused by flow interactions between building-generated corner vortices and localised turbulence around the modules. Additionally, it was found that the southerly wind direction was able to introduce large module loads due to reattachment of the flow separation bubble as shown in Figure 8.4. At the flow

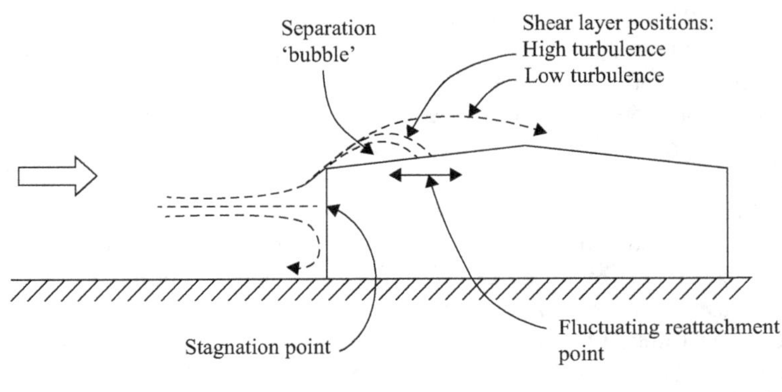

Figure 8.4 Flow separation and reattachment at the windward side of a building, Holmes [11]

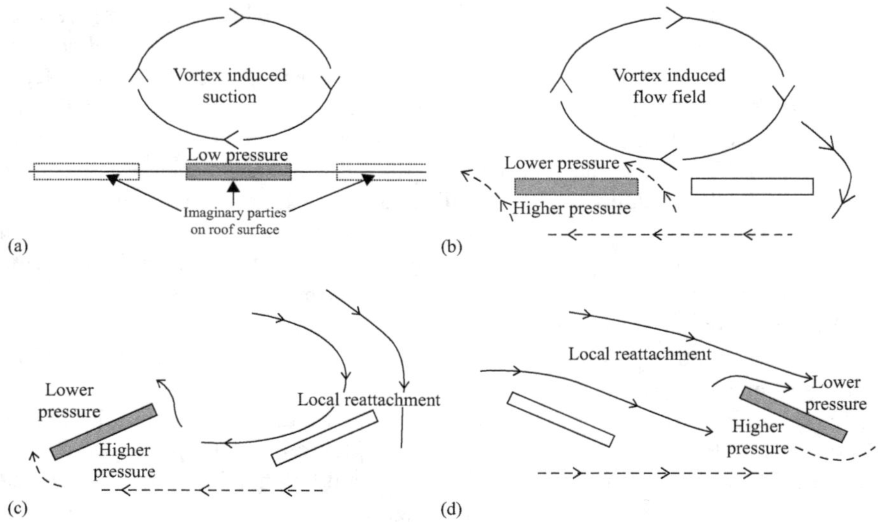

Figure 8.5 PV module uplift flow diagrams for (a) bare roof; (b) 2° module tilt; (c) 20° module tilt under southerly wind; (d) 20° module tilt under northerly winds, Pratt and Kopp [45]

reattachment point, the recirculating wind can change direction and blow against the back of the modules, causing uplift loading.

Pratt and Kopp [45] examined these effects in a further wind tunnel study using the same modelling and wind flow characteristics. It was shown that building corner vortices drive critical uplift loads for modules with low tilt angles, whilst flow reattachment and local turbulence effects introduce large uplifts for high tilt angles, as shown by the wind flow diagrams in Figure 8.5.

Another unexpected finding from this wind tunnel simulation was that the presence of a PV array reduced the turbulent kinetic energy, k, above the roof. This ran counter to the authors' expectations, as it was thought that any roof-mounted obstacle should increase turbulence. A CFD simulation by Toja-Silva *et al.* [32] produced a reduction in k, which the authors attributed to damping induced by recirculation vortices forming between array rows (see Figure 8.6). It is noted, however, that considerable scepticism should be applied to turbulence models for building flow, as detailed in Section 8.6.

It is evident from these studies that wind loading on flat roof-mounted PV arrays is determined by complex wind flow interactions, building and module aerodynamics, wind direction and array geometry and positioning. These effects are studied by Banks [47] in wind tunnel tests undertaken to develop a standard approach for calculating solar array wind loads, the results of which have been used to develop a design guide published by the Structural Engineers Association of California [13], which has a wealth of useful information for wind load estimation

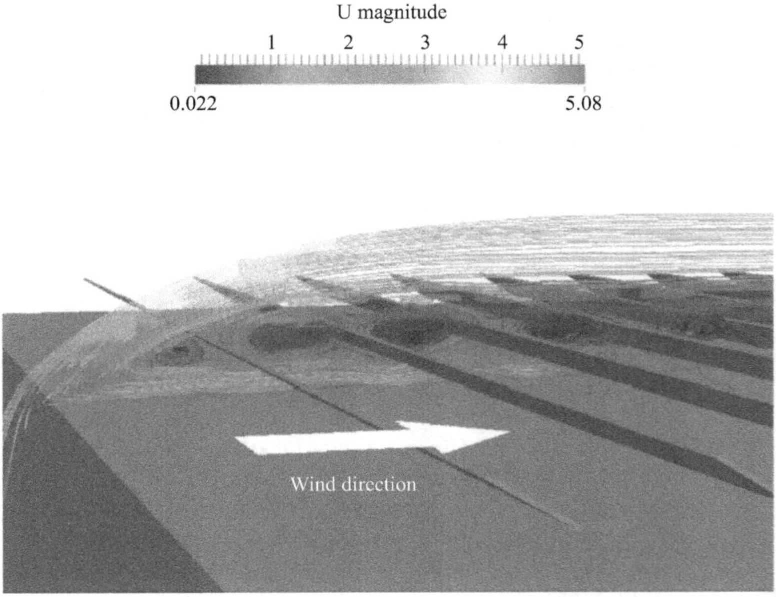

Figure 8.6 Inter-row recirculation vortices forming in a solar array during a CFD simulation, Toja-Silva et al. [46]. Wind speed, U, in m/s

and design. A 10 × 10 array of tilted modules was modelled in nine positions on the flat roof of a large, square, low-rise industrial style building subjected to an open terrain wind profile and varying wind directions. The model was created at 1:50 scale, and high frequency spectrum matching was used to provide realistic wind flow characteristics, as described above.

The findings of Banks [47] emphasise the importance of building-generated corner vortices in dictating peak wind loads on a solar array. It is claimed that, due to their dominance, any design method must primarily consider these vortices in the calculation process. Figure 8.7 illustrates the module uplift pressure contours on the modelled array, revealing the locations of peak wind loads across the roof in the darker areas, particularly in part (a) of the figure.

The critical northerly corner wind direction is clearly seen in Figure 8.7, with secondary critical zones produced by southerly corner winds. Regions of peak loading are shown propagating inwards from the windward roof corners, corresponding to the path of the building-generated corner vortices travelling over the array. The dashed black lines indicate the approximate positions of the vortex cores, whilst the solid lines show the positions of vortex flow reattachment. It is seen that peak module uplift occurs at the region of vortex reattachment, and that the highest loads are usually seen by the first windward row of modules. When the array is positioned at the building edge, the first row sees lesser loads due to its location within the flow separation bubble.

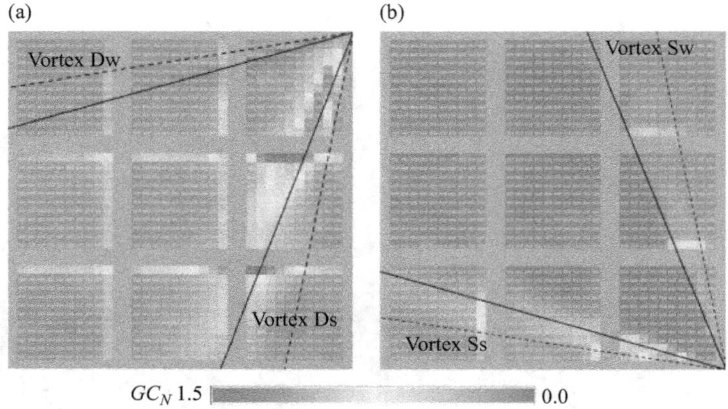

Figure 8.7 Pressure coefficient, GC_N, contours for a roof-mounted PV array under wind loading for wind at (a) 50° and (b) 140°. 0° is north, directly down the page, Banks [47]. The lines labelled 'Vortex …' show the trajectory of the vortices formed at the corners of the array

Other significant findings include:

- Southerly winds cause uplift loading for modules located in the windward third of the roof due to reattachment of the flow separation bubble.
- Highest cornering wind loads are seen by modules located a distance of one building height from the edge of the roof, corresponding to the reattachment zone of corner vortices and flow separation bubble.
- The presence of roof parapets can increase peak module loads by up to 30% as they increase the elevation of corner vortices above roof height, preventing them from being weakened by the windward edge of the array prior to reattachment. This agrees with the findings of Browne *et al.* [48].
- Based on the contour diagrams in Figure 8.7, suitable roof zones can be proposed for array positioning.

8.5.2 Sloped roof experiments

Comparatively, few studies have investigated PV module wind loading on sloped roofs. As previously discussed, Stenabaugh *et al.* [21] and Ginger *et al.* [22] derived C_p values for sloped roof residential buildings for the development of design standards. Later wind tunnel studies investigated the effects of altering multiple array and building variables on resultant wind loads on sloped-roof PV module installations. Aly and Bitsuamlak [49] and Stenabaugh *et al.* [21] modelled small arrays mounted parallel to sloped roofs of residential style buildings at 1:15 and 1:20 scale, respectively, under open terrain wind characteristics. Stenabaugh *et al.* [21] describes the choice of large model scale as a compromise between obtaining adequate data resolution and correct ABL scaling, noting a subsequent requirement for full-scale validation of test results.

Results of these tests are mostly consistent, indicating that highest wind loads occur for modules mounted close to the corners of the roof edge and in the vicinity of the roof ridge line, with maximum uplift occurring for wind directed parallel to the ridge. Interestingly, Stenabaugh *et al.* [21] found that a higher roof pitch (30° compared to a flat roof) results in slightly higher wind loads. It is evident that further wind tunnel testing of solar installations on sloped roofs is required to clarify the complex interaction of roof slope, building geometry, array spacing and location and orientation.

8.6 CFD simulations

CFD simulations of the wind flow over PV installations offer the prospect of reducing the cost of determining the wind loads, especially with the increased use of open-source software, particularly OpenFOAM [50], e.g. Toja-Silva *et al.* [32]. CFD can also be used for preliminary siting and investigating the effects of proposed building developments on the wind regime of an existing installation. Finally, CFD can model the deposition of dust on modules, e.g. Lu *et al.* [51], which reduces power output, and the convective heat transfer from modules. The latter is important as PV efficiency decreases with increasing cell temperature. This research is in its infancy, but some promising work has been done, e.g. Jubayer *et al.* [52]. Many areas of engineering use CFD of urban wind flow, and considerable expertise has developed in its implementation, e.g. Blocken [53,54]. It is often recommended that wind-related studies include both experimental (wind tunnel) and numerical (CFD) investigations e.g. Blocken [53]. The two methods have different advantages and drawbacks, which tend to offset each other. The issues regarding power spectra of the velocity discussed in Section 8.5 are a potential problem for wind tunnel testing. Further the model scales, typically from 1:200 to 1:500 mean that there can be a large discrepancy in Reynolds numbers. CFD can, in principle, avoid both these problems but has its own challenges.

Because of the geometric complexity of the urban landscape, most CFD has used the Reynolds Averaged Navier–Stokes (RANS) equations. It is necessary to model at least some of the terms in the RANS equations, and the accuracy of these 'turbulence models' is a controlling factor on the overall accuracy of the simulation. The next level of sophistication is large eddy simulation (LES), which has had limited application in urban simulations, e.g. Kono *et al.* [55], but is demonstrably more accurate than RANS modelling. To date, LES has been restricted to simple cases such as single buildings. A good introduction to RANS modelling and LES is given by Durbin and Reif [56]. Examples of RANS models include those for the production and turbulent transport terms for k, and the dissipation, ε in the well-known k–ε model. This model is probably the most used in wind engineering CFD because of its speed, stability and general lack of appreciation of its shortcomings. The worst of these shortcomings is the so-called stagnation point anomaly, as shown in Durbin [57] and Durbin and Reif [56], whereby k is greatly *over-predicted* in the stagnation region ahead of a body such as a rectangular building or a PV module. There is also evidence of an *under-prediction* of k and the mean velocity recovery in the wake region immediately downstream of the building or PV module, e.g. Mohammed and Wood [58]. These authors, along with Toja-Silva *et al.* [46] and Kono *et al.* [55], compared their simulations to the measurements of Meng and Hibi [59] of the simulated ABL flow over a generic rectangular building. Durbin [57] modified the eddy viscosity equation in the k–ε model to reduce the over-prediction of k, but this was found to be not entirely successful by Mohammed and Wood [58]. Those authors presented another correction that appears to be more accurate than Durbin's in the 3D flow of Meng and Hibi [59] but is inaccurate in 2D stagnation flows.

Very little generic flow study has been done that is relevant to PV modules. Lasher [60] showed that the k–ε model under-predicted the drag coefficient (1.5 instead of 1.98) for a two-dimensional thin flat plate, normal to a uniform flow, and significantly over-predicted the recirculation length, which is likely to have consequences for the modelling of downwind modules. The effect of aspect ratio, AR, defined as the span:width ratio, on flat plates has been extensively studied experimentally for uniform flow away from a roof or the ground, e.g. Ortiz *et al.* [61]. For sharp edged bodies like PV modules, Reynolds number effects should be of second order, making $AR = 1$ the smallest aspect ratio of practical interest. A PV module has $AR \sim 1.6$ for which the drag coefficient is about 1.1 when normal to the flow, suggesting a significantly different flow from high ARs. Dalhqvist [62] compared a number of RANS models to direct numerical simulations of finite AR flow and measurements of the drag. She reproduced Lasher's [60] results and found that RANS predictions of drag and recirculation length improved as AR decreased. The reason appears to be the increasing three-dimensionality of the vortex shedding from the module edges as AR decreases: this improves the accuracy of turbulence models developed in classical turbulent flows which are all instantaneously three dimensional. The consideration of generic flows applicable to PV wind loads seems to be a fruitful area of further basic research to establish the accuracy of various turbulence models and improve them.

Different module orientations, inclination and azimuth, were studied in the CFD simulations of Shademan and Hangan [63] using a Reynolds stress model (RSM) that solves for all Reynolds stresses. In principle, RSM should be more accurate than simpler models like k–ε, but there is a significant computational cost. They also compared their simulations for a 2D plate at various incidences and showed a similar, but smaller, under-prediction of the drag than found by Lasher [60]. The flow over a single PV array was modelled using the k–ω turbulence model by Jubayer and Hangan [64]. Using the same models, Jubayer and Hangan [65] analysed multi-component arrays. Generally, k–ω models handle the stagnation point anomaly better than the k–ε model (see Mohammed and Wood [58]) but do not perform significantly better in predicting the drag for normal flat plates, as shown in Dalhqvist [62]. Surprisingly, little study has been made of the effects of turbulence models on simulated urban wind flow or flow over modules and arrays. They can be significant both for the mean wind speed and the turbulence, e.g. Mohammed and Wood [66]. This section has shown the limited comparison of CFD results to laboratory measurements, and none at full-scale.

8.7 Discussion and analysis

A general goal of wind tunnel tests was to derive C_p values for array wind loading. As revealed by these studies, maximum wind pressures result in uplift, corresponding to negative net pressures under the commonly adopted sign convention. Discussions with Sunlock indicated that negative uplift pressures dictate their system design, with the most common failure of residential installations being pull-out of the mounting structure roofing screws due to module suction. For this reason, many studies are concerned with uplift aerodynamics only. The maximum net uplift pressure coefficients identified in these studies are displayed in Figure 8.8 for both flat roof and sloped roof models, with model scales shown for comparison.

These results show a wide range of maximum C_p values with no correlation with model scaling. According to Stathopoulos *et al.* [42], it is likely that variations in results between studies are caused by incorrect scaling of the ABL when simulating structures at large scales, as described previously. In addition, there is little consistency between studies relating to modelling conditions. Typical variations include building size, array size, module configuration, roof positioning, module tilt angle, choice of wind simulation standard and inherent differences between wind tunnels. It is therefore unsurprising that these studies yield differing results.

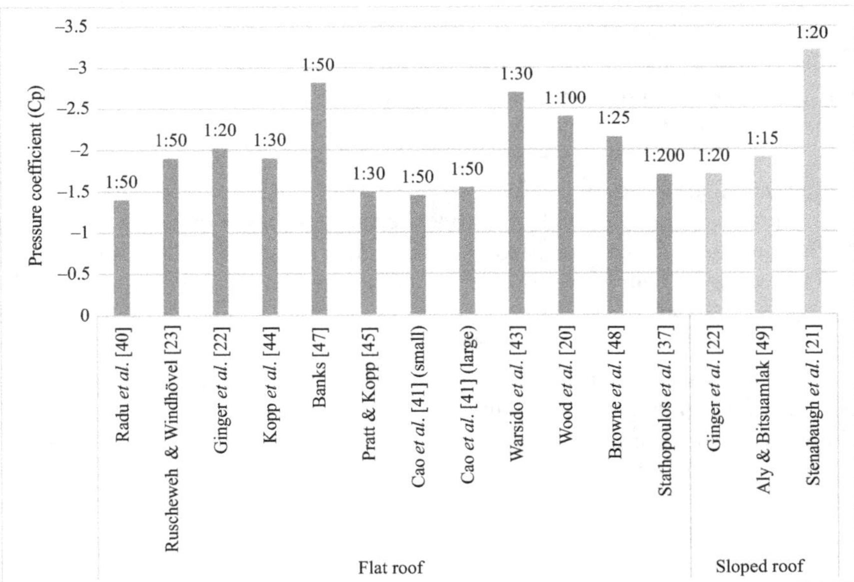

Figure 8.8 Net measured C_p values for various wind tunnel tests, corresponding to maximum module uplift. The maximum value for C_{fig} presented in AS 1170.2 is equal to the maximum C_p value given by Ginger et al. [22] for sloped roof buildings. The model scale is indicated above each bar

A need for greater consistency is discussed by Banks [36], arguing that accurate wind tunnel testing techniques, combined with correct assumptions relating to the aerodynamics of roof-mounted PV arrays, must be used in order to obtain realistic results.

Almost all studies modelled wind characteristics according to open terrain conditions, including those that formed the basis for *AS 1170.2* design data. Given that open areas result in the highest wind speeds, it is expected that this approach has resulted in conservative wind loads. The exception is the study carried out by Browne *et al.* [48] that considered suburban terrain wind conditions, resulting in a maximum C_p value of -2.15. In contrast to expectations, this is both higher than the flat roof study average of -1.96 and the maximum uplift value provided in *AS 1170.2* of -1.7, again displaying the inconsistencies between studies.

A recent series of proprietary wind tunnel tests by Kostas *et al.* [67], conducted with support from Sunlock, investigated the effects of modelling a flat roof-mounted solar array under suburban wind conditions. Outcomes of this test were compared to *AS 1170.2* to establish if the pressure coefficients given in the standard provide conservative results for built-up terrain categories. It was found that maximum C_p uplift values were reduced by as much as 30% for central portions of the array when compared to the standard. Values only approached those given in *AS 1170.2* for modules at the corners and windward edges of the array where high uplift loads are expected, as per [35], indicating the potential for design refinements under urban wind loads.

The high frequency spectrum matching methodology developed by Banks [36] represents the best approach for wind tunnel testing of rooftop PV arrays, satisfying requirements for adequate model sizing and accurate ABL modelling. Using this technique in conjunction with an urban wind profile would provide the most accurate indication of rooftop solar array wind loads. The peak pressure coefficient found in [35] by using this method for open terrain was larger than that of any other flat-roof study tested using traditional means and, at a value of -2.82, is 66% larger than the maximum value provided in *AS 1170.2*. It is therefore possible that any reductions in design C_p values achieved by altering wind approach characteristics may be offset by increased loads under accurate wind modelling.

Despite the rigorous testing carried out in [35], the SEAOC guide discusses limitations for adapting results of any particular wind tunnel test to other building and array configurations. The reason for this is explained by Kopp and Banks [68]: 'generalisation of data to account for roof and array zoning cannot be performed from a single test configuration because of variations in the wind field acting on the array caused by the building and thus the aerodynamic loads'. Therefore, attempts to refine design data and eliminate conservative assumptions will require careful testing of the effects of multiple parameters.

Significant potential for design optimisation exists in relation to module layout. Due to building-specific aerodynamic effects found throughout the surveyed research, it is vital to include the impact of array positioning in minimising wind loads. Studies commonly show large C_p reductions by moving an array into a more favourable location on a roof.

8.8 Conclusions

This literature review has investigated the suitability of modern code for wind loading, as exemplified by Australian Standard *AS 1170.2*, for the estimation of wind loads for the design of urban rooftop PV mounting structures. These structures contribute an increasingly significant proportion of the cost of a solar installation; therefore, it is critical to optimise their design in order to reduce costs and increase the penetration of solar energy systems.

AS 1170.2 was found to provide limited guidance for design of PV arrays under wind loads. The studies that formed the basis for module design pressures were reviewed, indicating the use of potentially inaccurate wind tunnel simulation methods caused by incorrect modelling of the ABL. Furthermore, these simulations used open terrain wind characteristics, presenting higher wind speeds than those expected at urban sites. A common finding was that critical wind loads for arrays on flat roofs occur for wind directed at a building's north corner. Such peak loads are the result of building-generated corner vortices acting on the modules in the array, and reattachment of the wind flow separation bubble near the edge of the building. The highest wind loads caused module uplift in all cases. The few published studies investigating arrays on sloped roofs revealed peak module uplift loads for wind directed parallel to the roof ridge line. Highest module loads were consistently found to occur at array edges. Aerodynamic effects caused by specific features of each building present difficulties in the generalisation of solar array wind loading data. However, module loads can be minimised by positioning an array in a central roof location.

The reviewed studies showed large variations in array wind design pressure coefficients. It is likely that these differences occurred as a result of incorrect boundary layer modelling, in conjunction with significant differences in modelling configurations. Advancements in test methodologies developed by Banks [36] enable satisfactory testing of large scale models in ABL wind tunnels, as required for modelling of PV arrays. A standard rooftop solar array wind tunnel test protocol has also been proposed by Banks [36] in order to eliminate inconsistencies and assumptions in the system design process. Use of Banks's methods represent the most accurate means for estimating wind loads on roof-mounted PV modules via wind tunnel testing and should be considered for inclusion in future editions of wind loading standards.

Significant research gaps exist. The few studies found in the literature commonly simulated wind loading on such systems using incorrectly scaled ABLs, with large variations shown in results. To develop an accurate understanding of wind loading on sloped-roof arrays, future wind tunnel testing should be carried out using high frequency turbulence spectrum matching to eliminate compromises between model scaling requirements and wind flow characteristics.

The majority of studies examined in this survey used open terrain wind models to determine PV module wind loads, likely yielding conservative results. Further, wind tunnel tests combining high-frequency spectrum matching and urban terrain wind profiles should be conducted for a variety of both flat-roof and sloped-roof solar array installations. Studies should use a standard test protocol to eliminate inconsistencies between tests in order to develop a realistic database of wind loads on various solar array configurations.

Stenabaugh *et al.* [21] have noted the requirement for full-scale validation of wind tunnel simulations. Studies aimed at measuring solar array wind loads on existing installations are scarce but would greatly benefit any future effort to standardise wind loads and optimise design practices.

The situation for computational modelling of wind loads is not significantly better. The complexity of the urban topography means that simplified turbulence models must be used for the simulations, and many of these models are known to be inaccurate for the generic flows that are applicable to PV wind load estimation. Due to the increasing availability of open-source software, and the need to investigate individual cases, however, it is likely that simulations will be increasingly used. It is recommended that the combined use of wind tunnel and computer modelling that has been advocated at several places in this review be augmented by wind load measurements on actual arrays.

Acknowledgements

David Wood's work has been supported by the NSERC Industrial Research Chairs programme in collaboration with Enmax Corporation. Jesse Brydges is supported by a Mitacs award in conjunction with RWDI Inc.

References

[1] California Energy Commission & California Public Utilities Commission, The Go Solar California!, 21 May 2017. [Online]. Available: http://www.gosolarcalifornia.org/.

[2] National Solar Repository of Singapore, [Online]. Available: http://www.solar-repository.sg/pv-systems-database. [Accessed 25 May 2017].

[3] Australian PV Institute Solar Map, 21 May 2017. [Online]. Available: http://pv-map.apvi.org.au/.

[4] IRENA (International Renewable Energy Agency), "The Power to Change: Solar and Wind Cost Reduction Potential to 2025," IRENA, 2016. [Online]. Available: http://www.irena.org/DocumentDownloads/Publications/IRENA_Power_to_Change_2016.pdf.

[5] M. Chang, C. H. Hsueh, H. Chen and C. Chen, "Evaluating the influence of typhoons on PV module reliability," in *32nd European Photovoltaic Solar Energy Conference and Exhibition*, Munich, 2016.

[6] S. Fortuin, M. Hermann, G. Styri-Hipp, P. Nitz and W. Platzer, "Hybrid PV-thermal collector developments: concepts, experiences, results and research needs," *Energy Procedia*, vol. 48, pp. 37–47, 2014.

[7] K. Pickerel, "What are the mechanical loading tests for solar panels?," Solar Power World, 5 July 2016. [Online]. Available: http://www.solarpowerworldonline.com/2016/07/mechanical-loading-tests-solar-panels. [Accessed 26 May 2017].

[8] ASCE/SEI, "Minimum Design Loads and Associated Criteria for Buildings and Other Structures," ASCE, Reston, Virginia, 2017.
[9] Australian/New Zealand Standards, "Structural Design Actions: Wind Loads AS1170.2 (Reconfirmed 2016)," Homebush, ANS, 2011.
[10] J. D. Holmes, C. S. Kwok and J. Ginger, "Wind Loading Handbook for Australia and New Zealand," Melbourne, Australasian Wind Engineering Society, 2012.
[11] J. D. Holmes, "Wind Loading of Structures," 3rd edition, Baton Rouge, LA, CRC Press, 2015.
[12] Australian Building Code Board, "National Construction Code 2016," Canberra, Council of Australian Government, 2016.
[13] Structural Engineers Association of California, "Wind Design for Low-Profile Solar Photovoltaic Arrays on Flat Roofs," Structural Engineers Association of California, Sacramento, 2012.
[14] American Society of Civil Engineers, "Minimum Design Loads for Buildings and Other Structures," Reston, VA, ASCE, 2013.
[15] Canadian Commission on Building and Fire Codes, "National Building Code of Canada," Ottawa, Canadian Commission on Building and Fire Codes, 2010.
[16] Osvaldo, "Extreme Wind Speed Estimation for Calgary," MSc thesis, University of Calgary, Calgary, 2017.
[17] C.-M. Cheng, M.-S. Tsai, Y.-L. Lo and C.-H. Wang, "Investigation on atmospheric boundary layers: field monitoring and wind tunnel simulation," in *6th European and African Conference on Wind Engineering*, Cambridge, 2013.
[18] A. Schellenberg, J. Maffei, K. Telleen and R. Ward, "Structural analysis and application of wind loads to solar arrays," *Journal of Wind Engineering and Industrial Aerodynamics*, vol. 123, pp. 261–272, 2013.
[19] P. Blackmore, "Wind Loads on Roof-based Photovoltaic Systems," BRE Bookshop, London, 2004.
[20] G. S. Wood, R. Denoon and K. C. Kwok, "Wind loads on industrial solar module arrays and supporting roof structure," *Wind and Structures*, vol. 4, pp. 481–494, 2001.
[21] S. E. Stenabaugh, Y. Iida, G. A. Kopp and P. Karava, "Wind loads on photovoltaic arrays mounted parallel to sloped roofs on low-rise buildings," *Journal of Wind Engineering and Industrial Aerodynamics*, vol. 139, pp. 16–26, 2015.
[22] J. Ginger, M. Payne, G. Stark, B. Sumant and C. Leitch, "Investigation on Wind Loads Applied to Solar Modules Mounted on Roofs," James Cook University, Townsville, 2011.
[23] H. Ruscheweyh and R. Windhövel, "Wind Loads at Solar and Photovoltaic Modules for Large Plants," in *Proceedings of the 13th International Conference on Wind Engineering*, Amsterdam, 2011, pp. 10–15.
[24] S. L. Walker, "Building mounted wind turbines and their suitability for the urban scale – a review of methods of estimating urban wind resource," *Energy and Buildings*, vol. 43, pp. 1852–1862, 2011.
[25] J. T. Millward-Hopkins, A. S. Tomlin, L. Ma, D. Ingham and M. Pourkashanian, "The predictability of above roof wind resource in the urban roughness sublayer," *Wind Energy*, vol. 15, pp. 225–243, 2012.

[26] F. C. Emerjeamara, A. S. Tomlin and J. T. Millward-Hopkins, "Urban wind: characterisation of useful guest and energy capture," *Renewable Energy*, vol. 81, pp. 162–172, 2015.
[27] E. R. Marciotto and G. Fisch, "Wind tunnel study of turbulent flow past an urban canyon model," *Environmental Fluid Mechanics*, vol. 13, pp. 403–416, 2013.
[28] A. A. Razak, A. Hagishima, N. Ikegaya and J. Tanimoto, "Analysis of airflow over building arrays for assessment of urban wind environment," *Building and Environment*, vol. 59, pp. 56–65, 2013.
[29] Encraft, "Warwick Wind Trials – Final Report," 2009.
[30] B. Tabrizi, J. Whale, T. Lyons and T. Urmee, "Extent to which international wind turbine standard IEC 61400-2 is valid for rooftop wind installation," *Journal of Wind Engineering and Industrial Aerodynamics*, vol. 139, pp. 50–61, 2015.
[31] IEC, "IEC 61400-2 Wind Turbines Part 2: Small Wind Turbines Edition 3.0 2013-12," 2013.
[32] F. Toja-Silva, A. Colmenar-Santos and M. Castro-Gil, "Urban wind energy exploitation systems: behaviours under multidirectional flow conditions – opportunities and challenges," *Renewable and Sustainable Energy Reviews*, vol. 24, pp. 364–378, 2013.
[33] B. R. Karthikeya, P. S. Negi and N. Srikanth, "Wind resource assessment for urban renewable energy application in Singapore," *Renewable Energy*, vol. 87, pp. 403–414, 2016.
[34] Flextronics Australia, "Sunlock Solar Panel Mounting System," Flextronics Australia [Online]. 2017. Available: http://www.sunlock.com.au/. [Accessed 18 January 2017].
[35] Flextronics Australia, "Rooftop Installation Manual," 2013.
[36] D. Banks, "Measuring peak wind loads on solar power assemblies," in *13th International Conference on Wind Engineering*, Amsterdam, 2011.
[37] T. Stathopoulos, I. Zisis and E. Xypnitou, "Local and overall wind pressure and force coefficients for solar modules," *Journal of Wind Engineering and Industrial Aerodynamics*, vol. 125, pp. 195–206, 2014.
[38] D. Banks, T. K. Guha and Y. J. Fewless, "A hybrid method of generating realistic full-scale time series of wind loads from large-scale wind tunnel studies: application to solar arrays," in *14th International Conference on Wind Engineering*, Porto Alegre, 2015.
[39] M. Aly and G. Bitsuamlak, "Aerodynamics of ground-mounted solar modules: test model scale effects," *Journal of Wind Engineering and Industrial Aerodynamics*, vol. 123, pp. 250–260, 2013.
[40] A. Radu, E. Axinte and C. Theohari, "Steady wind pressures on solar collectors on flat-roofed buildings," *Journal of Wind Engineering and Industrial Aerodynamics*, vol. 23, pp. 249–258, 1986.
[41] J. Cao, A. Yoshida, P. K. Saha and Y. Tamura, "Wind loading characteristics of solar arrays mounted on flat roofs," *Journal of Wind Engineering and Industrial Aerodynamics*, vol. 123, pp. 214–225, 2013.

[42] T. Stathopoulos, E. Xypnitou and I. Zisis, "Wind loads on rooftop solar module systems: a contribution to NBCC 2015," in *7th Workshop on Regional Harmonization of Wind Loading and Wind Environmental Specifications in Asia-Pacific Economies (APEC-WW 2012)*, Hanoi, 2012.

[43] W. P. Warsido, G. T. Bitsuamlak, J. Barata and A. G. Chowdhury, "Influence of spacing parameters on the wind loading of solar array," *Journal of Fluids and Structures*, vol. 48, pp. 295–315, 2014.

[44] G. A. Kopp, S. Farquhar and M. J. Morrison, "Aerodynamic mechanisms for wind loads on tilted, roof-mounted, solar arrays," *Journal of Wind Engineering and Industrial Aerodynamics*, vol. 111, pp. 40–52, 2012.

[45] R. N. Pratt and G. A. Kopp, "Velocity measurements around low-profile, tilted, solar arrays mounted on large flat-roofs," *Journal of Wind Engineering and Industrial Aerodynamics*, vol. 123, pp. 226–238, 2013.

[46] F. Toja-Silva, C. Peralta, O. Lopez-Garcia, J. Navarro and I. Cruz, "Roof region dependent wind potential assessment with different RANS turbulence models," *Journal of Wind Engineering and Industrial Aerodynamics*, vol. 142, pp. 258–271, 2015.

[47] D. Banks, "The role of corner vortices in dictating peak wind loads on tilted flat solar modules mounted on large, flat roofs," *Journal of Wind Engineering and Industrial Aerodynamics*, vol. 123, pp. 192–201, 2013.

[48] M. T. L. Browne, M. P. M. Gibbons, S. Gamble and J. Galsworthy, "Wind loading on tilted roof-top solar arrays: the parapet effect," *Journal of Wind Engineering and Industrial Aerodynamics*, vol. 123, pp. 202–213, 2013.

[49] A. M. Aly and G. Bitsuamlak, "Wind-induced pressures on solar modules mounted on residential homes," *Journal of Architectural Engineering*, vol. 20, 2014. https://doi.org/10.1061/(ASCE)AE.1943-5568.0000132.

[50] OpenFOAM CFD Ltd. (ESI Group), "OpenFOAM," OpenFOAM CFD Ltd. (ESI Group), 20 January 2017. [Online]. Available: http://openfoam.com/. [Accessed 18 January 2017].

[51] H. Lu, L. Lu and Y. Wang, "Numerical investigation of dust pollution on a solar photovoltaic (PV) system mounted on an isolated building," *Applied Energy*, vol. 180, pp. 27–36, 2016.

[52] M. Jubayer, K. Siddiqui and H. Hangan, "CFD analysis of convective heat transfer from ground mounted solar panels," *Solar Energy*, vol. 133, pp. 556–566, 2016.

[53] B. Blocken, "50 years of computational wind engineering: past, present and future," *Journal of Wind Engineering and Industrial Aerodynamics*, vol. 129, pp. 69–102, 2014.

[54] B. Blocken, "Computational fluid dynamics for urban physics: importance, scales, possibilities, limitations and ten tips and tricks towards accurate and reliable simulations," *Building and Environment*, vol. 91, pp. 219–245, 2015.

[55] T. Kono, T. Kogaki and T. Kiwata, "Numerical investigation of wind conditions for roof-mounted wind turbines: effects of wind direction and horizontal aspect ratio of a high-rise cuboid building," *Energies*, vol. 9, pp. 907–937, 2016.

[56] P. A. Durbin and B. P. Reif, "Statistical Theory and Modeling for Turbulent Flows," Chichester, John Wiley & Sons, 2011.

[57] P. Durbin, "On the k–ε stagnation point anomaly," *International Journal on Heat and Fluid Flow*, vol. 17, pp. 89–90, 1996.

[58] M. A. Mohammed and D. H. Wood, "Modifications to RANS turbulence models for wind flow over buildings," *International Journal of Sustainable Energy*, vol. 36, pp. 225–241, 2015.

[59] T. Meng and K. Hibi, "Turbulent measurements of the flow field around a high-rise buildings," *Journal of Wind Engineering*, vol. 76, pp. 55–64, 1998.

[60] W. C. Lasher, "Computation of two-dimensional blocked flow normal to a flat plate," *Journal of Wind Engineering and Industrial Aerodynamics*, vol. 89, pp. 493–513, 2001.

[61] X. Ortiz, D. Rival and D. H. Wood, "Forces and moments on flat plates of small aspect ratio with application to PV wind loads and small wind turbine blades," *Energies*, vol. 8, no. 4, pp. 2438–2453, 2015.

[62] E. Dalhqvist, "The Performance of CFD RANS Models in Predicting Wind Loads on Flat Plates: A Comparative Study with DNS," KTH Royal Institute of Technology School of Mechanical & Industrial Engineering, Sweden, 2016.

[63] M. Shademan and H. Hangan, "Wind loading on solar panels at different azimuthal and inclination angles," in *The Fifth International Symposium on Computational Wind Engineering*, Chapel Hill, 2010.

[64] C. M. Jubayer and H. Hangan, "Numerical simulation of wind effects on a stand-alone ground mounted photovoltaic (PV) system," *Journal of Wind Engineering and Industrial Aerodynamics*, vol. 134, pp. 56–64, 2014.

[65] C. M. Jubayer and H. Hangan, "A numerical approach to the investigation of wind loading on an array of ground mounted solar photovoltaic (PV) panels," *Journal of Wind Engineering and Industrial Aerodynamics*, vol. 153, pp. 1–11, 2016.

[66] M. A. Mohammed and D. H. Wood, "Computational modelling of wind flow over the university of Calgary campus," *Wind Engineering*, vol. 40, pp. 228–249, 2016.

[67] J. Kostas, Y. Padayatchy, E. Chong, J. Tan and W. H. Melbourne, "Wind Tunnel Measurements of Pressures on Solar Panel Arrays on Industrial Warehouse Buildings," *MEL Consultants Report* 73/13, 2014.

[68] G. A. Kopp and D. Banks, "Use of the wind tunnel test method for obtaining design wind loads on roof-mounted solar arrays," *Journal of Structural Engineering*, vol. 139, pp. 284–287, 2012.

Chapter 9

Design optimization of multi-energy hubs for community energy projects*

Azadeh Maroufmashat[1], Sean B. Walker[1], Ushnik Mukherjee[1], Michael Fowler[1], Ali Elkamel[1] and Sourena Sattari[2]

Abstract

In this chapter, the authors study the effect of accounting for the hourly emission factors on the optimal primary energy consumption, natural gas, and electricity, for a network of energy hubs. Each hub is representative of a part of an urban area such as a school, residential complex, or hospital, in Ontario. In order to accomplish this, a mathematical model is developed based on two objectives, reducing the total annual cost of the system and the greenhouse gas emissions. Different technologies such as combined heat and power systems, solar photovoltaics, solar-thermal collectors, boilers, and heat and electrical storage systems are considered. Different scenarios are defined to investigate the effect of energy storage systems and energy exchange between buildings on the optimal configuration of the system. The results are compared for different scenarios based on total annual cost, CO_2 emission, as well as natural gas and electricity consumption for constant and hourly emission factors.

9.1 Introduction

In this chapter, the authors examine the use of a network of energy hubs to improve the energy efficiency and reduce the greenhouse gas emissions of an energy system. Such a network of hubs, called a Smart Energy System (SES), is a means to meet energy demands more effectively by integrating multiple resources and technologies. An SES can connect power sources to end-users through energy conversion technologies and transmission utilities and from the demand side, through the

*Presented in *Proceedings of 2016 Energy and Water Symposium*, Windsor, Ontario, Canada, 22–23 June 2016.
[1]Department of Chemical Engineering, University of Waterloo, Canada
[2]Sharif Energy Research Institute, Department of Energy Engineering, Sharif University of Technology, Iran

market, operators, and service providers. The effective consumption of energy sources along with the generation of renewable energy are tied to issues of economic growth, climate change, and the energy security. In this analysis, a network of energy hubs in a region is created to conserve and store energy, and, thereby, provide a cleaner environment. Moreover, the introduction of distributed combined heat and power (CHP) systems as a component of an SES can provide an improvement in energy efficiency through the poly-generation of electricity and heat. The integrated nature of SES can provide a number of advantages including greenhouse gases emission reductions, energy demands satisfaction, more efficient use of fossil fuels, the development of renewable energy technologies, and a reduction in the need for new energy generation and transmission technology. These advantages are achieved through the integration of renewable energy to the grid and the natural gas system, the incorporation of energy storage, and more efficient use of energy generation technologies to meet a variety of electrical and heat demands. When hydrogen is used as an energy vector, it is possible to accommodate industrial demands for hydrogen, for lift trucks or refinery operations, or future fleets of hydrogen fuel-cell vehicles.

Distributed energy systems provide a new way of meeting the demands of energy users over a large geographic area compared to central energy systems [1,2]. In central generation systems, power plants are very large and centrally located units that involve a significant transmission infrastructure. Distributed energy generation systems are a new and progressive design in which the energy generation units are close to the consumers. Gradually, in the implementation of a distributed energy system, small power generation units are replaced by larger ones [2]. In other words, distributed energy systems can be defined as buildings which can completely supply their energy requirements for electricity, heat, and cooling [1]. Dunn *et al.* state that, in the future, the energy supply chain will be near the consumer's buildings and applications [3]. Although these systems enjoy a lot of advantages, the allocation of power generation technologies amongst various consumers is a competitive challenge economically, environmentally, and technically [4].

Energy hubs, developed by Geidl *et al.* [5], are defined as interconnected energy systems which include energy generation, conversion, and storage systems. They are interfaces between different energy generation and loads. Considering this definition, the synergies among different forms of energy represent an opportunity for system improvement and the use of emerging energy conversion technologies, including cogeneration systems and fuel cells. Syed *et al.* [6] develop a simulation model for the operational study of a fleet of plug-in fuel-cell vehicles and a commercial building based on novel energy hub concept. Moeini-Aghtaie *et al.* [7] develop a probabilistic economic dispatch optimization model for the optimal operation of multi-energy carrier systems. However, the studies cited above do not consider the use of energy storage systems in the energy hubs.

A mathematical model is developed by Koltsaklis *et al.* [8] and Bracco *et al.* [9] for the design and operation of several buildings, based on the use of CHP systems. Ashouri *et al.* develop a framework for the optimal design and operation of building services [10], while Wakui *et al.* develop a mixed integer linear programming (MILP) model for the operation of residential complexes by considering hot water network among the building cogeneration systems by the objective function is the minimization of the primary energy consumption [11].

In this chapter, the authors build on the previous examinations of energy hubs [12,13] by studying the effect of hourly emission factors on the optimal primary energy consumption, both natural gas and electricity, for a network of energy hubs. The variable hourly emissions factor is determined using published emissions factors for power generation technologies from the GREET database and production values provided by the Independent Electricity System Operator [14,15]. The operation of the energy hub using a constant emissions factor is also used for comparative purposes.

9.2 Methodology

The two main objectives of this optimization study are to minimize the total annual cost and CO_2 emissions as illustrate by the following equations:

$$Z_1 = \sum_{s} (Cost^s_{Cap} + Cost^s_{op} + Cost^s_{fuel} - Income^s) + Cost_{pipe} \tag{9.1}$$

$$Z_2 = \sum_{s} \sum_{j} \sum_{h} \sum_{m} CI_j \times P_{s,j,m,h} \times b_j \tag{9.2}$$

The cost objective comprises of the capital, operational, and maintenance, fuel and piping cost of all of the hubs existing within the system.

The number of energy conversion and storage technologies installed in the hubs are multiplied by their rated capacities and unit capital costs to determine the total capital investment. The total capital investment is then amortized by setting a project life time and interest rate. The Chemical Engineering Plant Cost Indexes have been used to update the cost of the individual components from their respective base year to the year in consideration in this study [13]. The components of each hub have a fixed and variable operating-and-maintenance (O&M) cost associated with them. In general, the fixed cost is taken to be as a fraction of the capital cost, whereas the variable cost includes the operational regime that the components are subjected to over the course of its lifetime. The term fuel cost represents the cost incurred in using the different primary energy vectors. The revenue earned by each energy hub is estimated based on the exchange of surplus energy between energy hubs. This term is added to the objective function as the income [12].

Equation (9.3) illustrates the governing energy balance equation associated with each hub, where $C_{s,i,j,m,h}$ is a coupling factor denoting either the individual component efficiencies, or a product of distributed energy system efficiencies. The balance of energy stored within the hub is given by (9.4) where $A^{ch}_{s,i}$ and $A^{dis}_{s,i}$ denote charging and discharging efficiencies, while $M^{std}_{s,i,m,h,q}$ denotes the standby losses associated with a particular energy storage technology.

$$L_{s,i,m,h} = \sum_{j} \left[C_{s,i,j,m,h}\right].\mathrm{I}.b_{s,j}.P_{s,j,m,h} - \sum_{q} (Q^{ch}_{s,i,m,h,q} - Q^{dis}_{s,i,m,h,q}) \tag{9.3}$$

$$M_{s,i,m,h,q} = M_{s,i,m,h-1,q} + A^{ch}_{s,i,q} Q^{ch}_{s,i,m,h,q} - A^{dis}_{s,i,q} Q^{dis}_{s,i,m,h,q} - M^{std}_{s,i,m,h,q} \tag{9.4}$$

Different constraints regarding the energy flow limit, energy conversion, and storage technologies are considered based on [13]. Moreover, it is assumed that the number of operating energy conversion and storage technologies cannot be more than installed.

The set of equations (9.5a) and (9.5b) show the governing equations employed for relating the exchange of energy vectors between hubs ($Tr_{sk,i,m,h}$).

$$T_{s,i,m,h} = \sum_{k \neq s} Tr_{sk,i,m,h} \quad \text{for } i = elec, heat, H_2 \tag{9.5a}$$

$$P^{network}_{s,j,m,h} = \sum_{k \neq s} Tr_{ks,i,m,h} f(d_{ks}) \quad \text{for } i,j = elec, heat, H_2 \tag{9.5b}$$

The term $P^{network}_{s,j,m,h}$ shows the transferred energy to other hubs and $f(d_{ks})$ is energy loss based on the distance [12]. A constraint also considers for the energy flow rate between energy hubs to be within the upper and lower values.

Different energy conversion technologies including PV, CHP systems are modelled according to [16].

The proposed mathematical model is carried out in general algebraic modelling software (GAMS) based on the MILP approach and solved by CPLEX solver [17].

9.3 Illustrative case study

The urban area considered for this optimization study is located within the province of Ontario, Canada. The urban area includes four buildings that can be categorized into four different groups, namely residential complex (7,765 m^2), office building (1,000 m^2), commercial building (75,000 m^2), and a restaurant (1,000 m^2). Each building is representative of a hub. The optimization model chooses the number of energy conversion and storage technologies to be installed within each hub. The problem selects the number of fixed unit size distributed energy systems as a way to optimally meet the energy demands throughout the hub network. The schematic of the hubs comprised within the urban energy system considered in the study and the energy network links between different hubs has been illustrated in Figure 9.1. The different energy vectors that are considered to be potential inputs to each hub in this study include natural gas, electricity (from grid and local solar panels), and thermal energy (solar). The distance separating the fours hubs from each other has also been shown in Figure 9.1.

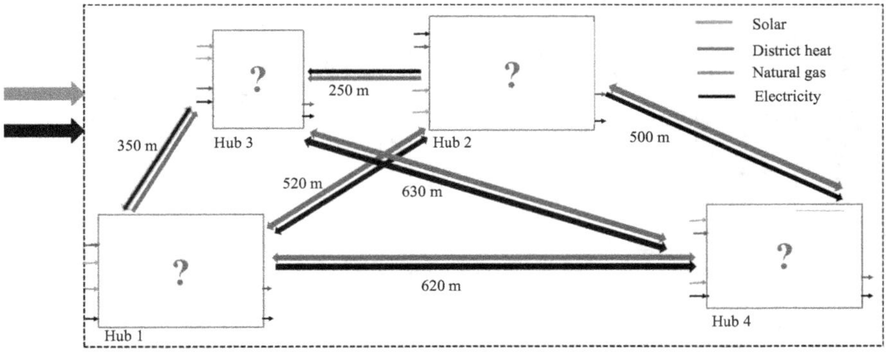

Figure 9.1 The schematic view of energy hub network

Tables 9.1 and 9.2 highlight the techno-economic information associated with the different distributed energy systems including energy conversion and storage technologies. The type, rated capacity, associated efficiencies, and heat-to-power ratio of the technologies have been included in the technical information. The economic information includes the capital and O&M costs of the individual components. Energy distribution network costs have also been listed in Table 9.1. Pipelines will be established for energy hubs exchanging heat between them. Electricity is proposed to be exchanged by means of the utility grid network. Energy demands of each individual energy hub are obtained from [12,13]. For the sake of simplicity in data representation, a sample day from each season is illustrated; however, the model is run for each hour of the year.

The O&M cost of technologies are usually mentioned as a percentage of investment cost ($/kW), while those of some technologies are based on the operating hours of the technologies. The system fixed O&M cost is the recurring annual cost that occurs regardless of the size or architecture of the power system. For some technologies, the O&M cost is based on the hourly value, and it is multiplied by the operating hours per year to calculate the annual O&M cost.

Electricity is one of the fuels existing within the urban energy system. The time of use price for electricity used by small-scale users within Ontario is employed in this study as a cost associated with the purchase of electricity [19]. The cost associated with delivering electricity to end users in the form of transmission and distribution cost has also been used in this study. Transmission cost of electricity is taken to be 1.21¢/kW h. Natural gas pricing includes commodity price, and also storage and transportation charges and is taken to be 22¢/m^3 for this study [20].

Solar irradiation data for southern Ontario has been used in the modelling study to assess the feasibility of using solar panels within the energy hubs. In order to account for incurred emissions while using natural gas and grid electricity, a natural gas emission factor of 0.187 kg/kW h and an average Ontario electricity grid emission factor of 0.177 kg/kW h has been used [21,22]. A discount rate of 8%

Table 9.1 Technical and economic information of the energy hub technologies and network infrastructure information [12,18]

	Rated capacity (kW)	Heat-to-power ratio	Max. number in each hub	Electrical eff.	Thermal eff.	Unit investment cost ($/kW)	Fixed O&M cost ($/kW year)	Variable O&M cost ($/kW h)
Boiler	100–530	–	3	–	0.82	100	–	0.027
Micro turbine	30–250	1.45–1.89	3	0.23–0.35	0.38–0.46	2,440–2,970	–	0.013–0.016
Internal combustion engines (ICE)	60–300	1.43–1.81	3	0.26–0.35	0.44–0.56	900–1,200	–	0.0111–0.016
PV	100	–	–	0.13	–	3,910	21	–
Solar-thermal collector	100	–	–	–	0.35	1,520	1% Cap	0.01
Piping	–	–	–	–	–	200 ($/m)		4.4% Cap

Table 9.2 Technical and economic information of the storage technologies [18]

	Rated capacity (kW h)	Max. number in each hub	Unit investment cost ($/kW h)	Fixed O&M cost ($/kW h year)	Variable O&M cost ($/kW h)	Charging/ discharging eff.	Standby loss (%/h)	Depth of charge (D.OD) %
Heat storage	50–300	1	23	–	0.0015	1/1	0.414	
Battery pack	30–50	2	300	2.9		0.9/0.85		80

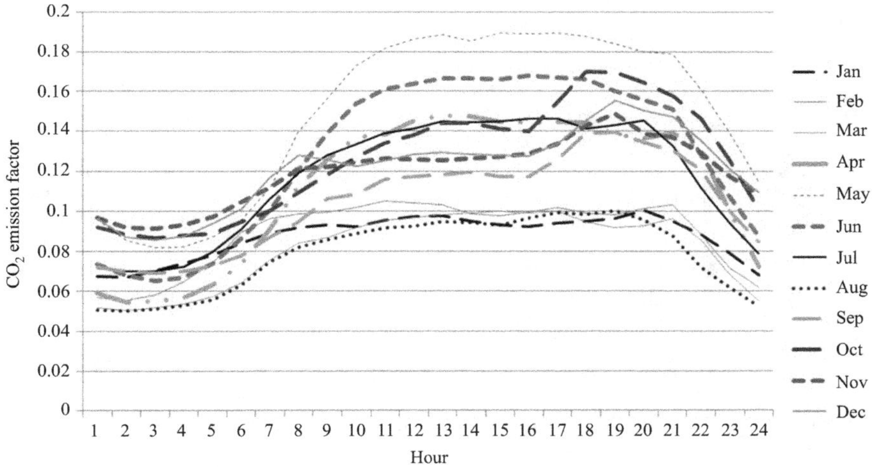

Figure 9.2 Hourly CO_2 emission factor for electricity generation in Ontario for 2014

and a project life time of 20 years are used to amortize the capital cost associated with the technologies used within the energy hub.

The hourly emission factor of electricity grid for Ontario is shown in Figure 9.2 [14,15].

9.4 Results and discussion

Using the methodology described in Section 9.2, the performance of the system is compared across the two scenarios as follows:

Scenario 1: Each building is considered to be as one hub. Energy exchange between hubs is not allowed. Distributed energy systems (DES) including different energy conversion and energy storage systems can be installed in each hub (no interaction/DES/storage).

Scenario 2: In contrast to Scenario 2, the energy exchange between hubs is allowed (interaction/DES/storage).

Figure 9.3 shows the optimal configuration of each energy hub including the selection of the technologies and their numbers of both scenarios.

A number of heat and electrical storage systems are selected in Scenario 1. Considering both DES and energy network in Scenario 4 results in more CHP installation.

In the subsections below, the performance of the SES is discussed for the case of the variable emissions factor and the constant emissions factor.

In Figure 9.4, the costing of the system, under a constant emissions factor is given in thousands of dollars per annum. As can be seen in Figure 9.4, the operation of hub 4 makes up the majority of the costing in each scenario. The CO_2 emissions

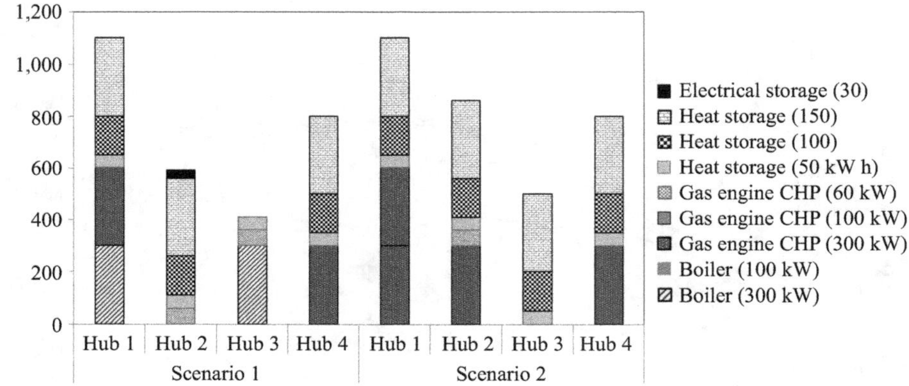

Figure 9.3 Optimal configuration of each energy hub for different scenarios

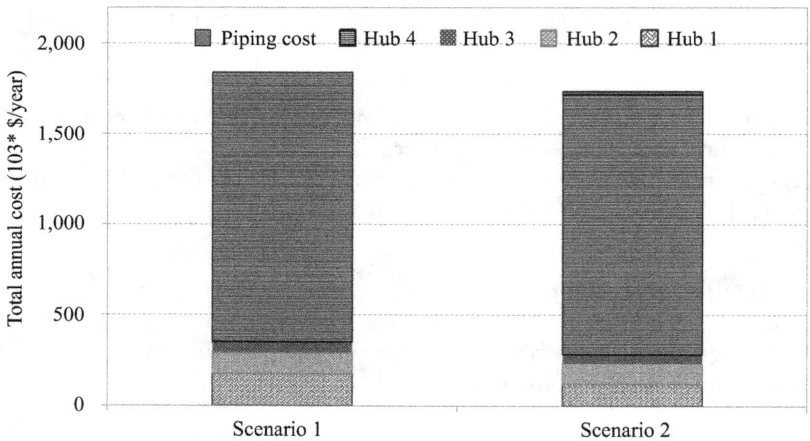

Figure 9.4 Annual cost for two scenarios based on the cost objective

are compared for the two scenarios based on the constant and variable emission factor as shown in Figure 9.4. In Scenario 2, the total annual cost decreases to 6% compared to Scenario 1. Results shows that considering DES and energy exchange between hubs can lower the total annual cost of the system (Figure 9.5).

Since the emission factor of the electricity grid is quite low in Ontario, as over 75% of Ontario's electricity comes from carbon neutral nuclear and hydro power generation, employing distributed energy systems can potentially increase greenhouse gases emissions [23]. The conventional energy systems, where the electricity is supplied by the grid, have the lowest greenhouse gas emission when compared to the other scenarios. For example, in Scenario 2 where the exchange of energy is not restricted, the carbon emissions increase significantly compared to the first

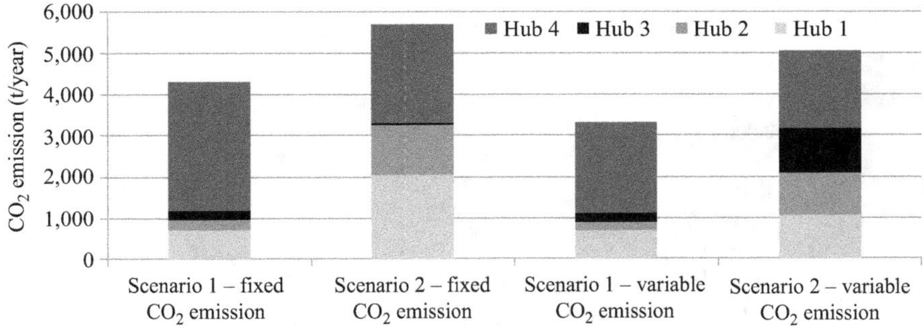

Figure 9.5 Greenhouse gas emissions for two scenarios using constant and variable emissions factor based on the cost objective

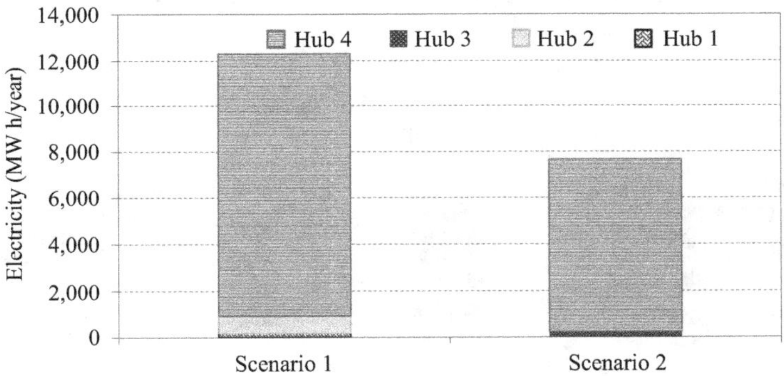

Figure 9.6 Electricity consumption from the grid for different scenarios based on the cost objective

scenario. Considering the dynamic emission factor can reduce the greenhouse gas emission of the total systems by 11%–22% for both scenarios, compared to those of constant emission factor. It means that, in order to precisely calculate the greenhouse gas emission for the integrated systems, one should consider the hourly emission factor, which is rarely considered in the previous literature.

A comparison of the primary energy consumption including electricity and natural gas in each scenario is provided in Figures 9.6 and 9.7.

The electricity consumption is cut in half when energy exchange is allowed between the hubs, in Scenario 1 compared with Scenario 2. However, the consumption of natural gas doubles, which indicates that the use of distributed energy storage decreases electricity consumption but favours the use of natural gas.

Examining Scenario 2, where full interaction occurs between the energy hubs, in detail it can be seen that energy hubs 1 and 2 meet most of the demand through the use of the DES, while the heat demand and 10% of the electricity demand of hub 3 is met by the neighbouring hubs.

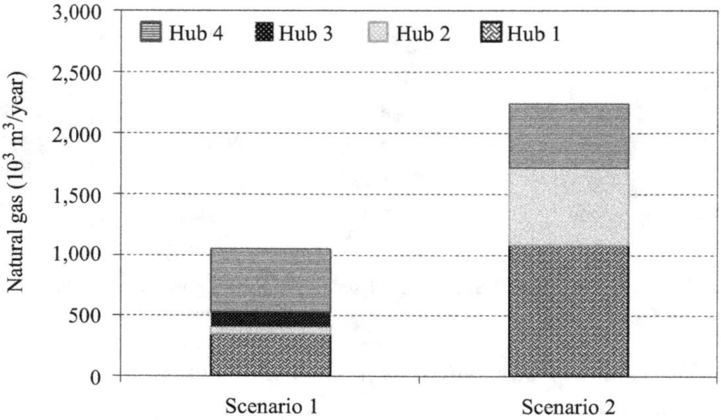

Figure 9.7 Natural gas consumption for two scenarios based on the cost objective

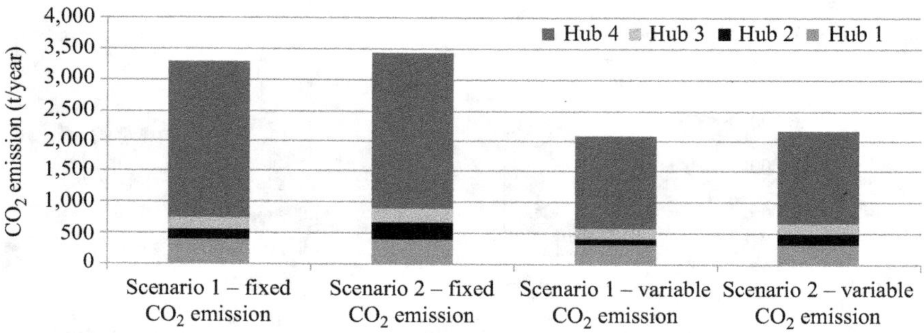

Figure 9.8 Greenhouse gas emissions for two scenarios using constant and variable emissions factor based on the CO_2 emission objective

Considering the optimization model by the objective of the minimization of greenhouse gas emission is shown in Figure 9.8. It can be seen that using the variable emission factor can reduce the greenhouse gas emission of Scenario 2 to 37%.

9.5 Conclusions

This work presents the application of energy hubs and its network model for the optimal design of urban energy systems using constant and variable emission factors. Unique to this research is the effect of a dynamic emission factor on the optimal design of the integrated systems, which is not considered in the prior literature. The generic mathematical model is developed in GAMS. The optimal results show that using dynamic emission factor can reduce the greenhouse gas emission of the system. In other words, using the average emission factor may overestimate the emissions of the integrated energy system. Moreover, comparing the scenario results demonstrates that considering the energy network between the energy hubs can reduce the total cost of the systems; however, due to the clean grid electricity production in Ontario, the CO_2 will increase. Furthermore, the electricity consumption will decrease when considering the network between energy hubs, while that if natural gas increases due to the use of energy systems such as CHP systems. The results regarding the minimizing the objective function of CO_2 emission show that the greenhouse gas emission decreases around 36% for both scenarios compared to those of constant emission factor.

References

[1] Ackermann, T., Andersson, G., and Södera, L. Distributed generation: a definition. Electric Power Systems Research, 2001;57:195–204.
[2] Helsinki, E. *Energy visions 2030 for Finland. VTT Energy*. 2001: Finland.
[3] Dunn, S. Hydrogen futures: toward a sustainable energy system. International Journal of Hydrogen Energy, 2002;27(3):235–264.
[4] Mallikarjun, S., and Lewis, H.F. Energy technology allocation for distributed energy resources: a strategic technology-policy framework. Energy, 2014;72:783–799.
[5] Geidl, M. Integrated modeling and optimization of multi-carrier energy systems, PhD thesis, ETH Zurich; 2007.
[6] Syed, F., Fowler, M., Wan, D., and Maniyali, Y. An energy demand model for a fleet of plug-in fuel cell vehicles and commercial building interfaced with a clean energy hub. International Journal of Hydrogen Energy. 2010;35 (10):5154–5163.
[7] Moeini-Aghtaie, M., Dehghanian, P., Fotuhi-Firuzabad, M., and Abbaspour, A. Multiagent genetic algorithm: an online probabilistic view on economic dispatch of energy hubs constrained by wind availability. IEEE Transactions on Sustainable Energy. 2014;5(2):699–708.
[8] Koltsaklis, N.E., Kopanos, G.M., and Georgiadis, M.C. Design and operational planning of energy networks based on combined heat and power units. Industrial & Engineering Chemistry Research, 2014;53(44): 16905–16923.
[9] Bracco, S., Dentici, G., and Siri, S. Economic and environmental optimization model for the design and the operation of a combined heat and power distributed generation system in an urban area. Energy, 2013;55:1014–1024.
[10] Ashouri, A., Fux, S.S., Benz, M.J., and Guzzella, L. Optimal design and operation of building services using mixed-integer linear programming techniques. Energy, 2013;59:365–376.
[11] Wakui, T., Kinoshita, T., and Yokoyama, R. A mixed-integer linear programming approach for cogeneration-based residential energy supply networks with power and heat interchanges. Energy, 2014;68:29–46.

[12] Maroufmashat, A., Fowler, M., Khavas, S.S., Elkamel, A., Roshandel, R., and Hajimiragha, A. Mixed integer linear programing based approach for optimal planning and operation of a smart urban energy network to support the hydrogen economy. International Journal of Hydrogen Energy, 2016;41 (19):7700–7716.

[13] Maroufmashat, A., Elkamel, A., Fowler, M., *et al.* Modeling and optimization of a network of energy hubs to improve economic and emission considerations. Energy, 2015;93:2546–2558.

[14] Argonne National Laboratory, GREET Database, 2016. [Online]. Available: https://greet.es.anl.gov/. [Accessed 5 June 2016].

[15] Independent Electricity System Operator, Data Directory, 2016. [Online]. Available: http://www.ieso.ca/Pages/Power-Data/Data-Directory.aspx. [Accessed 5 June 2016].

[16] National Renewable Energy Library, Distributed Generation Renewable Energy Estimate of Costs, Retrieved July 2014, from http://www.nrel.gov/analysis/tech_lcoe_re_cost_est.html.

[17] GAMS Development Corp., GAMS the solver manuals, Cplex 12., Washington, DC, 2015.

[18] Maroufmashat, A., Sattari, S., Roshandel, R., Fowler, M., and Elkamel, A. Multi-objective optimization for design and operation of distributed energy systems through the multi-energy hub network approach. Industrial & Engineering Chemistry Research, 2016;55(33):8950–8966.

[19] Ontario Electricity Price, Independent Electricity System Operator (IESO), Retrieved May 2014, from http://www.ieso.ca/.

[20] Historical Natural Gas Rates, Ontario Energy Board, Retrieved June 2014, from http://www.ontarioenergyboard.ca/OEB/Consumers/Natural+Gas/Natural+Gas+Rates/Natural+Gas+Rates+-+Historical.

[21] The Climate Registry, CO2 Emission Factor, Retrieved June 2014, from http://www.theclimateregistry.org/downloads/2014/02/2014-Climate-Registry-Default-Emissions-Factors.pdf.

[22] Environmental Protection Agency, Average Annual Emissions and Fuel Consumption for Gasoline-Fuelled Passenger Cars and Light Trucks Office of Transportation and Air Quality, 2008: EPA420-F-08-024.

[23] Miller, L., and Carriveau, R. Balancing the carbon and water footprints of the Ontario energy mix. Energy, 2017;125:562–568.

Chapter 10
Battery-based storage for communities

Boyuan Zhu[1], *Junwei Lu*[1], *Wayne Water*[1], *Markos Katsanevakis*[1], *Mojtaba Moghimi*[1], *Domagoj Leskarac*[1] *and Sascha Stegen*[1]

Abstract

Instantaneous load demand variations and fluctuations introduced by renewable energy sources (RESs) to the system, pose various challenges to the operation of the grid. Significant support to the grid can be provided by energy storage systems (ESSs) strategically placed and sized. In the presence of ESSs combined with RESs proximate to load centres, transmission losses are minimized and grid infrastructure upgrades are deferred, due to local generation and consumption of electric energy. This is without the need to transfer energy from remote power plants. Furthermore, ESSs introduce considerable flexibility to electric energy, which can now be consumed when it is needed the most (e.g. during the on-peak period). As a result, the loading levels of the grid components are reduced and the terminal voltages are maintained within the allowable limits. This chapter presents an in-depth study of battery storage for communities and an analysis of major battery-storage technology and application. This includes another special form of distributed energy storage – the electric vehicle.

10.1 Introduction

Renewable energy is no longer novel to those involved in the evolution of power, especially solar power and wind power due to their dramatically increased penetration. There is still some way to go in the take-up of solar energy as a means of powering our world. An incredibly large amount of small-scale photovoltaic (PV) installations in residential or commercial buildings are acting as virtual power stations, which decentralize power generation against the conventional utility. However, this introduces the problem of excess energy and unstabilized power, which affects existing electricity infrastructure. The concept of community energy storage (CES), as shown in Figure 10.1, has captured the interest of energy suppliers.

[1]Queensland Micro and Nanotechnology Centre (QMNC), Griffith School of Engineering, Griffith University, Australia

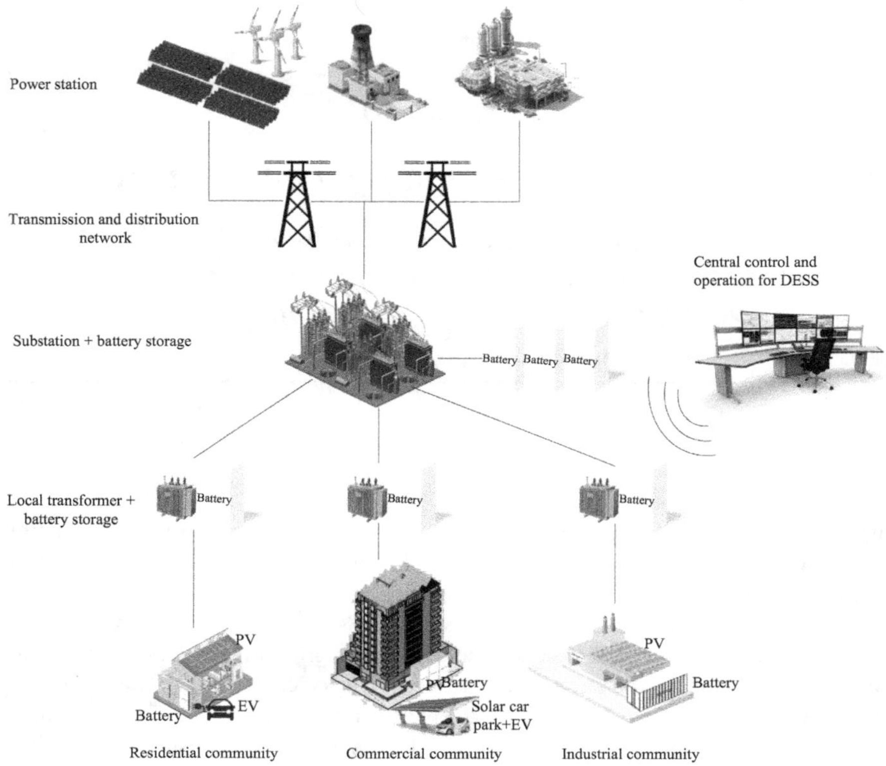

Figure 10.1 Concept of community energy storage

It operates distributed energy storage systems (DESSs) at end-user homes and businesses rather than using one or a few large-scale units at the utility side. These decentralized storages are acting as buffers between intermittent power resources and the utility grid, which significantly eases the pressure on local transformers.

- CES is a young market but it is being driven quickly by multiple trends: (1) the increased penetration of intermittent renewable resources in the grid urgently requires buffer technology to smooth power. (2) The continued advances in power electronics provide solid technique support. (3) The rapid proliferation of electric vehicles (EVs) significantly stimulates battery development and production, which makes battery storage economically viable. Therefore, CES is believed to provide numerous benefits with many possible combinations. Such benefits are as follows: it eases the burden on the centralized grid and increases the overall resilience of the power supply.
- Less transmission and distribution (T&D) equipment is needed for local demand in new development areas as it provides power locally, which reduces the need for T&D capacity.

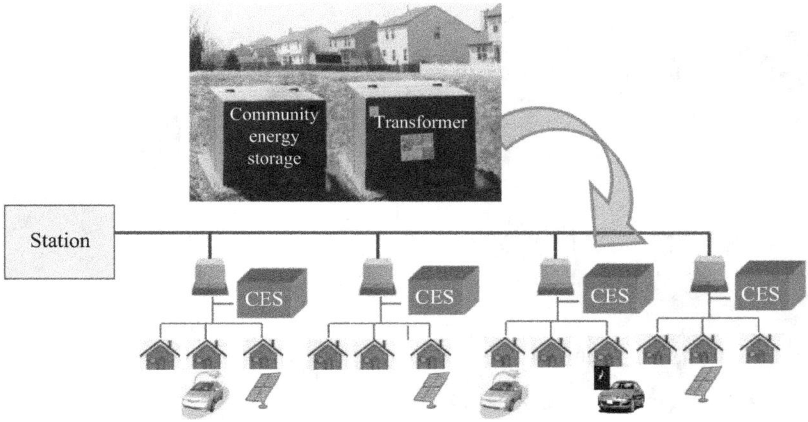

Figure 10.2 American electric power distributed CES [1]

- It extends the life of existing infrastructure by reducing the need to upgrade the transformer and network to meet local peak demand. The cost of the battery is more attractive compared to upgrading the grid.
- It serves as a robust, fast-responding and flexible alternative source of power generation. It can maintain the stability and reliability of the grid to overcome the integration challenge of renewable energy access.
- Low price energy can be stored during off-peak periods and used when the price is high. This creates a very economic environment benefiting energy users.

In 2005, American Electric Power (AEP) located a 2-MW system with advanced sodium–sulphur (NaS) battery storage at a substation as a pilot project to evaluate the prospects and merits of the CES concept [1]. Unfortunately, the battery didn't perform to its stringent standards and the full-scale programme was not resumed. Eventually around 2009, AEP deployed 80 individual distributed units of smaller scale systems, with 25 kW rated power and 25–75 kW h batteries, at or near end-user homes (see Figure 10.2). It aimed to apply CES at small scale of battery compared to megawatt scale. The specification of the CES was published as an open-source document for public use and feedback (as a guideline for further standardization).

Another trial project in 2010, led by DTE Energy, is another example of the concept of an aggregated CES system in utility territory [2]. There were twenty 25 kW/50 kW h battery-storage systems on local transformers and another 500 kW/ 250 kW h integrated to the substation, as shown in Figure 10.3. It demonstrated CES's capabilities of voltage/volt-ampere reactive (VAR) support, integration renewable generation, islanding during power outages and frequency regulation. The project used secondary EV batteries as storage, which identified an alternative application for the continuous life of EV type batteries.

Alkimos Beach community in Western Australia is currently leading the largest community-scale battery-storage project in Australia and will enable Synergy

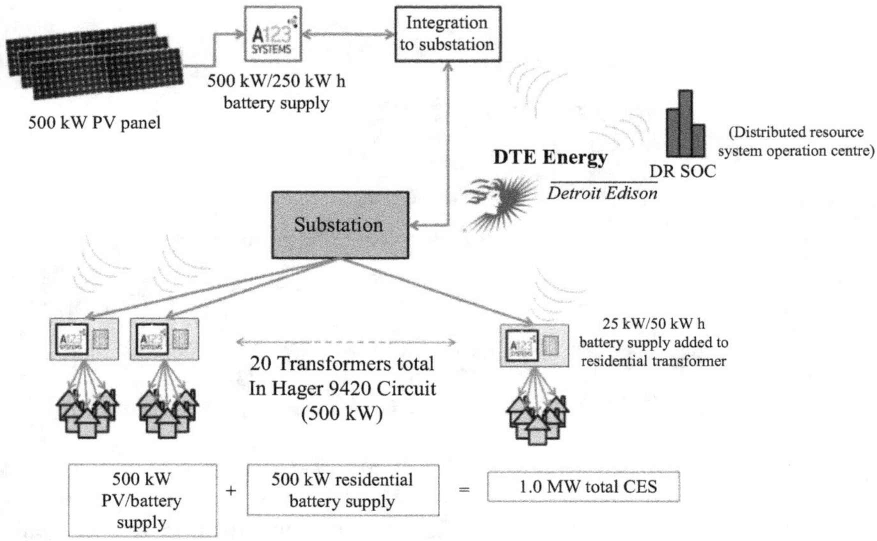

Figure 10.3 DTE Energy aggregated CES in utility territory [2]

to be on the leading edge of this technology [3]. The project installs a 300 kW/ 1.1 MW h fully contained lithium-ion (Li-ion) ESS connected to the distribution network, as shown in Figure 10.4.

Sonnen, a German company who takes a step further in smart storage systems, launched Sonnen Community in 2015 to link all its customers' solar and storage devices into a virtual grid (as demonstrated in Figure 10.5). This allows subscribed members to trade electricity peer to peer and sell excess power into the wholesale market [4]. This aggregation creates a new economy of electricity trading that challenges conventional utility modes.

CES raises a new transformation in the form of energy storage that aggregates a large amount of DESS. However, a new transformation is always accompanied by challenges. The government policy of renewable energy and market regulation are barriers to CES development and standardization. Some markets have low-cost electric power or are dominated by vertically integrated utilities. Selection of the right battery chemical and deployment of the right capacity size of DESS are key trade-offs in the economic benefit to users and the cost of the system. The battery is still not cheap enough, especially when the entire community is considered to go off-grid. The proliferation of EVs can be considered as alternative storage but there is high demand for them when rapid charging is conducted. Forecast and control algorithms are extremely challenging due to the unpredictable nature of intermittent renewables, various customer behaviours and load patterns, strict operation logic and conditions to prolong battery life and reliable communication methods.

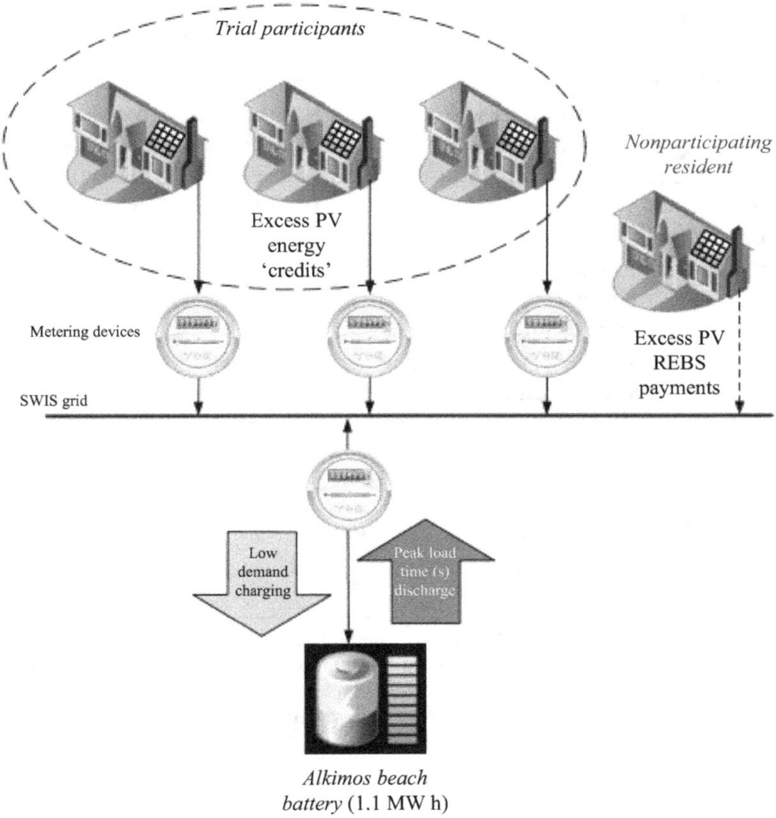

Figure 10.4 Alkimos beach energy storage trail concept diagram [3]

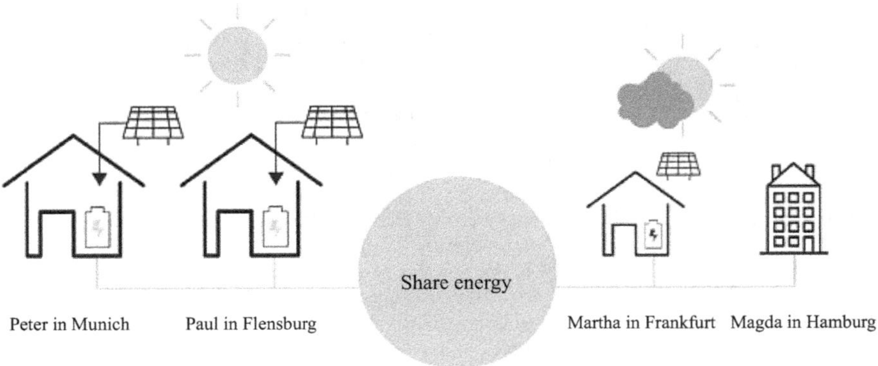

Figure 10.5 Sonnen virtual grid across Germany [4]

10.2 Technology of battery storage

Battery storage is one of most significant technologies that change way of using energy. For a large upfront cost, solar panels can generate 'free' electricity but don't work at night. Alternatively, off-peak power is much cheaper but only available in the middle of the night. Battery storage helps the grid to adopt variable forms of renewable energy and provides more flexible, reliable and efficient energy to customers. It is fully controllable and fast responding that plays an important role in maintaining a consistent energy balance within the renewable network by storing excess of generation and supplying additional energy to loads to compensate the deficit. A variety of pilot projects by means of different battery-storage technologies are conducted over the world that demonstrates capability and maturity throughout grid and community.

Battery technology varies according to different criteria. Table 10.1 introduces some major concepts that have to be understood before looking into battery storage.

Table 10.1 Major concepts of battery storage

Name	Unit	Description
Chemistry		It refers to the chemistry in a battery operation
Total capacity	A h	It refers to the total amount of electric charge battery can deliver at the rated voltage
Usable capacity	A h	Limited by battery characteristic and operation condition, not all battery can healthily deliver the full amount of energy it stores
State of charge	%	The current state of a battery in use
Depth of discharge	%	An alternate method to indicate a battery's state of charge
Charge/discharge rate	C	Also known as C rate which indicates a rate at which a battery is charged or discharged relative to its maximum capacity
Operating temperature	°C	A temperature range that a battery can provide reasonably good performance
Nominal voltage	V	It is measured at the midpoint between fully charged and fully discharged based on a 0.2 C rate
Operating voltage range	V	Open-circuit voltages when a battery is fully charged and fully discharged
Battery management system		An electronic control system that monitors and controls the charging and discharging of the battery
Hazards		Explosion, fire, leakage, toxic materials, gas emission, etc.
Self-discharge	%	Percentage of charge losses when stored at room temperature
Cycle life	Cycle	It is the number of complete charge/discharge cycles that a battery is able to perform before its end of life
Calendar life	Year	It indicates how long a battery is expected to last in terms of calendar years
End of life		When a battery's capacity fade under a certain percentage of its original capacity, typically 80%

10.2.1 Conventional and advanced lead-acid batteries

Lead-acid (PbAc) battery is the first rechargeable battery for commercial use which was invented in 1859 and is still in wide use today. The maturity of the conventional PbAc battery is dependable that it makes battery simple to manufacture and cost per watt-hour is really low. The conventional PbAc battery is good for power-driven applications, commonly classified into three usages: automotive (starter or SLI), motive power (traction or deep cycle) and stationary (UPS), due to its high specific power, capable of high discharge currents and wide range of temperature performance. However, limitations of chemistry characteristic constrain its application in energy fields. This kind of battery has low-specific energy and poor weight to energy ratio. It has to be charged slowly with correct voltage limits to avoid buildup of sulphation on the negative plate which robs battery performance. A full charge normally takes 14–16 h, and the battery must always be stored in full state-of-charge to prevent sulphation condition. Repeated deep-cycling reduces battery life. For example, a deep-cycle battery has limited cycle life around 400–500 cycles by half depth of discharge (DoD) with 50% remaining [5].

It is believed that lead-based technology has significant unused performance potential that can make the battery viable in the renewable applications by overcoming the core limitation of utilization of lead. In 2004, The Ultrabattery by Commonwealth Scientific and Industrial Research Organisation of Australia combined the asymmetric ultracapacitor with the PbAc battery, which provided a fundamental structure for advanced lead-carbon (ALC) battery development. Followed by research and commercialization of Japanese researcher, ALC battery was applied and tested in the hybrid vehicle in 2008. The ALC solves sulphate accumulation, partial charge and ageing issues by adding carbon into the cathode plate which significantly improves the performance of charge and discharge, as well as cycle life. Figure 10.6 illustrates the classic PbAc cell with the lead-negative plate being replaced with a carbon electrode to benefit from the qualities of a supercapacitor.

Unlike a regular PbAc battery, ALC battery is capable of eight times faster charging and three times in discharge power. Its lifecycle increases around six

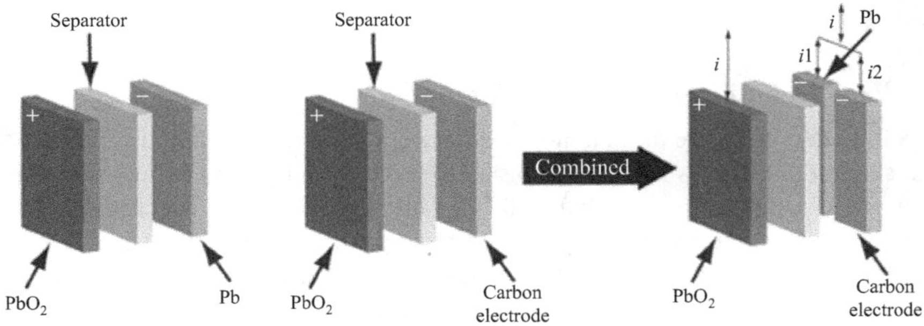

Figure 10.6 The classic PbAc develops into an ALC battery [6]

times up to 2,200 cycles with deep DoD, between 30% and 70%. It inherits the maturity of manufacture, wide temperature performance and dependence from PbAc battery that makes it more cost-effective. Today, the ALC is used as a replacement for the classic starter battery in start–stop applications and in 48 V micro and mild hybrid systems. Also, it is beginning to be widely used in energy storage of PV and wind application.

10.2.2 Lithium-ion batteries

Lithium is the lightest metal in catalogue and has the greatest potential that can provide the largest energy density for weight. Therefore, it is naturally ideal for energy storage but its inherent instability is standing in the way. The first rechargeable Li-ion battery was commercialized in 1991 by Sony Corporation. Li-ion is nonmetallic lithium that has slightly lower energy density than lithium metal but much safer. The battery has a congenital advantage that it does not have memory effect that requires scheduled cycling and also very low in self-discharge. But its overall drawback is also obvious. Li-ion is fragile which needs an extra circuit to protect the cell from under-discharge and over-charge. Thermal runaway is still a major concern in the safety aspect. Therefore, manufacturers keep making rapid improvement in enhancing different chemical combination with Li-ion to increase battery charge/discharge rate, cycle and calendar life and reduce the cost which is very important to make the battery economical viable. Current available types of Li-ion battery in the market are lithium cobalt oxide (LCO), lithium manganese oxide (LMO), lithium–nickel–manganese cobalt (NMC) oxide, lithium–iron–phosphate (LFP), lithium–nickel–cobalt–aluminium (NCA) oxide and lithium titanate (LTO), as compared in Table 10.2.

LCO is the original Li-ion which is average in all characters. It is widely used in mobile phones, cameras and laptops. An NCA battery is clearly the best in specific energy but shorter cycle life and less stability. It was widely used in Telsa electric powertrain for its EVs, produces by Panasonic. An NMC battery generally has a longer cycle life, more stability and less energy density and less cost than NCA. In terms of specific power and thermal stability, LMO and LFP are superior. Thus, NMC and LFP share large partitions in EV and energy storage market nowadays. LTO is the best in lifespan and safety aspect but low in capacity and expensive in cost which is main technique obstacle in commercializati on.

10.2.3 Sodium–sulphur batteries

A NaS battery is a type of molten-salt battery constructed from liquid sodium (Na) and sulphur (S). It uses molten salts as an electrolyte and gains conductivity which operates at a temperature of 250–350 °C. This type of battery has a high-energy density, high efficiency of charge/discharge and long cycle life and is fabricated from inexpensive materials. For example, the modern sodium–nickel–chloride battery is said to have an operating temperature from −40 to 65 °C with a cycle life of 3,500 at 80% DoD, a level comparable with Li-manganese and Li-phosphate. It can be deployed to support the electric grid, or for stand-alone renewable power

Table 10.2 Comparison of different Li-ion batteries

Chemistry	Lithium–cobalt oxide	Lithium–manganese oxide	Lithium–nickel–manganese
Abbreviation	LCO	LMO	NMC
Application	Mobile phones, tablets, laptops, cameras	Power tools, medical devices, electric powertrains	E-bikes, medical devices, EVs, Energy storage
Snapshot	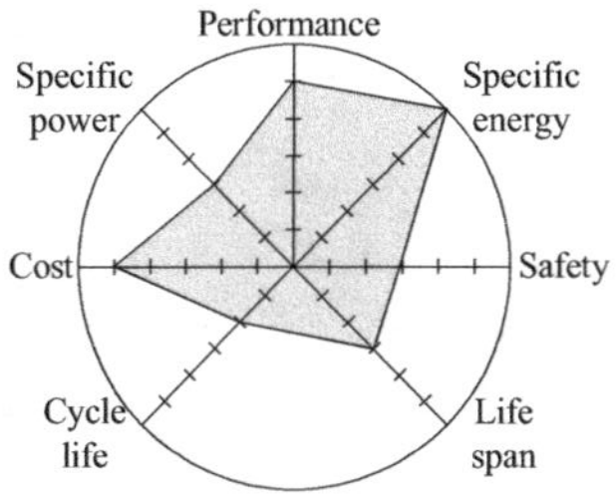	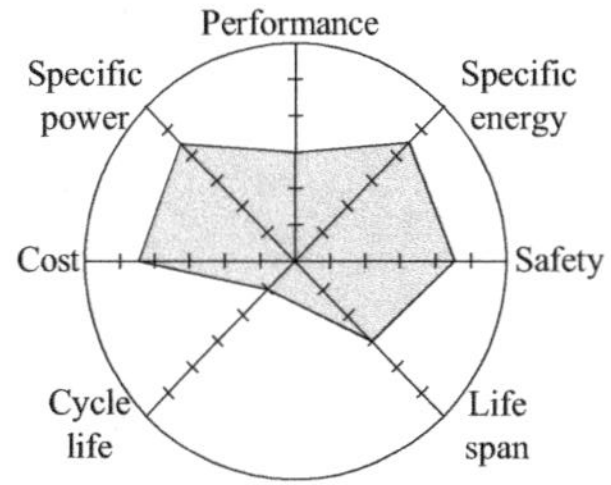	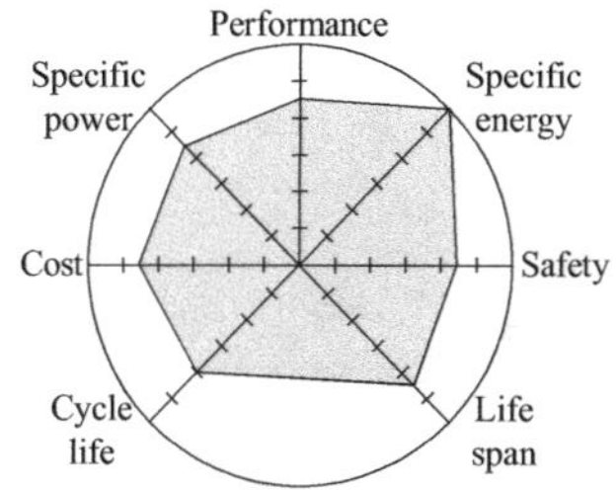
Nominal voltage (V)	3.60	3.70 (3.80)	3.60 (3.70)
Full charge (V)	4.20	4.20	4.20 (or higher)
Full discharged (V)	3.00	3.00	3.00
Minimal voltage (V)	2.50	2.50	2.50
Specific energy (W h/kg)	150–200	100–150	150–220
Charge rate	0.3 C (1 C max)	0.7–1 C (3 C max)	0.3–1 C
Discharge rate	1 C	1 C, 10 C possible	1–2 C
Cycle life	500–1,000	300–700	1,000–2,000
Thermal runaway (°C)	150 (full charge promotes thermal runaway)	250 (high charge promotes thermal runaway)	210 (high charge promotes thermal runaway)

(Continues)

Table 10.2 (Continued)

Chemistry	Lithium–iron phosphate	Lithium–nickel–cobalt–aluminium oxide	Lithium titanate
Abbreviation	LFP	NCA	LTO
Typical manufacturer	Energy storage, EVs	Medical devices, industrial, electric powertrain (Tesla), grid storage	UPS, electric powertrain (Mitsubishi i-MiEV, Honda Fit EV), grid storage
Snapshot	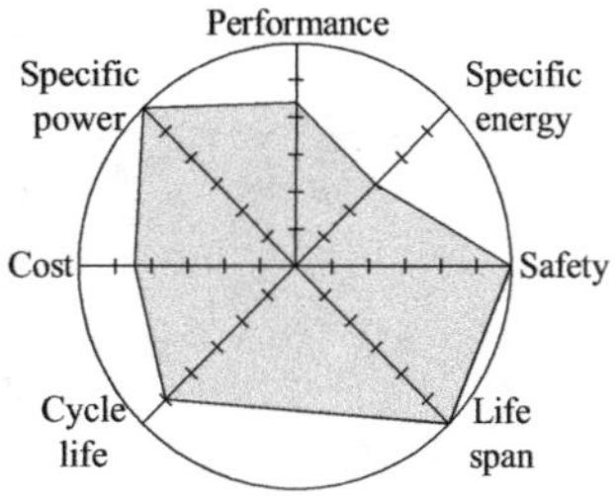	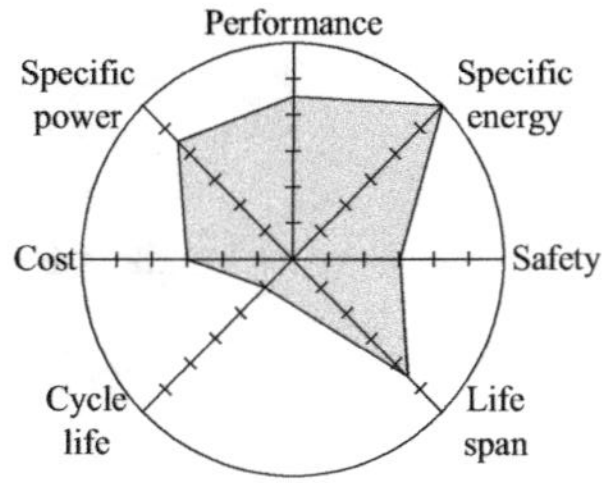	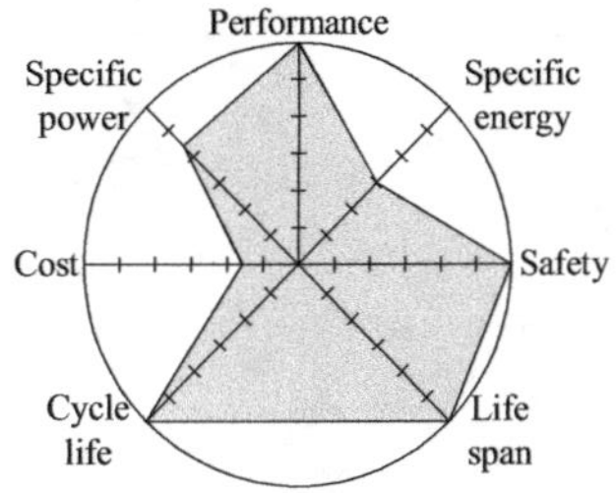
Nominal voltage (V)	3.20, 3.30	3.60	2.40
Full charge (V)	3.65	4.20	2.85
Full discharged (V)	2.50	3.00	1.80
Minimal voltage (V)	2.00	2.50	1.50
Specific energy (W h/kg)	90–120	200–260	70–80
Charge rate	1 C	0.5 C	1 C (5 C max)
Discharge rate	1–6 C	1 C	10 C possible
Cycle life	1,000–2,000	500	3,000–10,000
Thermal runaway (°C)	270 (very safe battery even if fully charged)	150 (high charge promotes thermal runaway)	One of safest

applications. However, pure sodium presents a hazard, because it spontaneously burns in contact with air and moisture, thus the system must be protected from water and oxidizing atmospheres.

10.2.4 Battery storage in power applications

Application of CES can demonstrate various capability under multi-mode operation via smart grid infrastructure, which includes frequency regulation, VAR support and voltage support, PV output shifting and levelling, demand response for peak shaving, islanding during outages, etc. Detail comparisons are listed in Tables 10.3 and 10.4.

Table 10.3 Comparison of battery storage in power application

Demand	**Battery storage installation locations**			
	Generation	**Transmission**	**Distribution**	**Consumer**
Charge and discharge arbitrage	√	√	√	√
Storage by seasons			√	√
Frequency regulation	√		√	
Voltage support			√	√
Load tracking			√	
Black start	√			
Transmission and distribution congestion alleviation		√	√	
Demand response for peak shaving			√	√
Off-grid		√	√	
Waste heat utilization	√		√	√
Heating and cooling			√	√

Table 10.4 Advantage and cost effect of battery storage in power application

Application	**Advantage**	**Cost effect**
Generation	• Reduce system capacity requirements; • Improve power plant utilization; • Improve renewable energy plans to generate electricity	• Reduce power plant investment and operating costs; • Reduce power generation costs; • Increase power generation efficiency
Transmission and distribution	• Delay investment in power transmission and distribution; • Improve asset utilization; • Improve the amount of renewable energy access	• Reduce investment in equipment and the loss of network
Consumer	• Demand-side management; • Improve the allocation of electricity equipment resources	• Reduced electricity costs; • Improved quality and reliability

10.3 Challenges of EV penetration in distribution grid

An EV refers to a vehicle that is entirely powered by an electric drivetrain or a vehicle that is powered by a series hybrid electric drivetrain that needs to be plugged into the electricity grid for battery recharging. An electric car uses electric motors and motor controllers in place of an internal combustion engine. Generally, the energy for the motor is stored chemically in battery packs (e.g. Li-ion battery) that are located in the vehicle. These batteries are then charged at the home, at the workplace, or via publicly accessible recharge stations.

In a plug-in hybrid EV (PHEV), the drive wheels are powered by an electric motor. A smaller petrol engine is fitted to the vehicle to generate power for the electric motor. The on-board batteries can also be charged by plugging the vehicle into the electricity grid. In addition to charging from the electricity grid, most fully-EV and PHEVs take advantage of regenerative braking systems that charge the on-board batteries.

A key challenge associated with the widespread market acceptance of fully-EVs is the current high cost of this technology relative to conventional vehicles and limited operating ranges between vehicle recharging. This issue is being progressively addressed by significant investment in battery technology and battery management systems (BMSs) with a view to extending the operating range of these vehicles in the future.

All forms of battery EVs (BEVs) now benefit from a gradual increase in range and a reduction in up-front costs. This gradual and substantial rise in plug-in EV (PEV) penetration will have an impact on distribution grids, as charging PEV can be more taxing on distribution infrastructure [7]. Charging methods, power capacities and vehicle-to-grid (V2G) integration are areas to consider within communities for the sustainable penetration of PEV. As the cost of energy storage continues to reduce, replacing the battery to restore range becomes more affordable. This brings about the market of second life EVs batteries, with potential for battery energy storage in communities.

10.3.1 PEVs in communities

There is a substantial growing trend with the uptake of EVs within numerous countries and communities including both the private and commercial sectors. There have been over a hundred times increase of BEV, from mere thousands in 2010 to over 700,000 in 2015. EVs have several benefits over traditional petrol vehicles with only a few drawbacks. Zero emissions and noise at the point of use, low-maintenance costs, low-energy costs and the potential for local energy production are all major benefits of EV over internal combustion engine vehicles. The primary disadvantage to date though has been the cost of the vehicles, stemming from high battery prices. This, in turn, limited the range of affordable EV and impeded market adoption. As the demand for lithium-based energy storage increased, the supply of energy storage became more predominant by reducing cost per kW h each year.

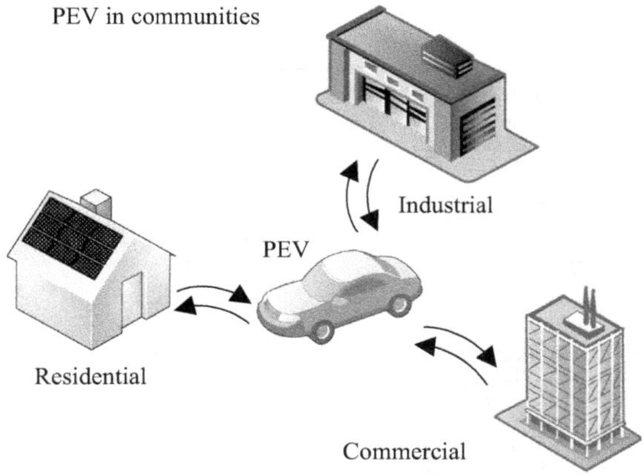

Figure 10.7 PEV relationship between residential, industrial and commercial load

BEV, hydrogen fuel cell EV and PHEV are the three classes of vehicles constituting PEV. All of these vehicle types have the ability to interconnect onto the distribution network with varying energy storage capacities [8]. BEV typically would contain the largest battery energy storage as HEV and PHEV hybridize energy storage with liquid or high-pressure fuels. Nevertheless, all PEV vehicles can charge or absorb energy from the grid and in some instances return that energy through V2G integration [9]. V2G integration transforms PEV into a substantial distributable load and potential energy source. The role of PEV in communities is illustrated in Figure 10.7.

Such mass storage has the capability to provide numerous benefits to both the energy supplier and local consumers of energy, residential, industrial and commercial. PEV owners would most likely have to charge facilities at their residential premises; some industries already include specified parking bays for PEV vehicles as do commercial buildings and parking lots. PEV integration is broadly seen as a future development in the Microgrid and Smart Grid.

10.3.2 EV charging technologies

One of the crucial challenges of EV penetration into the grid and increasing its usage in the communities is the charging capability and availability. The main concern of the consumers is to be able to charge EVs within the shortest possible time [10]. On the other hand, the EV charging stations are required to provide the charging capability for various EVs with different charging technologies. Therefore, it is necessary to develop EV charging technologies in parallel with the increasing EV usage in the communities.

To charge the EVs, two main charging technologies are available: contact-based charging and wireless charging. The contact-based charging includes alternating current (AC) and direct current (DC) charging methods. For the high power

Table 10.5 EV charging levels [12]

Charge method	Connection	Power (kW)	Voltage (V)	Location
Normal power	1-Phase AC connection	3.7	Single-phase 230 V AC	Domestic
Medium power	1- or 3-Phase AC connection	3.7–22	Single- or three-phase	Semipublic
High power	3-Phase AC connection	>22	Three-phase 400 V AC	Public
High power	DC connection	>22	480 V DC	Public

charging applications, AC on-board charging and DC off-board charging are currently used in the market. The benefit of DC over AC charging is mainly because of normalized voltage and current range, which is the same all over the world. Different sockets are in use to charge EVs' energy storage. The sockets are designed for DC and single- or three-phase AC connections. Power delivery is limited to the socket technology and the type of charging. The combined AC/DC sockets have the benefits of both charging methods and higher energy transfer range [8]. On the other hand, the contactless charging technology or wireless charging provides another solution to the EV charging. However, requiring complex infrastructure and limited power transfer are of disadvantages of wireless charging. Among different battery technologies employed in the EV charging, nickel metal hydride and lithium-ion have gained more popularity [7].

There are different standards for EV charging systems considering safety, reliability, durability, rated power and the cost of different charging methods [10]. Two of the main standards used in the literature are SAE J1772 [11] and IEC 61851-1 [12]. According to the standards, different charging levels are briefly explained [13]:

- Level 1 charging: It is mainly employed for slow single-phase on-board charging for domestic household applications or long-time EV parking.
- Level 2 charging: It is used for quick private and public EV charging.
- Level 3 AC and DC charging: It is for high power applications and utilized for fast public EV charging.

The summary of different levels of charging is presented in Table 10.5.

10.3.3 Infrastructure and control

One of the benefits of integrating a large number of EVs into the power grid is to provide a great potential energy storage for the community. This potential can be in the form of EVs injecting energy back to the grid or using second-life EV batteries as energy storage. The benefits of integrating a large fleet of EVs into the grid are mainly for two parties: EV owners and utility grid [9]. To be able to engage EVs as the energy generation resources, a bidirectional V2G infrastructure

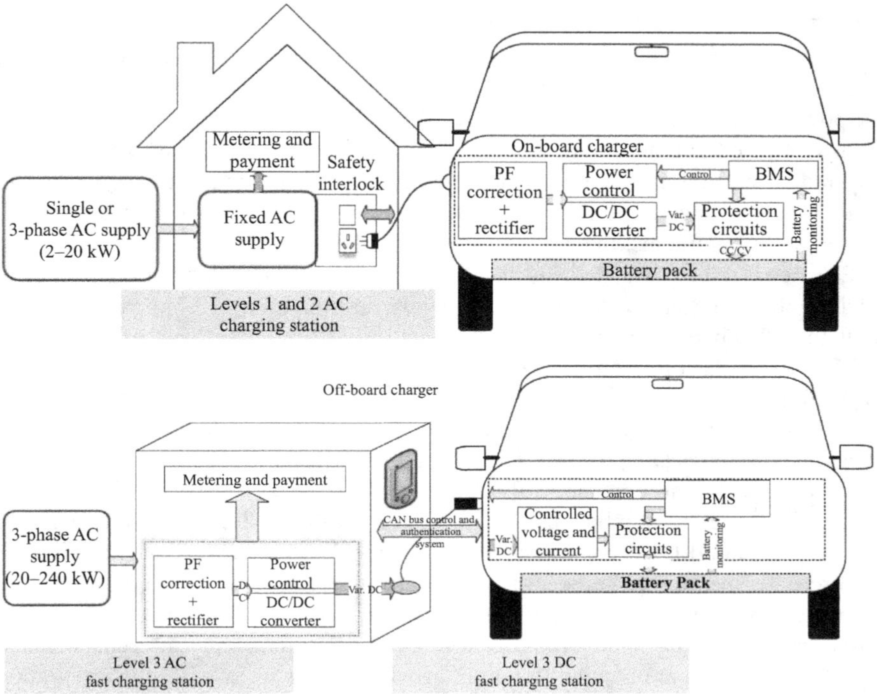

Figure 10.8 Charging station infrastructure for different EV charging levels

is required. V2G technology provides many services to the community such as frequency and voltage regulation, active power support, reactive power compensation, power factor regulation and supporting the integration of renewable energy resources [14].

One of the challenges in the V2G systems is the complexity of the bidirectional battery charger, which requires additional hardware and software to perform properly. The common components of a bidirectional battery charger are AC/DC converter, DC/DC converter, BMS and protection circuits. The DC/DC converter is employed as either buck or boost converter in the charging or discharging mode. The role of AC/DC converter is to rectify the AC power while charging the EV and invert the DC power to give the power back to the grid. The overall structure of the V2G system is illustrated in Figure 10.8.

The integration of EVs to the grid can be coordinated by intelligently scheduling EVs to discharge them at the periods that energy has higher prices (peak hours) and charge them when the price of energy is lower (off-peak hours). To implement such a system and infrastructure, an advanced control, communication and optimization system incorporated into the grid and EV management system is required.

10.3.4 Grid stability

One of the main challenges of EV penetration in distribution grid is to deal with grid stability especially in the case of large number of EVs. To fulfil the stability of the power grid, the impact of EVs on the load and generation side of the grid should be investigated. First, the generation units are required to meet EV charging demand. Therefore, it is important to control EV charging and mitigate the impact of EV loads as uncontrolled charging can lead to undesirable grid impacts such as power quality problems. In order to prevent extra EV load on the grid, the charging of EVs needs to be controlled and shifted to the off-peak hours and thus no need for additional power generation and consequently no limitation of EV penetration. To shift the EV charging to the off-peak hours, utilities employ time of use (TOU) tariff strategy. According to the TOU tariffs, the pay rate would be lower during the off-peak hours, which motivates the EV owners to charge their vehicle within this period. However, this should be taken into account that the load shifting to the off-peak hours requires an optimized scheduling; otherwise, it can cause an even higher increase in peak load demand [10].

To increase the number of EVs integrated into the grid, optimized charging/discharging algorithms that consider the benefits of EV owners and utilities are required. To maximize the benefits of the utility, the optimization algorithms are employed to minimize the impact of EVs on the grid, total electricity generation costs, energy losses and infrastructure costs. Furthermore, different demand response strategies are applied to flatten demand profile in case of high integration of EVs [13]. On the other hand, to provide the benefits of EV owners while considering grid constraints, optimization algorithms are employed to minimize the charging costs and maximize the discharging profits [14]. Additionally, charging rate and charging time are factors that should be taken into account in the design of optimization algorithms as they have an impact on the EV owners' benefits [15]. By fulfilling the requirements of EV owner and utility, the grid stability is less of a concern.

10.3.5 Limitations

As many benefits are imagined for EVs and V2G systems, there are challenges and limitations to overcome as well. Among many of the challenges, battery degradation, high investment cost and social barriers can be pointed out [14]. The batteries have a limited lifetime and many factors influence on the aging rate. One of the main reasons of aging is the charge and discharge rate, as higher rates would result in quicker battery degradation. However, a degraded battery may not be functional in EV and V2G system but does not mean it has no more benefits for the community. The EV batteries can be repurposed and reused for the stationary and utility applications after the end of their lifetime in the EV [16].

One of the other limitations is high investment cost of EV industry. The cost of EVs in the market and the required infrastructure is still high compared to that of combustion engine cars. In other words, there needs to be a high investment on the development and implementation of V2G infrastructure and different EV components such as batteries, battery chargers, and control systems. Energy losses and safety issues are also among the challenges, which need to be specifically considered for the EVs and V2G systems in the communities. One of the reasons of the high price of EVs is the low acceptance by the society. In order to increase the number of EVs and EV charging stations in the community, a comprehensive and reliable network of V2G is required to guarantee the requirements of EV owners at all the times [14].

10.4 Economic aspects of battery storage

Differentiating between price point and cost of battery storage is paramount to evaluate any battery storage with regard to performance and real value over the life of the project. Battery manufacturers commonly quote their up-front price based on per watt–hour (W h) on battery system only. However, the true cost of energy delivered during the usable lifetime of the battery is a more accurate and reliable method for estimating its economic value for projects and return on investment (ROI) of customers.

10.4.1 Cost metric

Apart from the environmental benefit, economic viability of battery storage is dominating the speed of its deployment. The upfront cost of battery storage is a critical factor that determines if it is economically viable as it takes a very large part of the overall cost. Power and energy are two basic orientations in battery-storage applications. The cost metric of '\$/W h', '\$/kW h' and '\$/MW h' are frequently used in measuring the normalized cost of energy storage devices. The other useful cost metric is based on power capacity that measured in watts, '\$/W'. They are very different from each other and have to be understood clearly when looking at the upfront cost of energy storage.

10.4.2 Effective cost of a battery

The effective cost of a battery is always hard to calculate while it is always a major concern for users when selecting and comparing different batteries for the application. Most common way is simplifying the figure based on the \$/kW h capacity value which is the price divided by the nominal storage capacity at full charge, as in (10.1). It seems this metric is useful in comparing batteries across various brands and models but actually very misleading. That is because the simplified calculation only considers the claimed battery capacity but not the real one during operation.

$$\text{Cost per capacity}(\$/\text{kW h}) = \frac{\text{Total cost of battery}(\$)}{\text{Battery capacity}(\text{kW h})} \tag{10.1}$$

The ultimate value of the battery is all about the total energy it can store and deliver. There are three important factors have to be taken into account: the cycle and calendar life of the battery, DoD, and round-trip efficiency (RE).

- The cycle life indicates the number of times that a battery can be charged and discharged. The calendar life tells how long a battery can last in storage, typically by calendar years. No matter whether a battery is in use or storage, degradation is irreversible that brings it to the end of life.
- DoD specifies what percentage of the battery capacity is actually used. Battery manufacturers normally specified a recommended maximum value of DoD for a nominal cycle life because the cycle life of a battery tends to decrease if DoD is increased.
- RE is dependent on the battery's DoD. It indicates the amount of energy that can be extracted from a battery as a percentage of total storage. The loss of energy is generally caused by heat or other inefficiencies within the system.

Therefore, a revised calculation of cost per storage can be derived from the following equation:

$$\text{Cost per storage}(\$/\text{kW h}) = \frac{\text{Total cost of battery}(\$)}{\text{Battery capacity}(\text{kW h}) \times \text{cycle life} \times \text{DoD} \times \text{RE}} \qquad (10.2)$$

Besides the performance profile of the battery itself, ancillary costs are not included in the equation. They are also very important, especially for those project developers and contractors. These costs vary case by case, which may include shipping cost, installation cost, ongoing maintenance cost and so on.

10.4.3 System cost breakdown

Most battery-based storage systems are DC power and interconnected to the utility grid or customer's load at AC side. Common component levels in a DC battery-storage system are

- Battery cells
  Normally, it refers to the smallest unit of battery that chemistry is packed into. They are a most basic unit of energy storage which is assembled into battery banks or modules.
- Battery banks/modules
  A group of battery cells are packed into a battery bank or module. It is generally standardized and mass produced by the manufacturer as an intermediate component of battery energy storage which can be easily replaced or repaired.
- Battery strings/racks
  Multiple battery banks or modules are connected in series or parallel to form battery strings/racks.
- Battery container
  Groups of battery strings/racks can be installed in a container to build large battery energy storage. It is normally used in commercial, industrial and grid application.
- Battery control unit (BCU)/BMS
  BCU/BMS cost is generally added into the price of the battery as a whole by the manufacturer.

However, to install and commission a complete AC ESS to a customer site, there are additional costs at AC side which can include:

- Power conversion system (PCS)
  Grid-tied energy storage mediums are predominately DC in nature. To effectively utilize the energy storage capacity on the present electric utility grid, the energy must be converted to a standard AC level and regulated through a converter. It bidirectional converts from AC to DC and DC to AC which interconnects and controls of multiple DC sources to provide regulated, stable and controllable power flow.
- Isolation transformer
  In certain circumstance, an isolation transformer is needed to transfer electrical power from AC power source to some equipment while isolating the powered device from the power source. For example, some PCS is delta connection and needs delta to star isolation transformer when connects to the four-wire network.
- Switchgear
  Electrical disconnect switches, fuses or circuit breakers are necessarily needed to control, protect and isolate battery and other electrical devices in the system.
- Cabling
  AC cable, DC cable and communication cable cost cannot be ignored. Normally, it is included in the price of battery or other equipment. However, additional cost has been counted among interconnections.
- Installation
  Mostly, it is the transportation, labour and machine cost for installation.

Therefore, it is important to clarify costs in different segments when faced with a value for battery storage cost in \$/W h. The most basic level is the cost of cells only which generally used by the battery manufacturer. However, ultimately customer would like to have the cost at most complete solution level which is also known as 'all-in-one' cost of an installed and commissioned battery-storage system.

10.5 Energy consumption pattern of a community

This is an analysis of energy usage data obtains from Ergon Energy Australia with effective measurements of 265 samples in a residential community throughout the year of 2012. It identifies the sufficient number of energy storage as well as the maximum power rating of power electronics used to get the rural household completely self-sustainable. Due to the privacy concern, details of the raw data cannot be disclosed, neither the designated household with regards to its address and the appliance information. However, the result shown can still be a very good reference for understanding the basic requirement of power electronics used for the off-grid ESS.

10.5.1 Regulated power supply

The energy usage of regulated power usage is temperature-dependent as shown in two example figures in Figure 10.9. Almost every household has the same energy

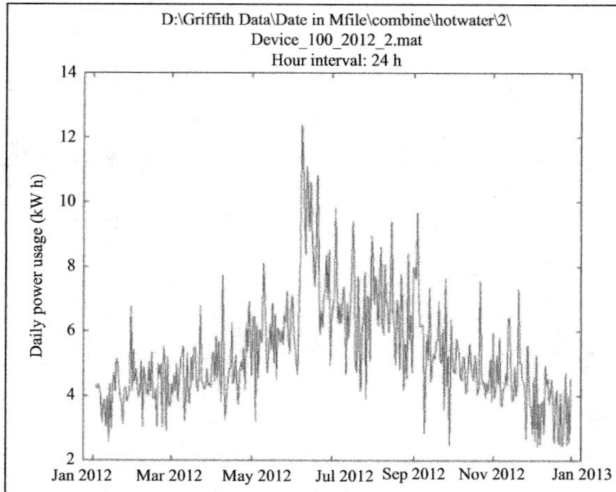

Device 100 – 5.33 kW h/day

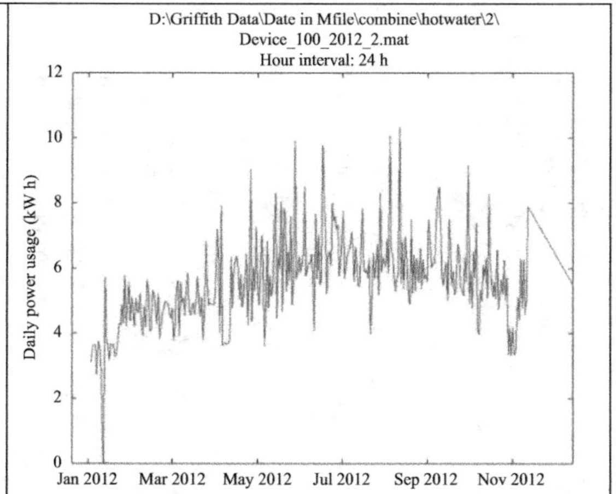

Device 93 – 5.55 kW h/day

Figure 10.9 Regulated power usage throughout 2012

usage pattern that the peak usage occurs during July period. The differences are the magnitude and the variation degree of the energy usage. Since it is a controlled-source power supply, the energy usage versus hour's pattern is not showing. On the other hand, the peak usage during a day is depending on the size of the hot water system (likely the biggest energy consumption device of a household). By looking at the top 10 peak energy usage within a period of time (e.g. 1 week), it can be justified whether there is a regulated power supply in the household if these readings are consistent.

10.5.1.1 Average daily power usage – all samples

In order to analyse the household energy usage, we first classify 265 samples into two groups: household with and without regulated power supply. In the following figures, the x-axis is the number of household samples and the y-axis shows the energy usage in kW h. 'Usage/day' is the average energy usage throughout a period, and '90% Usage' represents as the peak kW h energy usage required to cover 90% of days throughout the year. The '90% Usage' factor characterizes the peak energy usage required in the hourly scale, and this gives a good indication on the specification of power electronics required. For example, reading 30 kW h for '90% Usage' means the battery storage needs to be at least capable of exporting 30 kW h of energy. The actual requirement for the power electronic system will be higher. However, due to the details it involved, we will be focusing on the energy pattern in the hourly scale.

There are 182 samples for household without regulated power supply. The average power usage curve, as well as their daily usage ratio at 90%, is shown in Figures 10.10 and 10.11. To simplify the problem, we can focus our observations on 80% of people's energy usage behaviour (roughly for a number of household below 145) and following summaries are made:

- For general households, the average daily energy consumption is less than 19 kW h/day;

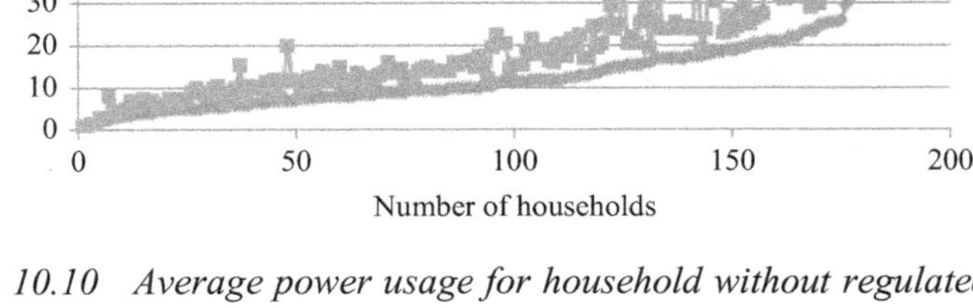

Figure 10.10 Average power usage for household without regulated meter

- 57% of households (among 182 samples) uses 5–15 kW h/day;
- The ratio of peak energy usage to the average daily usage is about 1.72.

There are 82 samples for household with regulated power. The average power usage curve, as well as the daily usage ratio at 90%, is shown in Figures 10.12 and 10.13. If we analyse 80% of people's energy usage behaviour, the following comments are made:

- The average daily energy consumption is less than 19.7 kW h/day
- 54.9% of people use 5–15 kW h/day
- The ratio of peak energy usage to the average daily usage is about 1.74.

We are also interested on the energy consumption behaviour of regulated power (controlled-power), and similar patterns are shown in Figures 10.14 and 10.15.

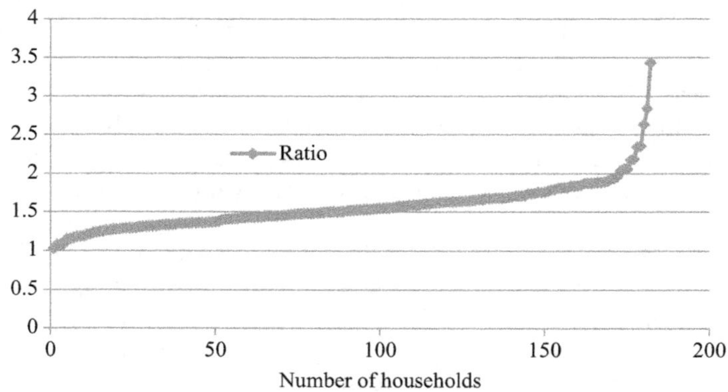

Figure 10.11 Ratio of average daily power usage versus 90% peak power usage

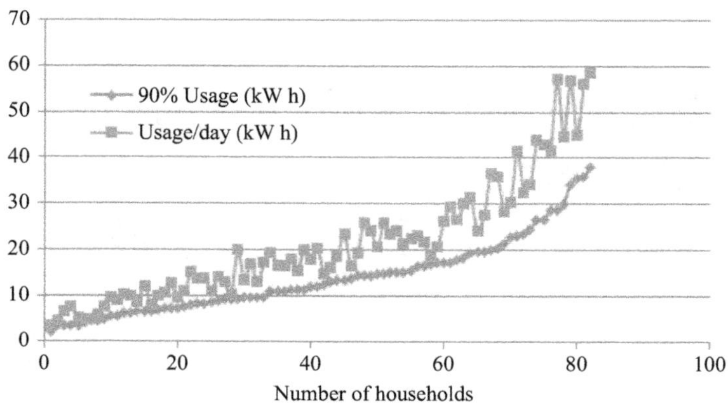

Figure 10.12 Average power usage for household with regulated power

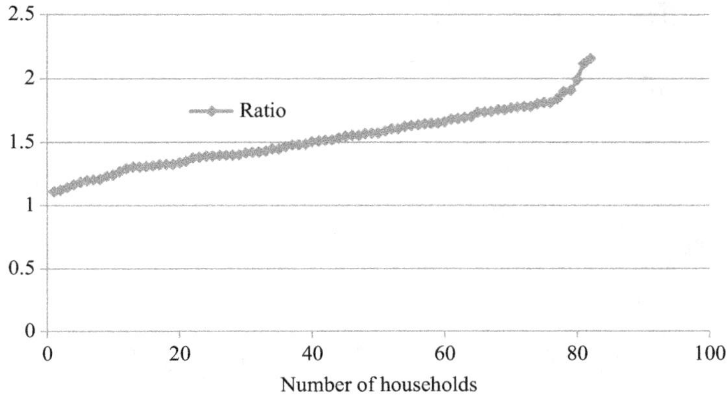

Figure 10.13 *Ratio of average daily power usage versus 90% peak power usage (household with regulated power)*

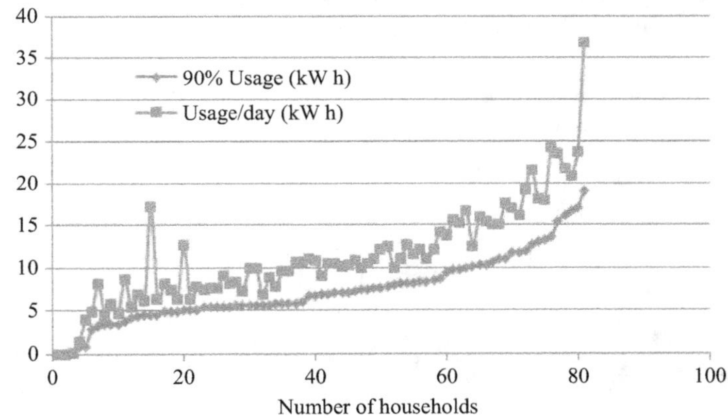

Figure 10.14 *Average power usage for regulated-power system only*

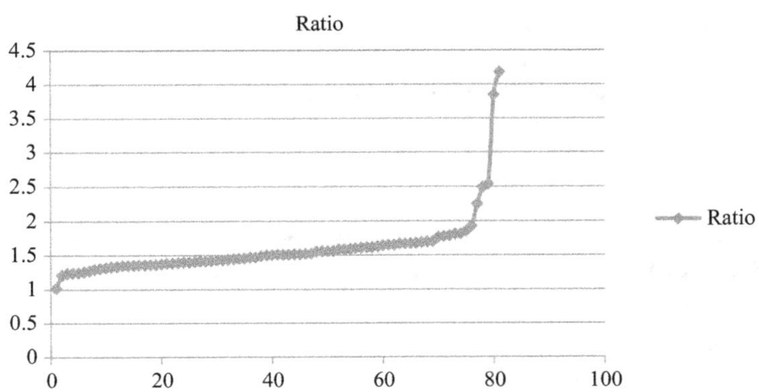

Figure 10.15 *Ratio comparison between daily energy usage and 90% of maximum usage for regulated-power system only*

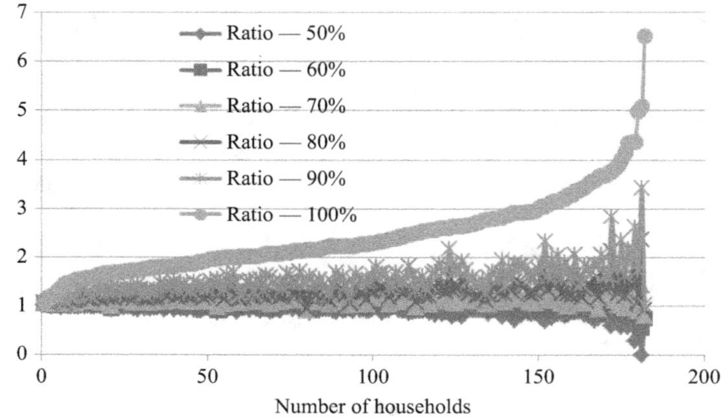

Figure 10.16 Comparison of level of energy coverage for no hot water households (50%–100% energy usage coverage), 186 samples

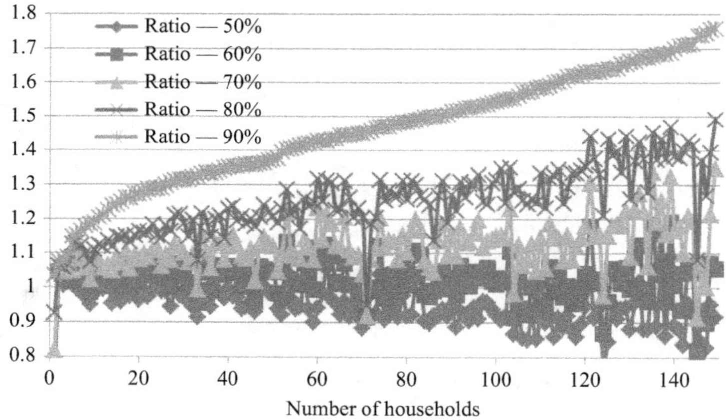

Figure 10.17 Comparison of the level of energy coverage for no hot water households (50%–100% energy usage coverage). 150 samples represent 80% of total sampling amount

If we use 80% of people's energy usage behaviour, the following comments are made:

- The average daily energy consumption is less than 10.24 kW h/day
- 72.8% of people use 5–15 kW h/day
- The ratio of peak energy usage to the average daily usage is about 1.68.

The above results are based on 90% coverage of energy usage days throughout a year. In other words, above analysis are used to justify what is the minimum energy consumption per day to cover 90% of days throughout a year. However, what happens when the degree of energy coverage varies? The comparison result is shown in Figures 10.16 and 10.17. The result showed here only focuses on

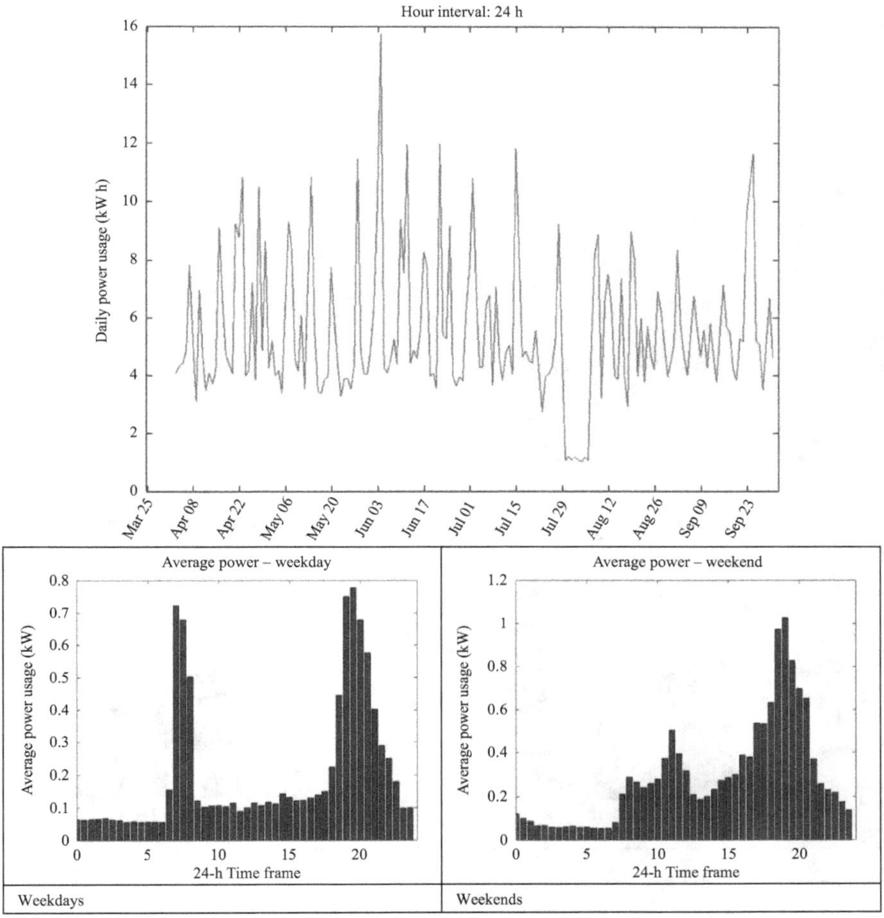

Figure 10.18 Double peak small usage (device 228 – 5.54 kW h/day)

household without controlled-supply. The household with or without controlled supply has very similar energy consumption figure.

As it can be observed, it is impractical to cover 100% days of energy usage as peak usage in some of days is three times greater than its average daily usage. If the coverage of days is below 80% of days, the ratio can be effectively reduced to 1.3–1.5 by looking at 80% of samples. The ratio is further dropped by bringing coverage days to 70%.

10.5.2 Energy usage pattern classification

The general household energy usage can be classified into four groups upon their energy usage pattern:

- Double peak type (Figures 10.18 and 10.19)
- Single peak type (Figure 10.20)

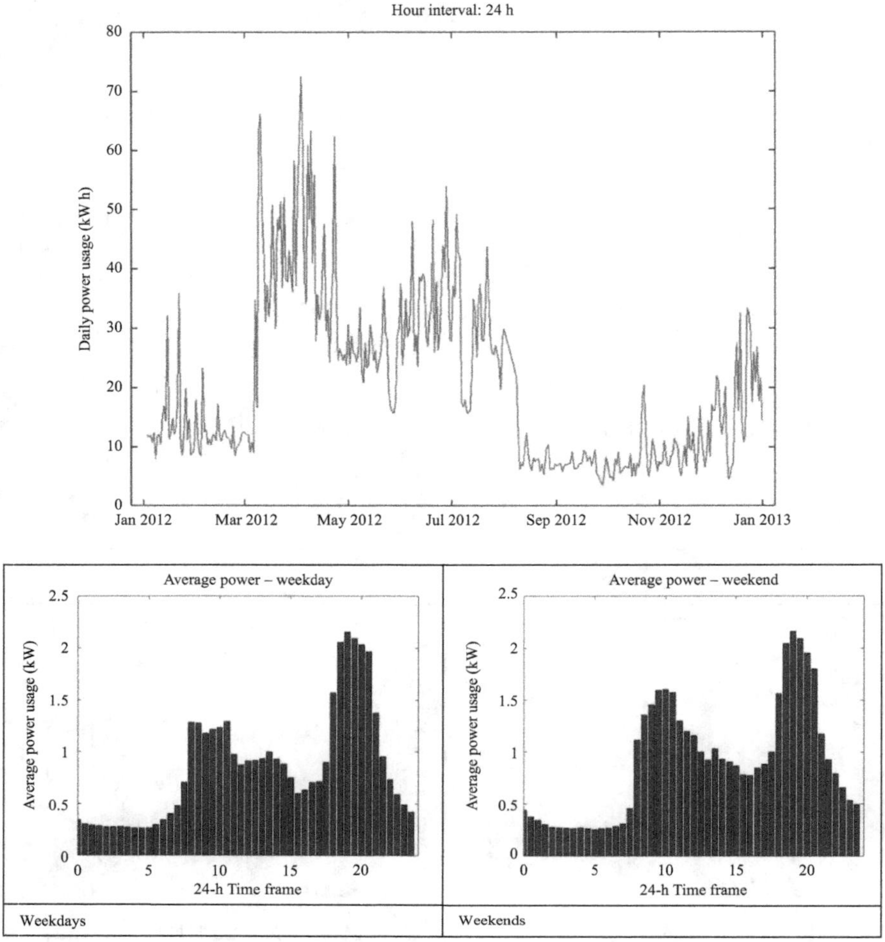

Figure 10.19 Double peak large usage (device 119 – 20.83 kW h/day)

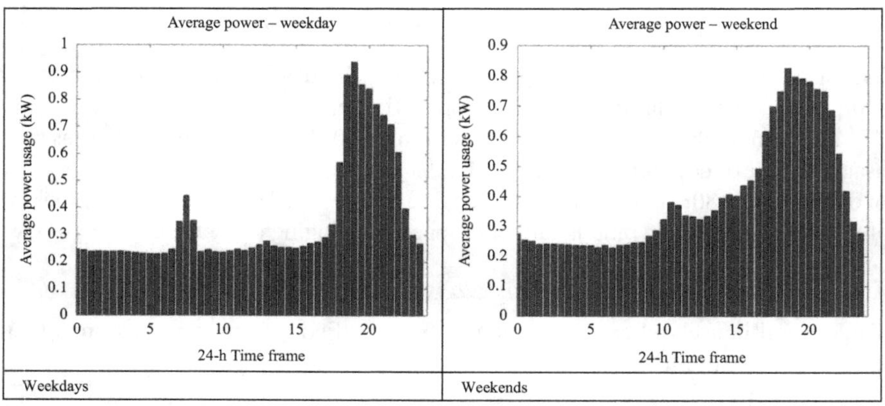

Figure 10.20 Single peak medium usage (device 139 – 8.92 kW h/day)

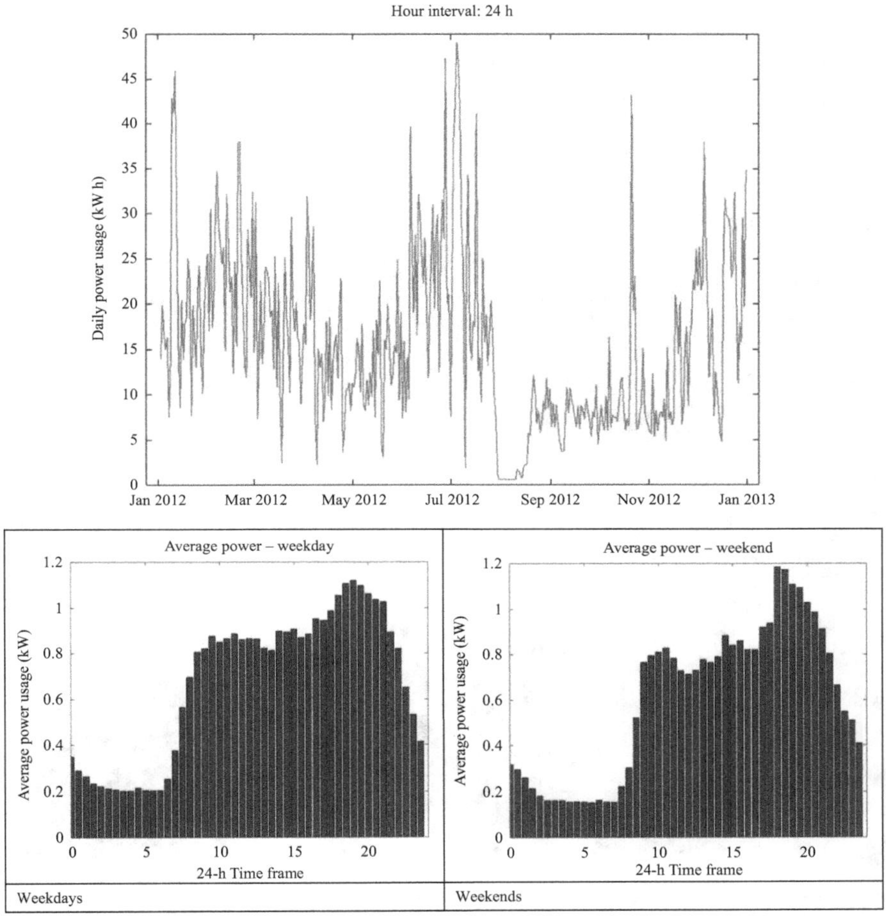

Figure 10.21 Mountain type (device 41 – 15.71 kW h/day)

- Mountain type (Figure 10.21)
- Valley type (Figure 10.22)
- Other types

10.5.2.1 Double peak type

The double peak energy usage pattern is the most commonly seen pattern form as shown in Figure 10.18. This type of household usually consisted by a single professional or a couple who has regular work routine from 09.00 to 17.00. During weekdays, the peak energy usage occurs from 07.00 to 08.30 in the morning and then from 19.00 to 20.30 in the evening. In weekends, energy usage pattern is distributed more evenly with additional daily energy usage compared to weekdays. This behaviour also reflects a repeated pattern in the daily power usage figure that

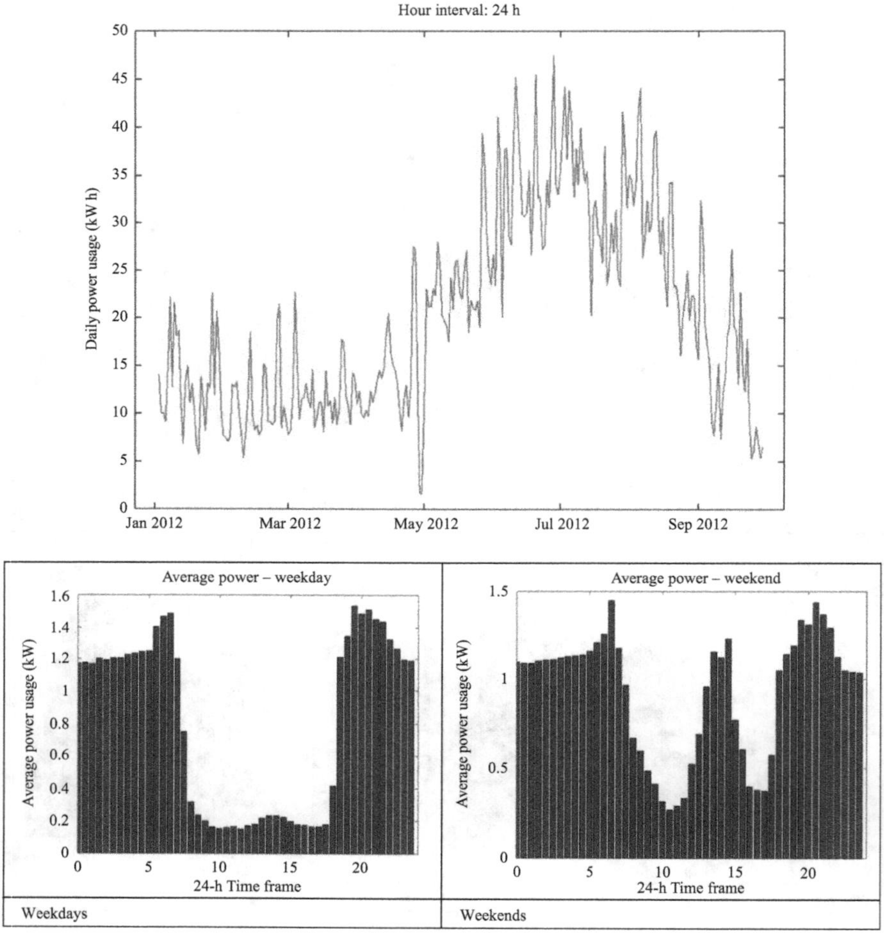

Figure 10.22 With solar system (device 229 – 20.28 kW h/day)

the regular peak energy usage occurring four times a month. Most of household energy usage pattern is very similar to the double peak type pattern, apart from that the peak time could be varied or the average daily usage could fluctuate upon the work type or the size of families.

For example, Figure 10.19 shows the double peak energy usage pattern of the household with greater family sizes. The average daily energy usage has increased by 15 kW h/day compared to the small-scale household example. Similarly, two peak energy usages occur between 08.00 and 10.00 in the morning and between 18.30 and 21.00 in the evening. Interestingly, households with greater daily energy consumption consume slightly less energy during the weekend. This could be attributed by some of the family members go out for the weekend, or the family weekend activities.

10.5.2.2 Single peak type

Another type of the energy consumption pattern is the single peak pattern as shown in Figure 10.20. One interesting observation was made on this type of energy consumption pattern. The household with this type of energy usage pattern tends to use less energy compared to other types. Rarely, household with this energy consumption pattern consumes greater than 10 kW h/day. Likely, this pattern is generated by young professionals or young independents who do not cook in the morning or does not have a regular work routine. Most of the single peak pattern has its peak energy usage at night, and only one or two households in samples perform differently.

10.5.2.3 Mountain type

Mountain type energy consumption pattern is shown in Figure 10.21. For energy usage patterns do not have an obvious peak usage during the day, it will be classified as the mountain type. For some households with the mountain type might still have the morning peak or the evening peak usage. However, the pattern will still be classified as mountain type as long as the afternoon energy usage is more than double of the early morning usage.

With this type of energy consumption pattern, it usually represents that the household size is greater and consequently represents as a higher daily energy consumption. Because of the diversity energy usage behaviour among the family members, the figure pattern is smoother without having an obvious peak usage compared to the double peak pattern and single peak pattern type. During the weekend, the main energy consumption period starts from 08.30 instead of 07.30 during weekdays. This could possibly due to people gets up late, and all appliances start up later.

10.5.2.4 Valley type

Energy usage pattern for households with solar system installation is very different from the previously introduced ones. Especially, the mountain type energy usage pattern can be easily observed when a large solar system is installed. Figure 10.22 shows a perfect example of energy usage pattern that the solar generation almost covers the energy usage during the daytime. As a result, the average daily energy usage decreases with increased solar generation and increases when the time comes to autumn and winter. During the weekend, daytime peak energy consumption occurred during the afternoon time between 13.00 and 15.00. This could be attributed to the operation of air-conditioning during the weekend period or due to heavy family activities.

As long as the afternoon power consumption is less than the early morning power consumption, the energy pattern is classified as valley type. The valley type energy pattern households will be excluded from the following analysis as we are more concern on the actual energy usage for the general household. The data of valley type pattern households are not valid for our following analysis. Furthermore, if the lower power consumption cannot be observed during the afternoon time, nor being classified into previous types, then it will be classified into other pattern types.

10.5.2.5 Analysis

Statistic number of different pattern types is shown in Figure 10.23. Double peak energy usage pattern attributes the largest portion due to lifestyle and people live in the area. Also, if taking the double peak and the single peak energy usage pattern into account, it attributes almost two-third of overall system samples. 6% was attributed to other pattern types as their energy usage behaviour is difficult to distinguish to any of the particular patterns.

The energy usage pattern of all samples is shown in Figure 10.24. As expected, the pattern is similar to the double peak type according to the result showing above. Peak energy usage occurs at 07.00–08.30 in the morning then 18.00–21.30 in the evening.

One interesting phenomenon in regards to the average energy usage upon on different energy usage pattern was observed. The comparison of the daily energy usage is tabulated in Table 10.6. The household with mountain type energy usage pattern consumes 20% more energy than average. This is can be attributed to a larger family size and consequently greater energy consumption accordingly. In

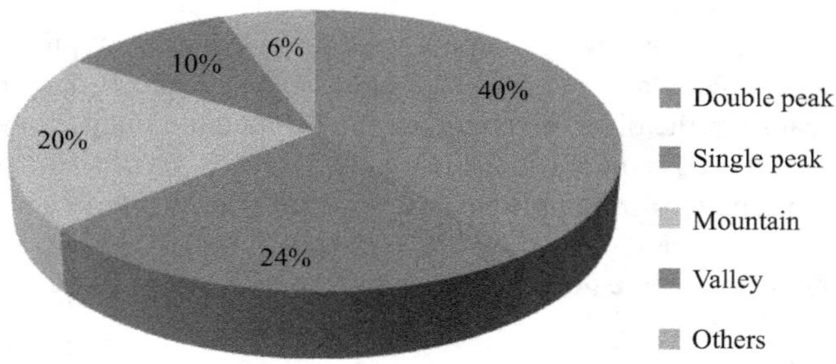

Figure 10.23 Percentage of each pattern type among 265 samples

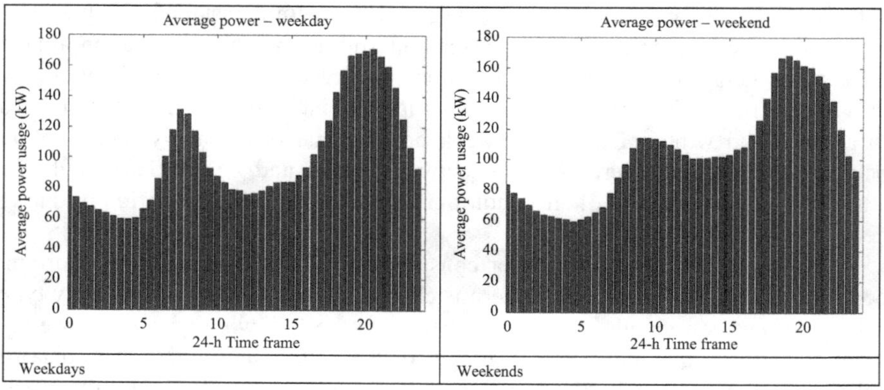

Figure 10.24 Total energy usage across the time frames (extracted from transformer raw data)

comparison, households with single peak energy usage pattern have the least energy consumption.

A similar outcome is obtained when plotting a generalized result upon on different energy usage pattern as shown in Figure 10.25. Keeping in mind that household with valley usage pattern has the solar system installed, and thus they consume less energy compared to users on other types.

Once the household samples are classified into groups upon on their energy usage pattern, result figure of the daily energy average usage to the 90% covering days ratio is shown in Figure 10.26. The ratio result does not make much difference compared to results obtained in earlier section. The ratio still varies between 1.6 and 1.8 when looking at 80%–90% of the overall household samples.

10.5.3 Peak apparent power (VA) identification

The reason we are looking into apparent power index is because it determines the size of the power electronics used in the ESS. The purpose of the above analysis has not only justified the energy usage pattern but also help to determine the size of the ESS. The index of real power only shows the actual energy usage of the general households. However, the actual power flowing in the system is determined by the apparent power and the rating of power electronics in the system needs to satisfy it. Looking into household's average daily energy usage (kW h) and generalize it with

Table 10.6 Comparison of average energy usage upon on the energy usage pattern

Type	Double peak	Single peak	Mountain	Valley	All
Average (kW h/day)	12.62	12.13	15.40	12.67	12.89

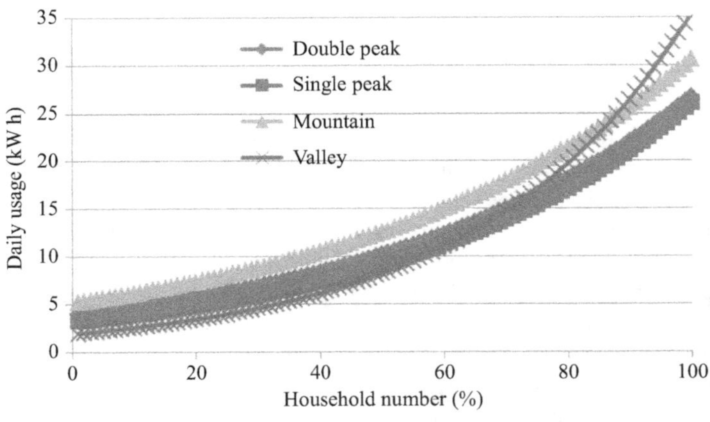

Figure 10.25 Generalized results of daily energy consumption upon on different energy usage pattern

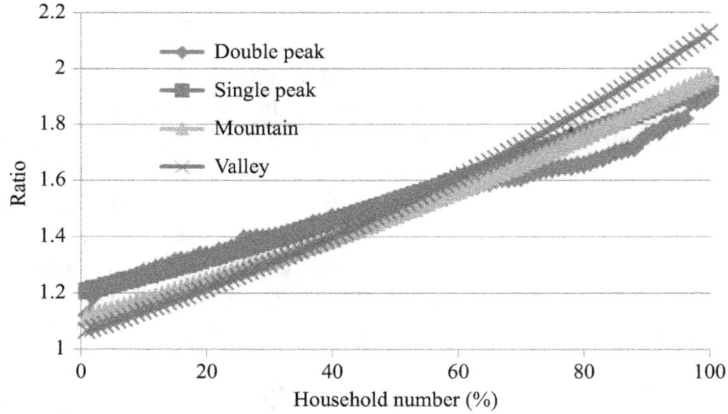

Figure 10.26 Generalized results of daily energy consumption ratio (90% days of coverage) upon on different energy usage pattern

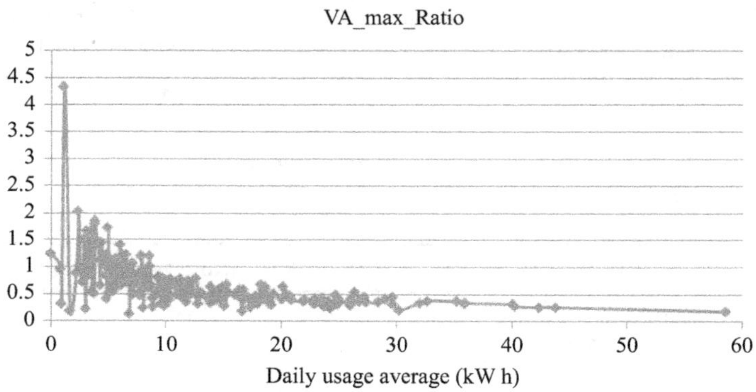

Figure 10.27 Ratio of peak VA usage versus average daily energy usage

its peak apparent power usage (VA), the relationship is shown in Figure 10.27. For households with daily energy usage less than 10 kW h/day, the ratio of peak VA to peak kW h is mostly greater than 1. However, with daily household energy consumption further increases, the ratio starts declining to as low as 0.5.

Also, it can be justified that most of the households have the peak power consumption at <10 kV A as shown in Figure 10.28. In other words, a battery-storage system with 10 kV A power electronics can satisfy >80% of household. Most of the households consume power from 4 to 10 kV A according to results shown in Figure 10.29. Thus, a 10-kV A power electronic product is probably the best size to suit general households for battery-storage systems.

Interestingly, the difference between peak real power and the peak apparent power has not much difference as shown in Figure 10.30. The ratio is almost

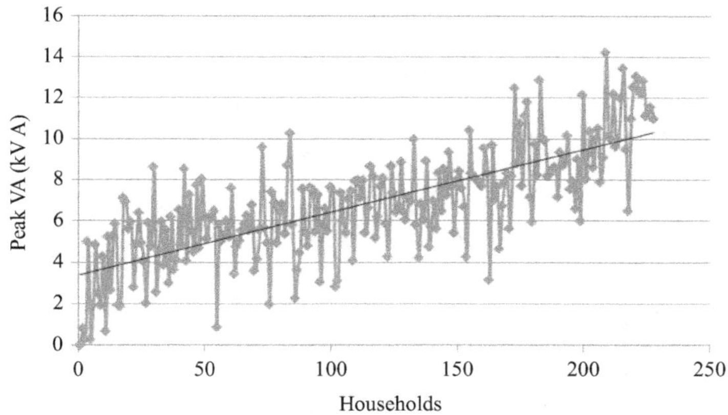

Figure 10.28 Peak VA usage sorted by average daily energy usage

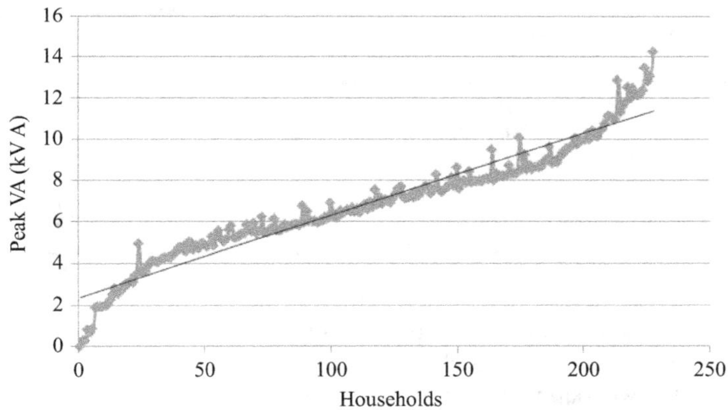

Figure 10.29 Peak VA usage sorted by peak VA usage

identical to each other for all the samples. Generally, a lot of appliances in the households are inductive and it was expected that the peak VA is higher than the peak kW h. If this is the case, then a simple solution of capacitor banks can be used to compensate the power factor. However, according to the result shown in Figure 10.30, the peak power consumed is almost identical to the peak apparent power required, then an active var compensator device (e.g. dStatcom) is recommended to install along with the system.

10.5.4 Summary

- The energy usage comparison under different household samples is tabulated in Table 10.7. This comparison is based on 80% of people's energy usage behaviour. Interestingly, household with regulated-power even consumes more

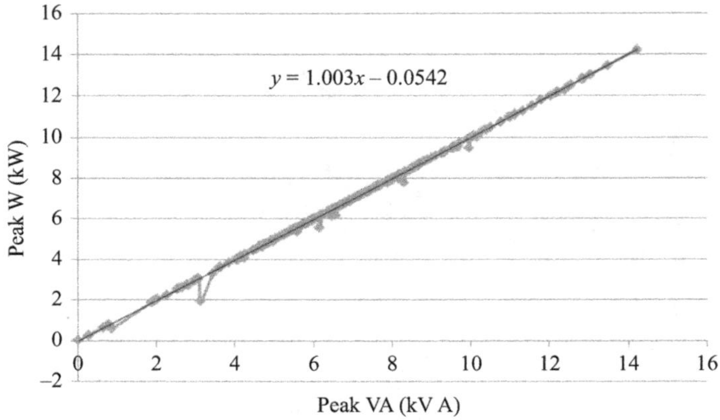

Figure 10.30 Peak VA versus peak W usage

Table 10.7 Comparison of energy usage when using 80% of people's behaviour as the sample

	Maximum usage (kW h/day)	**80% Ratio**	**Average usage (kW h/day)**
No regulated supply	19	1.72	5–15
With regulated supply	19.7	1.74	7–17
Regulated supply	10.34	1.68	4–11

power than household without regulated-power supply. This could be attributed to the factor that household with greater energy usage tends to utilize regulated-power supply on reducing their electricity costs.

- The peak energy required to cover 90% days of usage is roughly around 1.7 times greater than the average daily usage. However, this peak energy usage to daily energy usage ratio can be reduced to 1.3–1.5 with the coverage days reduced to 80%. When the coverage days decrease to 70%, the ratio is further reduced to 1.1–1.3.
- Peak instantaneous power required for a general household is roughly at about 10 kV A. The maximum power required is 14.23 kV A the worst case among all samples.

10.6 Selection process of battery storage

Applications to be performed by ESSs can be found across the levels of the electricity grid. ESS applications include energy shifting from the low demand period to the high demand period aiming to shave the peak load, shifting the energy produced by renewable energy sources (RESs) – such as PVs – from the morning to

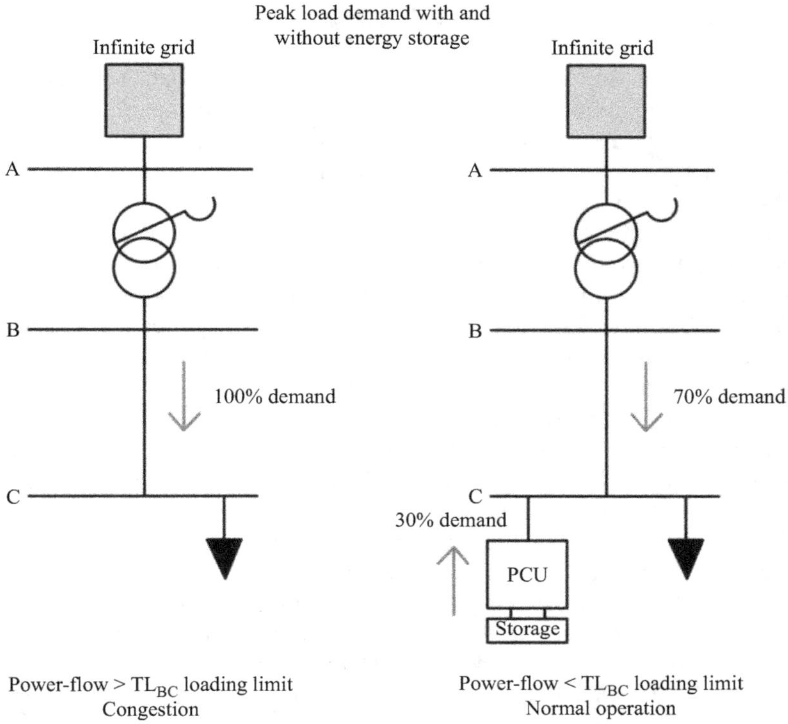

Figure 10.31 ESS for peak shaving and congestion relief

the evening hours where it is more favourable for the grid, frequency regulation, voltage support and congestion relief, to name a few. Considering that the above-mentioned applications require different time horizons and number of cycles per annum, and also the existence of constraints such as spatial limitations and storage mobility, the appropriate storage technology should be identified to perform the application effectively.

For example, an ESS employed for peak shaving and congestion relief, such as in Figure 10.31, should be placed downstream of the transmission line that is threatened with overloading, so that during the peak load period, the ESS supplies part of the load demand (e.g. 30%) while the residual load demand (e.g. 70%) is met by the power flow through the transmission line. Thereby transmission line congestion is avoided.

To identify the appropriate ESS technology for a specific application, 14 criteria are considered in the selection processes presented in this work. Prior to the deployment of the technology selection processes, a brief discussion of the traditional and the improved methods is realized [17–24].

A traditional ESS technology selection process includes generation of a table for each of the candidate energy storage technologies, where the table is filled with the characteristics the ESS should possess to perform a specific application (e.g. energy shifting). Further, the table is also filled with values to indicate the competitiveness of one ESS technology compared to another. The selection is realized by examining one by one the characteristics and comparing the recorded values. This comparison will eventually lead to the selection of the appropriate storage technology for the application under consideration. This process is demonstrated in [17]. A table is also employed in [18], where there, the pros and cons per ESS technology are recorded, and by comparing them, competent technologies are selected for further evaluation. In [19], a discussion-based comparison is realized, leading to the selection of the appropriate ESS technology. The author in [20] combines all of the above selection processes, by creating a table with specific characteristics the storage technologies should maintain, and a value per characteristic is assigned, representing the competitiveness of the storage technology for a certain characteristic. The table in [20] also includes advantages and disadvantages per ESS technology. In turn the author, by considering the tabulated values as well as the advantages and disadvantages per technology, selects the appropriate ESS for energy shifting.

Other selection processes that exist in the literature are more precise, and they can include radar charts (RCs) [21], decision matrices (DMs) [22,23] or the selection is realized by identifying the solution with the shortest distance from the ideal. The method that identifies the shortest distance is termed technique for order preference by similarity to ideal solution (TOPSIS) [24].

The RC in [21] values from 0 to 10 the extent to which the candidate storage technology meets each of the required criteria. The authors in [22,23] utilize DMs, where each candidate technology is compared with the others, and depending on which technology is better at a certain criterion, it receives higher total grades. One column chart is assembled per criterion, to compare the performance of one technology against the others, in turn the values of the column charts, serve as inputs to the TOPSIS method [24]. The selection processes continue, by assigning weights to each of the criteria, with the weights to be included in the calculations. The results establish a ranking of technology competitiveness factors (TCFs). For the RC and DM processes, the storage technology that presents the TCF with the highest percentage is perceived to be the best match for the application. Since the storage technology that maintains the shortest normalized distance from the ideal solution is sought in the TOPSIS method, the technology with normalized distance 1 is better than the technology with normalized distance 2, since the technology with distance 1 is closer to the ideal solution.

10.6.1 Criteria participating in the selection processes

The storage technologies that participate in the selection processes, as well as the 14 criteria (Ci) that are considered in this work, to shed light on the storage capabilities, are visualized in the RC of Figure 10.32 [21]. Normalized values from 0 to 10 are utilized in the RC of Figure 10.32, with the values prior to normalization to be available in [17,25,26].

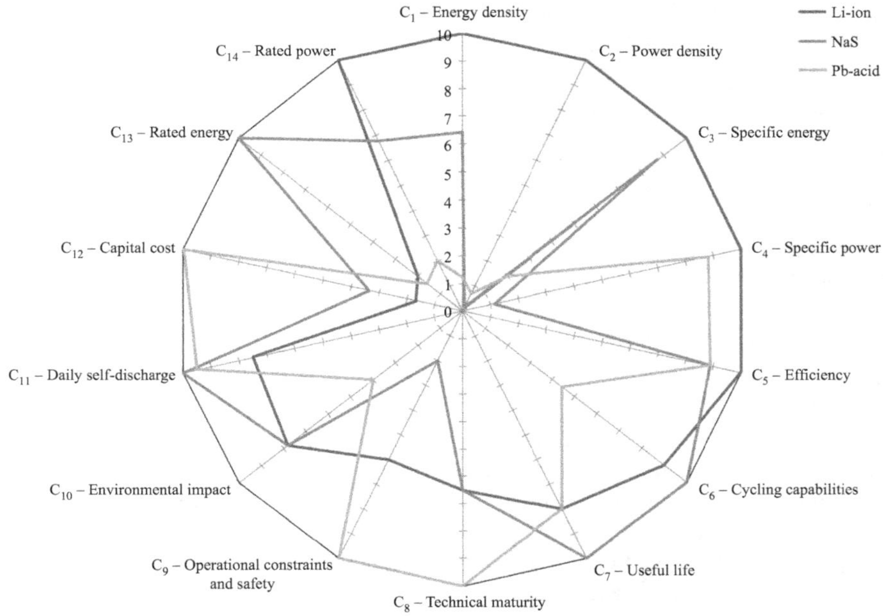

Figure 10.32 C_1–C_{14} utilized for the comparison between the storage technologies

Battery technologies mainly participate in the selection process, because of their consideration by many authors for a plethora of applications, such as energy shifting for peak shaving, RESs integration, frequency regulation, voltage support, congestion relief, etc. [27–32].

The storage technologies that are examined include Li-ion, NaS and PbAc batteries, because they maintain high efficiencies, long useful life and low daily self-discharge losses. The Li-ion technology in addition to the previous also maintains high energy and power densities among other competitive characteristics, while the NaS technology is well known for high-rated energy capabilities. The PbAc technology has been used extensively, and considerable experience is available, while it is also the lowest cost solution.

The comparison between candidate storage technologies, apart from RC utilization such as in Figure 10.32, can also be realized by assembling DMs [22], as shown in Table 10.8. For each of the selection criteria (i.e. C_1–C_{14}), another DM is generated, and one storage technology is compared with the rest in respect to the criterion associated with the DM. For example, in Table 10.8 is presented the comparison between Li-ion, NaS and PbAc battery technologies in respect to energy density (i.e. C_1), and it can be seen that the Li-ion technology receives the highest score (i.e. 9) resulting from the total grades (Gtotl) the technology received, when compared with the NaS and PbAc technologies.

Table 10.8 DM associated with C_1, for comparison between the candidate energy storage technologies

Grading scale	C_1–energy density	Li-ion	NaS	PbAc	$Gtot_1$
1 – Poor	Li-ion		4	5	9
2 – Diminished	NaS	2		5	7
3 – Similar	PbAc	1	1		2
4 – Enhanced					
5 – Superior					

Note: The grading scale is presented on the left of the table.

Column charts are another option to clearly identify the difference between the candidate storage technologies in respect to a specific criterion. For example in the column charts of Figure 10.33, it can be seen that the energy density of the Li-ion technology is superior to the energy density of the NaS and PbAc battery technologies. In terms of capital cost, the most competitive option is the PbAc battery, thereby this technology receives 10.

The numerical values of the column charts are utilized in the calculations of the TOPSIS method, thereby the TCFs of the TOPSIS method derive, as it is analysed in Section 6.2.

In the case the feeder is dominated by residential loads, the energy shifting application performed by the ESS, involves charging the battery in the morning and discharging it in the evening, for peak load shaving and for congestion relief. Alternatively, when commercial loads are the majority at the feeder, the ESS pattern includes charging at night hours, and discharging the next morning and afternoon. In both of the above cases (i.e. residential or commercial feeder), similar operating periods are anticipated ranging a few hours a day. Regarding the cyclic requirements of the ESS, they can reach 250 cycles per annum [33], which is proximate to one complete cycle per workday.

10.6.2 Weighting description and TCFs identification

The weighting presented in Table 10.9 is adapted to the energy shifting application. Criteria, such as operational constraints and safety (w9) and capital cost (w12), always receive high weights (i.e. w9 = w12 = 9) to highlight their importance. Other significant criteria include energy and power densities (w1 and w2), specific energy and power (w3 and w4) and efficiency (w5), among others. Energy and power densities receive the weights of w1 = w2 = 7.5, which is above the equitable weighting (i.e. 7.1428), because a storage technology with high energy to volume (i.e. kW h/m^3) and power to volume (i.e. kW/m^3) ratios is encouraged. This is because, higher amount of energy and power are provided by the technology (e.g. Li-ion), while the minimum volume is required. Similarly for specific power (kW/kg) and energy (kW h/kg), a storage technology, that meets these two criteria

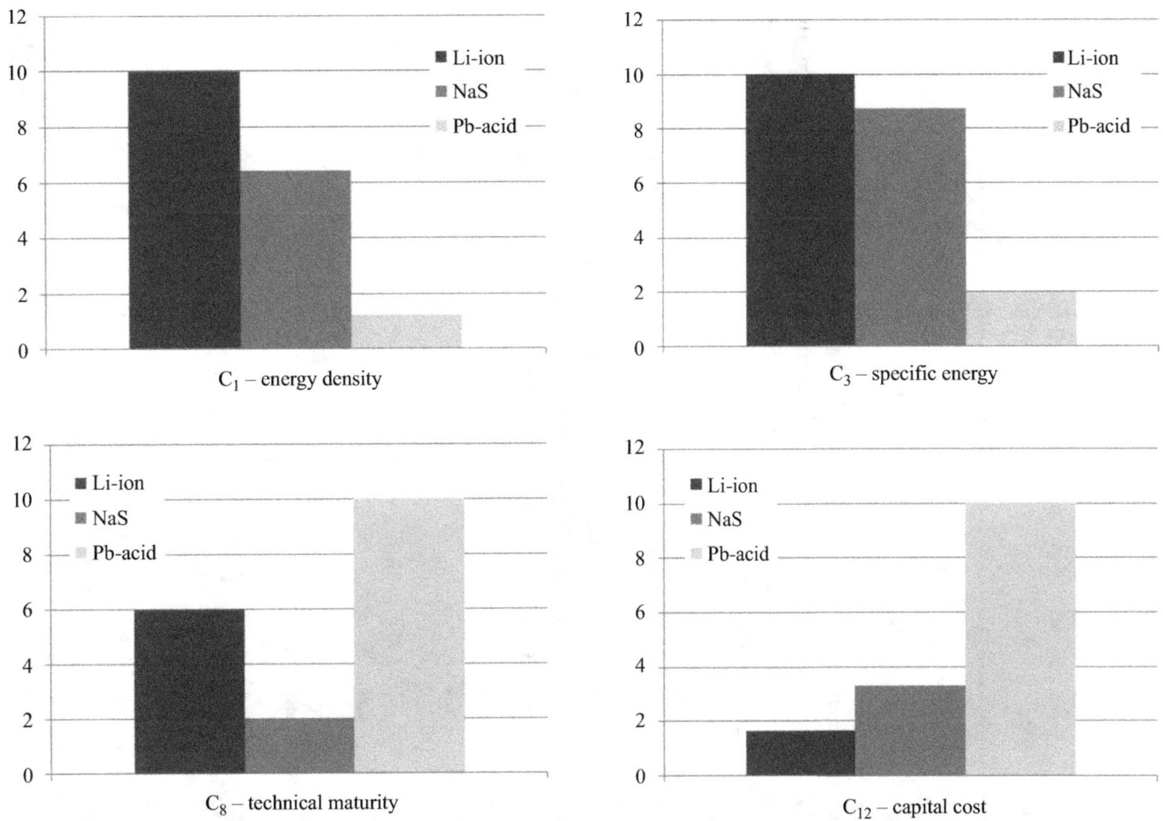

Figure 10.33 Column charts for comparison of the candidate storage technologies

Table 10.9 Criteria weighting for the application of energy shifting

Weight	Value	Weight	Value
w1	7.5	w8	5
w2	7.5	w9	9
w3	7.5	w10	7.5
w4	7.5	w11	2
w5	7.5	w12	9
w6	7.5	w13	7.5
w7	7.5	w14	7.5

to a high extent, is capable to provide high power and energy for the minimum weight requirements.

The storage characteristic of daily self-discharge receives a low weight (i.e. w11 = 2), because one complete charging-discharging cycle is performed by the ESS, almost every workday of the week (250 cycles pa [33]), therefore limiting the effect of the daily-self discharge.

For the calculation of the TCFs for each of the different selection processes discussed in Section 10.6.1, (10.3)–(10.6) are utilized. Values from the RC of Figure 10.32 are substituted into (10.3) and (10.4), and along with the weights of Table 10.9, a TCF per storage technology is identified. Values from the DMs, such as Table 10.8, are passed into (10.5), thereby the TCFs of the DMs method derive. For the calculation of the TCFs with the TOPSIS method, (10.6) is employed, with the criteria values to be supplied by Figure 10.33, and the criteria weights to be provided by Table 10.9.

$$\mathrm{App_{sum}} = \sum_{i=1}^{n} w_i C_i,\ n = 14 \tag{10.3}$$

$$\mathrm{TCF_{RC}} = \frac{\mathrm{App_{sum}}}{\sum_{i=1}^{n} w_i C_{\max}} 100,\ n = 14 \tag{10.4}$$

$$\mathrm{TCF_{DM}} = \sum_{i=1}^{n} w_i \frac{G_{\mathrm{tot}i}}{G_{\max}},\ n = 14 \tag{10.5}$$

$$\mathrm{TCF_{TOPSIS}} = \sum_{i=1}^{n} \left(\frac{w_i}{100}\right)^2 (C_i - C_{\max})^2,\ n = 14 \tag{10.6}$$

The TCFs, identified for the application of energy shifting, are presented in Table 10.10. From the table, it is clear that the Li-ion technology is leading the competition thereby it is selected for the application, with the NaS and Pd-acid technologies to follow. Particularly, the TCF of the RC method, and the TCF of the DM method, both for Li-ion technology (+), termed $\mathrm{TCF}^{+}_{\mathrm{RC}}$ and $\mathrm{TCF}^{+}_{\mathrm{DM}}$, maintain the percentages of 76.74% and 70.30% in accordance. The TOPSIS selection process, that determines the most appropriate storage technology, by identifying the one that maintains the shortest distance from the ideal solution, results in $\mathrm{TCF}^{+}_{\mathrm{TOPSIS}}$ =1.14. The value of 1.14 is the shortest normalized distance from the

Table 10.10 TCFs and suitability per energy storage technology, for the application of energy shifting

Technology	TCF_{RC} (%)	Suitability	TCF_{DM} (%)	Suitability	TCF_{TOPSIS}	Suitability
Li-ion (+)	76.74	1st	70.30	1st	1.14	1st
NaS (*)	62.52	2nd	58.95	2nd	2.07	2nd
PbAc (Δ)	56.13	3rd	53.95	3rd	2.45	3rd

ideal solution, compared to the longer distance of 2.07 for NaS, and the even longer distance of 2.45 for PbAc battery technology.

The results are somewhat expected, because the Li-ion technology outperforms the rest of the storage technologies in criteria 1–5 and 14, as it can be seen in Figure 10.32, while for the majority of the residual criteria, the Li-ion technology maintains competitive values. At this stage, it needs to be noticed that the ESS technologies included in the selection process were deemed competitive candidates for the application of energy shifting; therefore, they were included in the process. Among the competitive candidates, the Li-ion technology was eventually selected because it presented the highest TCFs as it can be observed in Table 10.10.

In the case, the selection process was intending to determine the most appropriate ESS technology for the application of frequency regulation, the candidate technologies would be different, and the selection process would include the technologies Li-ion, superconducting magnetic energy storage (SMES) and flywheel. Then, the selection processes analysed in this section would be re-applied, and the appropriate storage technology for frequency regulation would be identified. For example, because the effect of charging-discharging cycles on the SMES is negligible compared to the effect of the cyclic operation of the Li-ion battery, the SMES technology would be a more suitable solution compared to Li-ion for frequency regulation.

Regarding the utilized selection processes, the RC and TOPSIS processes are more objective compared to the process that employs DMs. This is because the inputs of the RC and TOPSIS methods are precise values obtained from [17,25,26], and they precisely indicate the extent to which the storage technology satisfies the criterion. Conversely, the selection process that involves DMs is based on a 1–5 grading scale that is applied subjectively. All the three methods discussed in this section are capable to identify the appropriate energy storage technology, with the RC and TOPSIS methods to be more objective.

10.7 Safety consideration

As important as efficiency and lifetime, safety is a non-negligible fact of the system throughout design, installation and operation phases. Each individual device in the system has to meet the design specification with sufficient evidence and record of testing conducted under the regular and worst case scenario. System integrator has to make sure all components meet factory specifications relative to their intended use cases. There is a lot needs to be considered regards to safety issues in battery-based community storage, apart from meeting prerequisite international standards.

10.7.1 Safety hazard of batteries and mitigation

Batteries, due to their intensive energy density and chemical nature, can be a serious safety risk if incorrectly installed and operated, potentially leading to electric shock, fire, flash burns, explosion or exposure to hazardous chemicals and released gases.

- Electric shock
  Banks of battery cells can induce a severe electrical shock if remain charged or other voltage applied to battery bank by another energy source in the system as many battery banks are likely operating at a hazardous voltage which must be electrically isolated. Battery terminals must be isolated with secure rated insulating materials and avoid direct contact. Minimum labelling for battery storage has to be permanently displayed at the prominent position as required.
- Energy hazard
  A battery has sufficient energy to cause an arc flash if it suffers a short circuit or fault. An arc flash can have temperatures above 12,000 °C, capable of melting metal or causing fires and explosions. Generally, higher battery energy storage capacities have a higher risk of arc flash. Arcing faults may cause catastrophic failure of battery cell enclosures unless the fault currents are removed quickly by correctly rated electrical protective devices. Batteries of all types and their associated DC cabling need to be protected from mechanical damage.
- Fire and explosion
  Gas emission and thermal runaway are major causes of fire and explosion. Most PbAc batteries generate hydrogen and oxygen while charging that needs adequate ventilation. The chemistry of Li-ion batteries makes them prone to 'thermal runaway' if they are damaged or overheated by overcharging. It can also result from component failure, a short circuit or loose connections. Temperature monitor and control are essential for batteries in operation and storage. Location of BESS needs to be carefully considered when installing on a customer premise, especially for those have elevated ambient temperatures.
- Hazardous chemicals
  Battery casings can degrade or be damaged by impacts. They can also rupture as a result of excessive temperatures and excessive pressure generated from a change in chemical reaction from over-charging or following a short circuit. Electrolyte (fluid or gel) can leak from a ruptured casing, resulting in toxic fumes, burns, corrosion or explosion. Some compounds produced during the failure of a cell can be extremely toxic. The cleanup, decontamination and disposal of damaged equipment may require specialized equipment and skills. Disposal of contaminated items or batteries at the end of their service life usually will require treatment as a hazardous waste as MSDS is essential to determine risk mitigation methods regarding chemical hazards.

10.7.2 Location of installation

The battery is preferred to be installed in a dedicated equipment room or area with restricted access to prevent access by unauthorized persons. It should also take into account by physical requirements of system characteristics, ventilation requirements, environmental factors and another risk of damage. Extreme ambient temperature or wide difference ambient temperature and humidity shall be avoided since battery life is shortened at a prolonged high temperature and battery capacity is degraded under low temperature; meanwhile, temperature and humidity change may cause condensation issue.

- Indoor installation
  For domestic dwellings, it is not recommended to install batteries in habitable rooms for health and safety. Typically, garages, storage rooms and plant rooms are suitable for batteries in these buildings, which comply with installation requirements as discussed. For nondomestic buildings, batteries need to be installed in a designated plant room or separate battery room which is separated from the remainder of the building. The separation shall be considered using construction with a fire resistance level material.
- Outdoor installation
  Outdoor installation requires more protection on the battery itself – a fenced-off section that creates a restricted area for authorized persons only. Shade should be considered over the battery enclosure.

10.7.3 Battery storage enclosure

A stand-alone battery enclosure needs well designed to protect the battery from damage. It usually has lockable doors or covering the box with lid. Steel fabrication with IP 55 or NEMA 3R rating seems like a typical requirement to protect the battery from physical damage, dust and water. The prolonged explosion under the sun, charging and discharging operation and limited interior space brings insulation, cooling and ventilation into a combinational concern on the enclosure. All battery terminals inside the enclosure have to be readily reachable for maintenance and service. Also, the battery enclosure should be insect and vermin-proof.

10.7.4 Safety policies and standards

Safety policies and standards in the growing energy storage are key challenges in keeping the industry on the right track, especially for Li-ion batteries. Thus, improving knowledge of safety hazards and renovating standards turns essential. In Australia and New Zealand, there are standards involving safety testing from cell level to system level already in place for battery installations that should be considered in the application of battery-based community storage.

- AS 1170.4 Structural design actions – earthquake actions in Australia
- AS 1319 Safety signs for the occupational environment
- AS 2676.1 Guide to the installation, maintenance, testing and replacement of secondary batteries in buildings – vented cells

- AS 2676.2 Guide to the installation, maintenance, testing and replacement of secondary batteries in buildings – sealed cells
- AS/NZS 3000 Electrical installation (known as the Australian/New Zealand Wiring Rules)
- AS 3011.1 Electrical installation – secondary batteries installed in buildings – vented cells
- AS 3011.2 Electrical installation – secondary batteries installed in buildings – sealed cells
- AS 4086.2 Secondary batteries for use with stand-alone power systems – installation and maintenance
- AS/NZS 4509.1 Stand-alone power systems – safety and installation
- AS/NZS 4509.2 Stand-alone power systems – system design
- AS/NZS 4777.1 Grid connection of energy system via inverters – installation requirements
- AS/NZS 4777.2:2015 Grid connection of energy systems via inverters – inverter requirements
- AS/NZS 5000.1 Electric cables – polymeric insulated – for working voltages up to and including 0.6/1 (1.2) kV
- AS/NZS 5000.2 Electric cables – polymeric insulated – for working voltages up to and including 450/750 V
- AS 60950.1 Information technology equipment – safety – general requirements
- AS 62040.1.1 Uninterruptible power system (UPS) – general and safety requirements for UPS used in operator access areas
- AS 62040.1.2 UPS – general and safety requirements for UPS used in restricted access locations
- NZS 4219 Seismic performance of engineering systems in buildings
- DR_AS 5139-2017 Electrical installation-safety of battery systems for use with power conversion equipment (revision of AS 4086.2-1997)

Several IEC standards are also developed.

- IEC 61427-1:2013 Secondary cells and batteries for renewable energy storage – general requirements and methods of test – Part 1 Photovoltaic off-grid application
- IEC 61427-2:2015 Secondary cells and batteries for renewable energy storage – general requirements and methods of test – Part 2 On-grid applications
- IEC 62109-1 Safety of power converters for use in photovoltaic power systems – Part 1: General requirements
- IEC 62109-2 Safety of power converters for use in photovoltaic power systems – Part 1: Particular requirements for inverters
- IEC 62619:2017 Secondary cells and batteries containing alkaline or other nonacid electrolytes – safety requirements for secondary lithium cells and batteries, for use in industrial applications.

Beside standards, there are guidelines for designers, installers and contractors. In 2014, German solar industry association, BSW Solar, created an extensive report

Safety of Guidelines – Li-ion Home Battery Storage Systems. In 2016, Australian Clean Energy Council issued Install Guidelines for Accredited Installers – grid-connected energy systems with battery storage that provided interim guidance to designers and installers of grid-connected energy systems with battery-storage systems. In the meantime, Energy Storage Council published the Australian Battery Guide – Guide for Energy Storage System that gave guidance to the energy storage industry and consumers in the interim while formal Australian Standards and being developed for the sector.

10.8 Conclusion

This chapter has analysed the CES in-depth focusing on battery technologies. Recent studies have shown growing interest in the concept of applying batteries in a distributed manner at localized applications. It has benefits that target shifting renewable generation, managing load peaks, providing uninterruptible power, local reactive power support and grid services when aggregated with many units. Most importantly, it easily adopts existing grid infrastructure which significantly reduced the cost of investment and eases the pressure of network upgrade.

Various battery technologies are viable for CES application. PbAc batteries were initially wide used for the industrial and residential UPS and automotive and provided lower capital cost solutions. However, its implementation was very limited due to low charging speed and short life cycles. Thus, new ALC batteries are developed with significant improvements and now raising in renewable energy applications which PbAc is not capable of. Li-ion has emerged as the technology of choice for short duration applications. Thanks to the popularity of EVs, Li-ion batteries are gaining rapid development enabled by fast price reduction and performance improvement. Li-ion batteries have been implemented in a wide variety of applications such as peak load management, renewable integration and diesel reduction. There are also many other battery technologies in competition with pros and cons targeting different applications in the field.

EVs, provide a special form of battery storage, potentially provided a significant amount of energy storage for the community. However, the gradual and substantial rise of EV penetration also has an impact on distribution grids, as large numbers of EVs on the load and generation sides challenge grid stability. V2G technology makes it possible to store surplus electricity generated from intermittent renewable solar and wind sources in EV batteries during nonpeak periods and feed power back to the grid when needed, enhancing grid stability and reducing electricity costs at peak hours. CES is also a feasible application of those after vehicle batteries for many grid support applications and ancillary services for the grid.

Due to the rapid uptake of energy storage, safety integration introduced significant technical and regulatory challenges during implementation of battery storage. There is no specific standard for battery-based CES, but relevant guidelines for battery storage of different chemistries are constantly improving. In general, battery hazards have to be investigated and identified to mitigate risks of application.

Requirements for battery installation location and enclosure vary according to different battery chemistry. Safety policies and standards in the growing energy storage sector is a key challenge in keeping the industry on the right track, especially for Li-ion batteries. Australia is a pioneer country in regulatory intervention to drive adoption of energy storage.

References

[1] Thomas, R P, 2011, *American Electric Power Community Energy Storage, IEEE Power Engineering Society Energy Storage – Super Session*, accessed 29 June 2017, ⟨https://www.smartgrid.gov/files/American_Electric_Power_Community_Energy_Storage_201109.pdf⟩.

[2] Asgeirsson, H, 2010, *Detroit Edison's Advanced Implementation of A123's Community Energy Storage System for Grid Support*, DTE Energy, accessed 29 June 2017, ⟨http://www.sandia.gov/ess/docs/pr_conferences/2010/hawk_dte.pdf⟩.

[3] Wearmouth, A, 2016, *Alkimos Beach Battery Storage Trail*, Synergy, accessed 29 June 2017, ⟨http://eeaust.com.au/images/stories/virtuemart/category/presentation-to-the2016api-summer-school-alkimos-battery-storage-project_(no-charge-curve)-andy.pdf⟩.

[4] Sonnen, 2017, *What is the sonnenCommunity*, accessed 29 June 2017, ⟨https://www.sonnenbatterie.de/en/sonnenCommunity⟩.

[5] Battery University, 2016, *BU-201: How Does the Lead Acid Battery Work*, accessed 4 July 2017, ⟨http://batteryuniversity.com/learn/article/lead_based_batteries⟩.

[6] Battery University, 2017, *BU-202: New Lead Acid Systems*, accessed 29 June 2017, ⟨http://batteryuniversity.com/learn/article/new_lead_acid_systems⟩.

[7] Dubey A, and Santoso S. Electric vehicle charging on residential distribution systems: impacts and mitigations. IEEE Access. 2015;3:1871–93.

[8] D. Leskarac, C. Panchal, S. Stegen, and J. Lu. PEV Charging Technologies and V2G on Distributed System and Utility Interfaces. In: Lu J, Hossain J, editor. Vehicle-to-Grid: Linking Electric Vehicles to the Smart Grid. London, United Kingdom: The Institution of Electrical Engineers; 2015. p. 157–221.

[9] Mwasilu F, Justo JJ, Kim E-K, Do TD, and Jung J-W. Electric vehicles and smart grid interaction: a review on vehicle to grid and renewable energy sources integration. Renewable and Sustainable Energy Reviews. 2014;34:501–16.

[10] Sbordone D, Bertini I, Di Pietra B, Falvo MC, Genovese A, and Martirano L. EV fast charging stations and energy storage technologies: a real implementation in the smart micro grid paradigm. Electric Power Systems Research. 2015;120:96–108.

[11] SAE. Electric Vehicle and Plug in Hybrid Electric Vehicle Conductive Charge Coupler. SAE J1772, 2016.

[12] IEC61851. Electric Vehicle Conductive Charging System. 2017.
[13] Maheshwari P, Tambawala Y, Nunna HSVSK, and Doolla S, editors. A review on plug-in electric vehicles charging: Standards and impact on distribution system. 2014 IEEE International Conference on Power Electronics, Drives and Energy Systems (PEDES); 2014 16–19 Dec. 2014.
[14] Tan KM, Ramachandaramurthy VK, and Yong JY. Integration of electric vehicles in smart grid: a review on vehicle to grid technologies and optimization techniques. Renewable and Sustainable Energy Reviews. 2016; 53:720–32.
[15] Moghimi M, Leskarac D, Nadian N, Stegen S, and Lu J. Impact of PEV behavior on peak demand reduction in a commercial microgrid. 2016 Australasian Universities Power Engineering Conference (AUPEC); 2016 25–28 Sept. 2016.
[16] Ahmadi L, Yip A, Fowler M, Young SB, and Fraser RA. Environmental feasibility of re-use of electric vehicle batteries. Sustainable Energy Technologies and Assessments. 2014;6:64–74.
[17] Chatzivasileiadi A, Ampatzi E, and Knight I. Characteristics of electrical energy storage technologies and their applications in buildings. Renewable and Sustainable Energy Reviews. 2013;25:814–30.
[18] Yang B, Makarov Y, Desteese J, *et al.* On the use of energy storage technologies for regulation services in electric power systems with significant penetration of wind energy. IEEE 5th International Conference on European Electricity Market (EEM). Lisboa, 2008.
[19] Suberu MY, Mustafa MW, and Bashir N. Energy storage systems for renewable energy power sector integration and mitigation of intermittency. Renewable and Sustainable Energy Reviews. 2014;35:499–514.
[20] Poullikkas A. A comparative overview of large-scale battery systems for electricity storage. Renewable and Sustainable Energy Reviews. 2013; 27:778–88.
[21] Saez-de-Ibarra A, Milo A, Gaztanaga H, *et al.* Analysis and comparison of battery energy storage technologies for grid applications. IEEE PowerTech. Grenoble, 2013.
[22] Ibrahim H, Dimitrova M, Dutil Y, Rousse D, Ilinca A, and Perron J. Wind-diesel hybrid system: energy storage system selection method. IEA ECES 12th International Conference on Energy Storage (InnoStock). Spain 2012.
[23] Ibrahim H, Ilinca A, and Perron J. Comparison and analysis of different energy storage techniques based on their performance index. IEEE Electrical Power Conference (EPC). Canada, 2007.
[24] Wang J-J, Jing Y-Y, Zhang C-F, and Zhao J-H. Review on multi-criteria decision analysis aid in sustainable energy decision-making. Renewable and Sustainable Energy Reviews. 2009;13:2263–78.
[25] Diaz-Gonzalez F, Sumper A, Gomis-Bellmunt O, and Villafafila-Robles R. A review of energy storage technologies for wind power applications. Renewable and Sustainable Energy Reviews. 2012;16:2154–71.

[26] Mahlia TMI, Saktisahdan TJ, Jannifar A, Hasan MH, and Matseelar HSC. A review of available methods and development of energy storage; technology update. Renewable and Sustainable Energy Reviews. 2014; 33:532–45.

[27] Laaksonen H, Saari P, and Komulainen R. Voltage and frequency control of inverter based weak LV network microgrid. 2005 IEEE International Conference on Future Power Systems, pp. 1–6, Amsterdam.

[28] Qi Y, and Jia H. Research on a new voltage control strategy for photovoltaic grid-connected system. 2011 IEEE International Conference on Electrical Machines and Systems (ICEMS), pp. 1–6, Beijing.

[29] Hill CA, Such MC, Chen D, Gonzalez J, and Grady WM. Battery energy storage for enabling integration of distributed solar power generation. IEEE Transactions on Smart Grid. 2012;3:850–7.

[30] Gao W-j, Jing T-j, and Yang M-h. A economic dispatch method of energy storage systems under time-of-use pricing in microgrids. 2012 IEEE China International Conference on Electricity Distribution (CICED), pp. 1–5, Shanghai.

[31] Li Y, Vilathgamuwa DM, and Loh PC. Design, analysis, and real-time testing of a controller for multibus microgrid system. IEEE Transactions on Power Electronics. 2004;19:1195–204.

[32] Pokharel B, Ojo O, and Balogun A. Standalone operation of a DFIG-based wind turbine system with integrated energy storage. 2015 IEEE 6th International Symposium on Power Electronics for Distributed Generation Systems (PEDG), Aachen, Germany.

[33] Eckroad S, Gyuk IP, Mears LD, Gotschall HL, Key T, and Kamath H. EPRI-DOE Handbook of energy storage for transmission & distribution applications. Electric Power Research Institute (EPRI) – US Department of Energy. Palo Alto, California 94304 and S.W. Washington, DC 20585. 1001834. 2003.

Chapter 11

Power-to-gas and power-to-power for storage and ancillary services in urban areas*

Ushnik Mukherjee[1], Sean B. Walker[1], Azadeh Maroufmashat[1] and Michael Fowler[1]

Abstract

In this study, power-to-gas, whereby hydrogen is generated electrolytically, and power-to-power, whereby electricity is used to produce hydrogen which is used, in turn, to generate electricity at a later time, are examined for their efficiency and emissions reductions in providing energy storage and ancillary services. Due to a large baseload of nuclear energy in the province of Ontario, and with wind generation added to the grid prior to 2014, the supply of electricity exceeds demand at certain times of the year during off-peak hours. To manage this excess supply, electricity is exported to neighboring provinces and states at a low, often negative price, due to the decreased demand and surplus generation. To curb these exports, the Independent Electricity System Operator (IESO) has switched renewable generators from nondispatchable to dispatchable energy sources that can be turned on or off or adjusted to output a different quantity of energy. In addition, due to the large baseload, the IESO has also shifted to allowing loads to offer demand response services previously only offered by generators. In this analysis, the rapid response of polymer electrolyte membrane electrolyzers, used to generate hydrogen in power-to-gas and power-to-power systems is also able to offer important and high value auxiliary and regulatory power services. In power-to-gas systems, the hydrogen produced is an alternative energy vector which can be contained within the natural gas infrastructure or other storage medium. For this analysis, the authors employ the General Algebraic Modeling Simulation to develop a simulation of a 2-MW power-to-gas and power-to-power system that produces hydrogen for energy storage and then uses this hydrogen to generate electricity when there is a peak in energy demand. This energy is then reintroduced into the electrical grid using a hydrogen turbine. The power-to-power scheme, although typically less energy efficient,

*Presented at *Proceedings of 2016 Energy and Water Symposium*, Windsor, Ontario, Canada, June 22–23, 2016.

[1]Department of Chemical Engineering, University of Waterloo, Canada

provides the flexibility to meet changing energy demands while generating hydrogen that can be used for industrial purposes and as a transportation fuel.

11.1 Introduction

With Ontario's increasing reliance on renewable energy, and preexisting nuclear facilities the power grid must manage periods of surplus energy generation. The presence of surplus energy creates an opportunity to use cheaper off-peak energy for energy arbitrage. The conversion of electricity to gas energy via power-to-gas is implemented through the use of electrolyzers and allows simultaneous grid stabilization, seasonal storage of bulk power, geographic transmission of energy and dispatchable regeneration of distributed renewable energy [1]. An advantage of power-to-gas is its ability to move energy between the electrical and natural gas infrastructures. As illustrated in Figure 11.1, power-to-gas is an energy solution which uses multiple energy streams to produce hydrogen gas which is then injected into the natural gas infrastructure, used on-site or stored in tanks or reservoirs. Power-to-gas systems can move energy from one location to another, where it can then be used to generate electrical or thermal energy. The only new infrastructure required for a power-to-gas system is electrolyzers and possibly compressed storage tanks. With this, the natural gas systems offer a large, distributed, energy storage network that can also transport energy from one place to another while providing energy storage for hours, days or months. In Figure 11.1, the system includes multiple sections: energy supply, energy conversion, the transmission and storage systems, distribution, conversion and final use. The hydrogen conversion technology, electrolysis, has the advantage of releasing no harmful emissions in comparison to steam methane reforming (SMR).

There are numerous available power-to-gas pathways to facilitate different applications [3,4]. In Figure 11.1, there are four pathways for power-to-gas including: power-to-pipeline, power-to-methanation, power-to-hydrogen and power-to-power. The power-to-pipeline pathway occurs when the hydrogen that is produced by electrolysis is injected into the natural gas infrastructure to create hydrogen-enriched natural gas, which is then fed to natural gas end-users via pipeline. Power-to-methanation refers to the production of renewable natural gas through the methanation of CO and CO_2 from hydrogen produced through power-to-gas. The renewable fuel produced through this stream can be safely injected into the natural gas pipelines. Power-to-power is an energy pathway whereby hydrogen is stored in a fuel cell on-site to produce power, on demand. One of the useful aspects of power-to-power is that it can provide useful backup power and be used to generate income through energy arbitrage. Under an energy arbitrage scheme, the system would purchase electricity at a low price to produce hydrogen and then use the hydrogen fuel cell to generate electricity when the price of electricity rises. This would allow for the hub to make a significant amount of revenue. In the power-to-hydrogen stream, hydrogen is either fed into the natural gas infrastructure and then separated out as needed, using a pressure swing absorber, or stored in

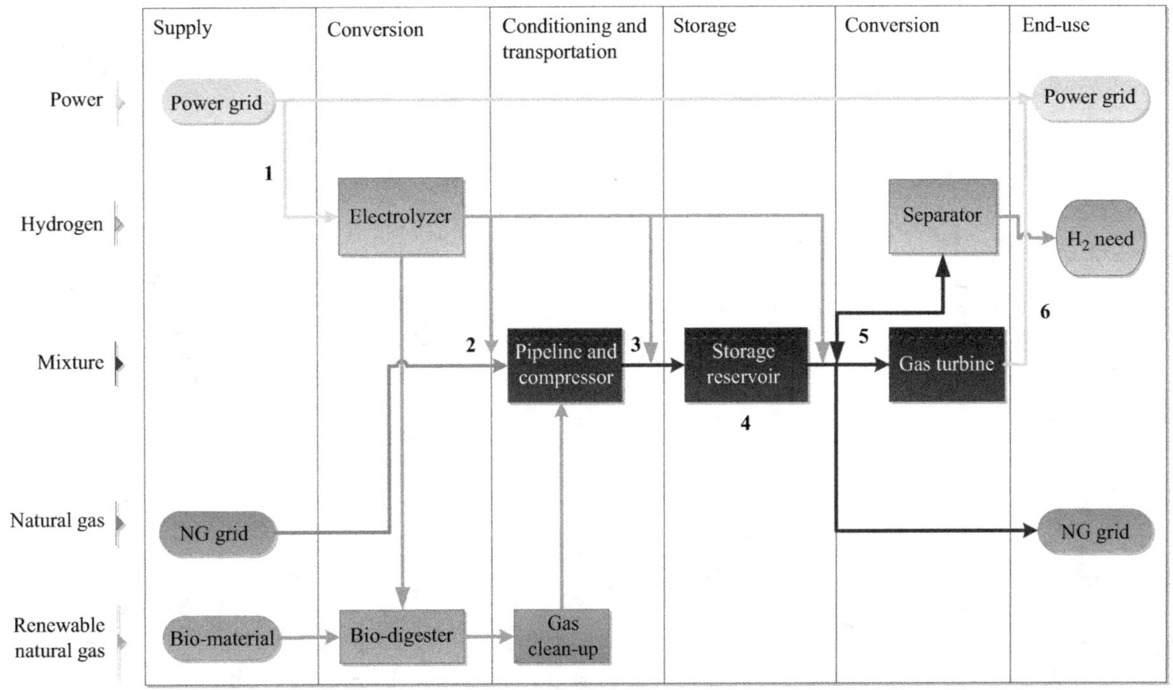

Figure 11.1 Basic concept illustration for power-to-gas [2]

tanks on-site for industrial hydrogen and mobility fuel. There is a high demand for hydrogen for both transport and industry, where hydrogen is used in the production of ammonia and petroleum. If the chemical industry seeks to reduce the carbon emissions of their operations, due to the prevalence of carbon taxes or, as in Ontario, Cap and Trade, power-to-hydrogen is an excellent technological choice. Power-to-hydrogen provides green hydrogen which can be delivered by pipeline or produced on-site to replace the use of SMR, a significant source of greenhouse gases. Of course, the power-to-hydrogen stream can also be used to power hydrogen fuel cell vehicles (FCVs). In this case, the hydrogen would be produced via electrolysis at a hydrogen refueling station and pumped into vehicles. This stream is potentially lucrative due to the many FCVs that are expected to enter the market by 2020 [5].

11.2 Methodology

The focus of the study is on optimally operating a power-to-gas hub capable of providing: (1) power-to-hydrogen service to a FCV refueling station and (2) power-to-power service to the power grid. The different technologies existing within the energy hub include:

- 2×1 MW polymer electrolyte membrane (PEM) electrolyzers developed by Hydrogenics Inc. [6];
- 1 ASME steel vessel capable of storing 89 kg of hydrogen at 172 bar [7];
- 1 prestorage reciprocating compressor module with a maximum flow capacity of 42 kg/h that compresses gas from a minimum inlet pressure of 3–310 bar [8];
- 1×3-stage reciprocating booster compressor capable of handling inlet pressures as low as 20 bar and a capacity of 87 kg/h. The booster compressor is used to complete the refueling cycle of FCV [8];
- 1 hydrogen dispenser and precooling system generally used in the modeling software Hydrogen Refueling Station Analysis Model developed by Argonne National Laboratory [9] and
- 1 heavy duty fuel cell power module rated at 33 kW developed by Hydrogenics Inc. [10].

A mixed integer linear programming (MILP) model is formulated to determine the optimal operation of the energy hub to provide power-to-hydrogen service and also sizes a fuel cell power module to provide power-to-power. Currently, Hydrogenics and Enbridge are under contract to independent electricity system operator (IESO) to develop a 2 MW power-to-gas hub in the Greater Toronto Area that provides ancillary services to the grid. In this paper, it is of interest to see the potential of a similar system to provide power-to-hydrogen. To ensure a simple and cost-effective design, only one of each the prestorage compressor, storage vessel and booster compressor have been chosen to be installed at the refueling station energy hub. However, the number of fuel cell power modules required to provide the power-to-power service is a decision variable in the optimization.

11.2.1 Mixed integer linear programming formulation

The primary objective in Model 1 is minimizing the overall cost of operation of the power-to-hydrogen energy hub. Equation (11.1) shows the objective function used in this case.

$$\begin{aligned} Minimize{:}\ Cost = {} & O\&M_{Electrolyzer} + Capital_{Electrolyzer} + (Capital_{Booster\ Compressor}) \\ & + (Capital_{Compressor,Prestorage}) + (Capital_{Tank\ Storage}) \\ & + Capital_{Dispenser} + Capital_{Cooling\ System} \\ & + (Capital_{Fuel\ Cell} \times N_{Fuel\ Cell}) + Net\ Expenses \end{aligned} \tag{11.1}$$

where $Capital_{Electrolyzer}$ denotes the amortized capital cost of the 2-MW PEM electrolyzer system. The annual operating and maintenance cost of the electrolyzers, $O\&M_{Electrolyzer}$ is taken to be 2.5% of the total capital investment. Similarly, the terms $Capital_{Booster\ Compressor}$, $Capital_{Compressor,Prestorage}$, $Capital_{Tank\ Storage}$, $Capital_{Dispenser}$, $Capital_{Cooling\ System}$, and $Capital_{Fuel\ Cell}$ are the amortized capital costs of the booster compressor, prestorage compressor, hydrogen storage system, dispenser, cooling system and fuel cell power module, respectively. Costs of each of the components in the energy hub are amortized over the 20-year lifetime of the project at an interest rate of 8%. $N_{Fuel\ Cell}$ is an integer decision variable that denotes the number of fuel cell power modules chosen. The amortized capital costs of the components are parametric inputs to the optimization problem and their values are given in Table 11.1, found in Section 11.3.

The term *t Expenses*, in (11.1) and (11.2), is the difference between the annual operating cost and the revenue earned by the energy hub.

$$\begin{aligned} Net\ Expenses = {} & \sum_{h=1}^{8,760} [(H_2\ Produced_h \times Water\ Consumption\ Rate \times Cost_{water}) \\ & + [\{E_h + (H_2\ Inflow_h \times Energy\ Consumption\ Factor_{Compressor,Prestorage})\} \\ & \times HOEP_h] \\ & + [\{E_h + (H_2\ Inflow_h \times Energy\ Consumption\ Factor_{Compressor,Prestorage})\} \\ & \times Transmission\ Charge_h] - (H_2\ Outflow_h \times H_2\ Premium\ Price) - (E_{FC,h} \\ & \times Offer\ Price)] - (Credit_{Emissions,existing} \times H_2\ Production\ Emission\ Offset) \end{aligned} \tag{11.2}$$

Equation (11.2) accounts for the various operating expenses incurred by the energy hub over the course of a year. Using electrolyzer specifications [6], the annual cost of water for producing hydrogen via electrolysis is calculated by multiplying the variable H_2 *Produced*, measured in kmol/h, with the parameters

Table 11.1 Component costing information of the energy hub

Component	Cost ($)
$O\&M_{Electrolyzer}$ ($), Annual operating and maintenance cost of electrolyzer	Confidential [6]
$Capital_{Electrolyzer}$ ($), Annual investment on total capital cost of electrolyzer	Confidential [6]
$Capital_{Booster\ Compressor}$ ($), Annual investment on total capital cost of booster compressor	$37,368 [8]
$Capital_{Compressor,Prestorage}$ ($), Annual investment on total capital cost of compressor prestorage	$25,442 [8]
$Capital_{Tank\ Storage}$ ($), Annual investment on total capital cost of tank storage	$30,422 [7]
$Capital_{Dispenser}$ ($), Annual investment on total capital cost of dispenser module	$10,898 [9]
$Capital_{Cooling\ System}$ ($), Annual investment on total capital cost of cooling system	$20,064 [9]
$Capital_{Fuel\ Cell}$ ($), Annual investment on total capital cost of fuel cell module	$28,104 [10]

Water Consumption Rate, in L H_2O/kmol of H_2 [6] and the cost of water $Cost_{water}$, in \$/L H_2O [11]. The Hourly Ontario Electricity Price (*HOEP*, \$/kW h) [12] is used to calculate electricity consumption charges for operating the electrolyzers and prestorage compressors. The energy consumed by the electrolyzers is denoted by *E* (kW h), whereas the energy consumed by the prestorage compressor is the product of the hydrogen flowing into the compressor (*H_2 Inflow*, kmol/h) and the energy (kW h) consumed by the compressor per kmol of hydrogen compressed ($Energy\ Consumption\ Factor_{Compressor,Prestorage}$, 2.5042 kW h/kmol) [8]. The hydrogen is fed to the prestorage compressor at the condition at which it is produced—30 bar and 21 °C. A fixed transmission charge for electricity is considered based on data provided by the local energy provider [13].

There are three potential revenue streams for the energy hub. The primary revenue is obtained from selling hydrogen to the FCV end users at a premium price of \$8/kg H_2 [14]. The secondary revenue stream is from the carbon credits earned from offsetting emissions in the production of hydrogen compared to the traditional SMR process. The emission factor of the process (EMF_{SMR}) in this study is assumed to be 18 kg of CO_2 to produce a kmol of hydrogen [15]. The power bought from the grid has an associated hourly emission factor having the unit kg CO_2/kW h of energy consumed. This data was developed based on the hourly generation output data made available by the IESO. The lifecycle emission factor (kg $CO_{2,e}$/kW h) of the different generation sources has been taken from a report prepared by Intrinsik Corp., Ontario for Ontario Power Generation [16]. This emission factor fluctuates with time as different types of power generation are brought online or offline. Based on the efficiency factor of the electrolyzer system existing within the energy hub, the emission factor unit is converted from kg $CO_{2,e}$/kW h to kg

CO_2/kmol of hydrogen produced (EMF_{H_2}) by multiplying it with the electrolyzer efficiency.

$$H_2\ Emioffset = \sum_{h=1}^{8,760} [(H_2\ Inflow_h \times EMF_{SMR}) - (H_2\ Inflow_h \times EMF_{H_2})] \qquad (11.3)$$

In (11.3), the potential CO_2 emission offset between the two hydrogen production technologies is calculated. This offset is used to determine the total carbon credit revenue. The third revenue source is the monetary benefits earned from selling power produced by the fuel cell system collocated at the refueling station. The selling price (*Bid Price*, \$/kW h) of the electricity produced by the fuel cell is determined by the ratio of the amortized capital cost of the fuel cell power module ($Capital_{Fuel\ Cell}$, \$/year) over the maximum power rating of the unit (33 kW). This fraction is then further divided by the number of hours in a year to obtain the *Offer Price*, 0.0972 \$/kW h. The product of E_{FC} (Energy produced by the fuel cell, kW h) and the *Bid Price* is the revenue earned from selling electricity back to the grid.

The ratio of PEM electrolyzer efficiency and the higher heating value of hydrogen yields a parametric coefficient (also called hydrogen production efficiency) having the unit kmol of hydrogen produced per kW h energy consumed.

$$H_2\ Produced_h = 9.675 \times 10^{-3} \times E_h \qquad (11.4)$$

In (11.4), the product of the hydrogen production efficiency and the energy consumed (E, kW h) is shown to be the hydrogen produced in an hour (H_2 *Produced*, kmol/h). Since the electrolyzer system size is fixed at 2,000 kW (E_{max}, 2 MW), the energy consumed in an hour cannot exceed this maximum limit. Therefore, the inequality constraint in (11.5) limits the hourly energy consumption.

$$0 \leq E_h \leq E_{max} \qquad (11.5)$$

The hydrogen produced at the electrolyzers is stored in the 89-kg tank storage unit at 172 bar. In (11.6), the hydrogen flow from the electrolyzers to the hydrogen in flow to the tank is equated. The variables H_2 *Inflow* and $H_2\ Produced_{Fuel\ Cell}$ represent the hydrogen produced for the refueling station and the hydrogen produced for the fuel cell system, respectively.

$$H_2\ Produced_h = H_2\ Inflow_h + H_2\ Produced_{Fuel\ Cell,h} \qquad (11.6)$$

The flow of hydrogen before entering the tank goes through a prestorage compressor to compress it from 30 bar, the electrolyzer hydrogen discharge pressure, to 172 bar, the storage tank pressure. Since the system uses a reciprocating compressor capable of handling a flow rate of 42 kg/h ($Max\ Inflow_{Compressor,Prestorage}$), the maximum hydrogen flow compressed into the tank cannot exceed this maximum capacity of the compressor. Therefore, (11.7) constrains the H_2 *Inflow* to less than or equal to the $Max\ Inflow_{Compressor,Prestorage}$.

$$H_2\ Inflow_h + H_2\ Produced_{h,\ Fuel\ Cell} \leq Max\ Inflow_{Compressor,Prestorage} \tag{11.7}$$

At any given hour (*h*), the amount of hydrogen stored within the tank is determined by doing a simple inventory balance as shown in the following equation:

$$H_2\ Inventory_h = H_2\ Inventory_{h-1} + H_2\ Inflow_h + H_2\ Produced_{Fuel\ Cell,h} - H_2\ Outflow_h - H_{2,Fuel\ Cell,h} \tag{11.8}$$

The index $h-1$ indicates the hydrogen inventory at the end of the previous hour in the tank. $H_2\ Inventory_h$ is the hydrogen inventory within the tank at the end of hour *h*. $H_2\ Outflow_h$ is the amount of hydrogen taken out of the tank and sent to the booster compressor before being sent to the refueling station. The variable $H_{2,Fuel\ Cell,h}$ is the flow of hydrogen withdrawn from the tank that is sent to the fuel cell system to produce electricity.

In this optimization, the power-to-power energy recovery pathway, the energy hub tries to achieve energy arbitrage. Here, electricity is used to produce hydrogen and store it in the refueling station tank when the *HOEP* is low. Later on, when the difference between the *Bid Price* and *HOEP* is favorable, hydrogen stored in the tank is withdrawn and sent to the fuel cell system to produce power. Therefore, at any hour, only one flow variable, $H_2\ Produced_{Fuel\ Cell}$ or $H_{2,Fuel\ Cell}$ can be greater than 0 while the other should be equal to 0. However, this does not imply that both of them cannot be 0 in an hour. Equations (11.9)–(11.11) have been used to constrain the model such that the two flow variables are not greater than zero at the same time. In (11.9), two binary variables *c* and *d* are defined.

$$c_h + d_h \leq 1 \tag{11.9}$$

Equation (11.10) is a constraint set on the maximum amount of hydrogen that can be produced for the fuel cell system in any given hour. Since the hydrogen produced for the fuel cell is stored in the tank, the product of the maximum flow capacity of the prestorage compressor and the binary variable *c* is used as a limiting constraint.

$$H_2\ Produced_{h,\ Fuel\ Cell} \leq c_h \times Max\ Inflow_{Compressor,Prestorage} \tag{11.10}$$

The hydrogen that is withdrawn from the tank and sent to the fuel cell system can be no greater than the maximum possible hydrogen that can be withdrawn from the tank, as calculated in the following equation:

$$H_{2,Fuel\ Cell,h} \leq d_h \times (Inventory_{Max} - Inventory_{Min}) \tag{11.11}$$

The maximum and minimum amount of hydrogen that can be stored in the tank at any instant is set by the upper ($Inventory_{Max}$, 45.4 kmol) and lower ($Inventory_{Min}$, 8.5 kmol) bounds, as shown in the following equation:

$$Inventory_{Min} \leq H_2\ Inventory_h \leq Inventory_{Max} \tag{11.12}$$

Equation (11.13) is used to determine the amount of electricity produced by the fuel cell system at any given hour (E_{FC}, kW h). E_{FC} is a function of the variable hydrogen fed to the fuel cell system ($H_{2,Fuel\ Cell}$, kmol/h). The terms $\eta_{Fuel\ Cell}$ and LHV_{H_2}, denote the efficiency of the fuel cell system (55%) and the lower heating value of hydrogen [10].

$$H_{2,Fuel\ Cell,h} \times \eta_{Fuel\ Cell} \times LHV_{H_2} = E_{FC,h} \tag{11.13}$$

In (11.14), the number of fuel cells is constrained by $E_{FC,max}$, the rated capacity of the a single fuel cell module (33 kW), and $N_{Fuel\ Cell}$ is an integer variable that decides the number of fuel cells that can be installed within the energy hub.

$$0 \times N_{Fuel\ Cell} \leq E_{FC,h} \leq E_{FC,max} \times N_{Fuel\ Cell} \tag{11.14}$$

The hydrogen outflow from the tank at every hour needs to be equal to the hydrogen demand placed on the refueling station. Therefore, (11.15) is used as an equality constraint to equate the tank out flow variable (H_2 *Outflow*, kmol) with the hydrogen demand, $Demand_{H_2}$, kmol, used as a parametric input to the model.

$$Demand_{H_2,h} = H_2\ Outflow_h \tag{11.15}$$

The FCV hydrogen demand curve used in this study is the default Chevron Demand Profile [8]. To account for variations in the total daily demand placed on the station over a period of 1 week, variability data from a feasibility analysis of a hydrogen fueling station in Honolulu carried out by Hill and Penev is considered [17]. In order to assess the maximum daily hydrogen production capacity of the 2-MW electrolyzer system, the default profile is incrementally scaled up until the optimization problem gives an infeasible solution. As the typical FCV refuels 2.64 kg/day, a 100 kg/day station can refuel approximately 38 cars every day [18]. An incremental analysis shows that the 2-MW system is able to meet a maximum daily hydrogen production capacity of 670 kg. The set number of storage vessels and prestorage compressors used as parametric inputs in Model 1 are determined by running a simple optimization model where the sole purpose was to size the individual components to meet the 670 kg/day hydrogen demand. As the hydrogen flow through the booster compressor is equivalent to the demand placed in an hour, carrying out a post-modeling run calculation in excel to estimate its annual operating cost simplifies the optimization problem. However, to elaborate in more detail, the following equations show how the operating cost of the booster compressor is added to the value of net expenses obtained from (11.2). The pressure of hydrogen going into the booster compressor is dependent on the pressure of hydrogen within the tank and is therefore variable. Equation (11.16) shows the variation of tank pressure as a function of the variables hydrogen inventory (H_2 *Inventory*, kmol), and compressibility factor (z) of hydrogen as a function of temperature and pressure within the tank. An empirical equation has been developed for tank pressure ranging from 30 to 172 bar in MATLAB® Simulink™ using a lookup table.

$$P_{tank,h} = \frac{H_2\ Inventory_h \times R \times T \times z_h \times 1{,}000}{V} \tag{11.16}$$

V, R and T are parameters which denote the tank volume (m^3), ideal gas constant (m^3 bar/K/mol) and the temperature (K) inside the tank. The pressure inside the tank derived from (11.14) is the pressure of the hydrogen flow going into the booster compressor. Based on the work by NREL [8], the theoretical work done by the booster compressor ($W_{Booster\ compressor,theoretical}$, kJ/kmol) is calculated using by the following equation:

$$W_{Booster\ compressor,theoretical,h} = \frac{\bar{z}R_{comp}Tk}{k-1}\left[\left(\frac{P_{out}}{P_{tank,h}}\right)^{(k-1)/k} - 1\right] \tag{11.17}$$

The parameters used in (11.17) include: R_{comp}(kJ/kmol/K), universal gas constant used for booster compressor; k, heat capacity ratio of hydrogen; P_{out}(bar), outlet pressure of compressor; $\bar{z}$, the variable compressibility factor of hydrogen going in to the booster compressor, as a function of $Pressure = (P_{tank} + P_{out})/2$; and, temperature, which is assumed to be a constant set at tank storage temperature.

$$Energy_{Booster\ compressor,h} = \frac{W_{Booster\ compressor,theoretical,h} \times H_2\ Outflow_h}{\eta \times 3{,}600} \times 1\ h \tag{11.18}$$

In (11.18), the ratio of theoretical work and the efficiency of the booster compressor (η, 65%) [8] is multiplied with the incoming hydrogen flow from the tank to estimate energy consumed in kJ/h. Therefore in order to have units of kW h, the kJ/h term is divided by 3,600, the number of s/h.

Equation (11.19) is used to estimate the net present value (NPV) of the project. The term C_0 is the initial investment at the beginning of the project. C_0 is calculated by summing the annual amortized capital cost of all of the components within the energy hub. Since the amortized payments are split over 20 installments, after the initial investment at period "0," there are 19 more installments. The expression within the summation term accounts for the time value of the net benefit earned by the project over the 19 years. The negative of the term *Net Expenses* calculated in (11.2) is expressed as *Net Income*. The difference between the annual *Net Income* and the annual amortized capital cost (C) as described above is used to estimate the net benefit each year. Since the amortized capital cost payments begin at period "0," the last term in (11.19) is the net benefit earned for the last year. An internal rate of return (i) of 8% is used in the study. It should be noted that the term *Net Income* is an annual value and is assumed to be the same for every year until the end of the project life. Therefore, it is necessary to account for the time value of money over the project lifetime.

$$NPV = -C_0 + \left(\sum_{n=1}^{19} \frac{(Net\ Income - C)}{(1+i)^n}\right) + Net\ Income \tag{11.19}$$

11.3 Results and discussion

The optimization is run for the 2014 year. The CPLEX algorithm in the General Algebraic Modeling System (GAMS 22.6) is used to solve the MILP problem. In Table 11.1, the capital costs of the individual components existing within the energy hub are given. The amortization of these individual costs is used in (11.1).

The economic feasibility of the project is assessed by calculating its NPV at the end of its 20 year lifetime. In this section, the NPV of the project is calculated for *price mechanism A*, given in Table 11.2.

Under *price mechanism A*, the *NPV* of the project after 20 years is \$10,313,448. The project achieves a positive NPV at the end of the first year, which denotes that the payback period for the energy hub is 1 year.

As the MILP problem is run for a period of 1 year, the total annual revenue earned from the three revenue streams is \$1,830,364. The proportion of each of the three revenue streams is given in Figure 11.2. The majority of the revenue is earned through the power-to-hydrogen energy recovery pathway because 66.8% of the total electrolyzer capacity is used for meeting the hydrogen demand of the refueling station. For a carbon tax value of \$15/t of CO_2, the system earns \$19,094 for offsetting 1,273 t of CO_2 over the course of a year [19]. This makes up 2.45% of the total revenue share. The introduction of the cap and trade market in Ontario will determine the price of allowances set on carbon emissions and at what price these allowances could be traded between different emitters. As the cap on emissions becomes tighter, the price of these allowances could rise. Concepts like power-to-gas energy hubs will only benefit from this due to their ability to link a relatively clean Ontario power grid to different energy sectors across the province through the use of clean hydrogen in different energy recovery pathways. The optimization problem chooses one 33 kW fuel cell power module. Therefore, the power-to-power energy recovery pathway has the smallest share of the total revenues at 0.93%. The total revenue earned by the power-to-power energy recovery pathway is \$7,233. This is determined from the unit electricity selling price (*Offer Price*) of 9.7 cent/kW h from the fuel cell.

11.3.1 Development of a premium price mechanism for the energy hub

A favorable selling price for hydrogen in *price mechanism A* enables the system to achieve a quicker payback. Therefore, to assess the economic feasibility of the energy hub under a more conservative price mechanism, a post-processing

Table 11.2 Price mechanism A

Power-to-hydrogen pathway: H_2 selling price (\$/kg)	Power-to-power pathway: electricity offer price (cent/kW h)	CO_2 emissions offset credit (\$/t)
8	9.7	15

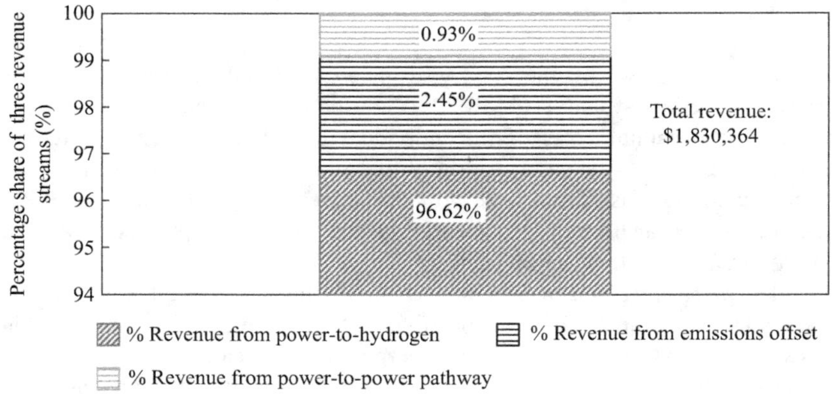

Figure 11.2 Percentage share of total revenue earned from the power-to-hydrogen, power-to-power and the credits earned from potential emissions offset while displacing hydrogen production from steam methane reforming with the water electrolysis process

calculation of the cost objective function in (11.1) is carried out for *price mechanism B*. The values of the variables that characterize the operating regime of the components within the energy hub when the MILP problem is run for *price mechanism A* are used here.

In comparison to *price mechanism A*, *price mechanism B* uses a hydrogen selling price that is equivalent to the hydrogen production price of the energy hub. The hydrogen production price of the system is determined using the following equation:

$$H_2\ Production\ Price = \frac{H_2\ Cost}{H_2\ Total} \tag{11.20}$$

The H_2 *Production Price*, determined from (11.20), is the levelized cost of hydrogen. In this equation, H_2 *Cost* is the annual cost of the energy hub, which includes the operating as well as the annual amortized capital cost payment. Equation (11.21) is used to estimate H_2 *Cost*.

$$\begin{aligned} H_2\ Cost &= O\&M_{Electrolyzer} + Capital_{Electrolyzer} + (Capital_{Booster\ compressor}) \\ &+ (Capital_{Compressor,Prestorage}) + (Capital_{Tank\ Storage}) + Capital_{Dispenser} \\ &+ Capital_{Cooling\ System} + (Capital_{Fuel\ Cell} \times N_{Fuel\ Cell}) \\ &+ (H_2\ Produced_h \times Water\ Consumption\ Rate \times Cost_{water}) \\ &+ \sum_{h=1}^{8,760}\Big(\Big[\big\{E_h + (H_2\ Inflow_h \times Energy\ Consumption\ Factor_{Compressor,Prestorage})\big\} \\ &\times HOEP_h\Big] \\ &+ \Big[\big\{E_h + \big(H_2\ Inflow_h \times Energy\ Consumption\ Factor_{Compressor,Prestorage}\big)\big\} \\ &\times Transmission\ Charge_h\Big]\Big) \end{aligned} \tag{11.21}$$

Table 11.3 Price mechanism B

Power-to-hydrogen pathway: H_2 selling price ($/kg)	Power-to-power pathway: electricity offer price (cent/kW h)	CO_2 emissions offset credit ($/t)
3.32	9.7	15

Table 11.4 Comparison of the total revenues earned under the two price mechanisms

Total revenue earned under *price mechanism A*	**Total revenue earned under *price mechanism B***
$1,830,364	$778,228

The total hydrogen produced by the electrolyzer system over the course of a year is estimated by summing the hydrogen produced every hour is determined in the following equation:

$$H_2\ Total = \sum_{h=1}^{8,760} H_2\ Produced_h \tag{11.22}$$

Equation (11.20) yields a levelized hydrogen production cost of \$6.64/kmol of H_2 (\$3.32/kg). In Table 11.3 the price of the three potential revenue streams for the energy hub in *price mechanism B* are given. Only the selling price of hydrogen is changed, the price values for the power-to-power energy recovery pathway and the emission offset credit remains the same.

As the payback period under *price mechanism B*, 22 years, exceeds the project lifetime of *price mechanism A*, 20 years, it is useful to determine a pricing structure that enables the energy hub to recover its costs. In this study, a new pricing structure for each of the three revenue streams with respect to payback periods of 8, 9 and 10 years has been proposed. For a fixed annual net income, the additional annual cash inflow ($Cash\ Flow_{Additional}$) required to achieve the three set payback periods is determined by solving the following equation:

$$-C_{Total} + (Net\ Income + Cash\ Flow_{Additional}) \times \left(\frac{(1+i)^n - 1}{i(1+i)^n}\right) = 0 \tag{11.23}$$

An NPV analysis of the energy hub operating under *price mechanism B* yields a negative NPV of −\$87,593. Therefore, the payback period for the project is greater than the project lifetime. Upon extending the calculation of the NPV, a positive value for the NPV is obtained at year 22. In Table 11.4 the total revenue earned by the energy hub when operating under *price mechanisms A* and *B* have been compared.

Table 11.5 Breakdown of additional cash inflow required from the power-to-hydrogen, and power-to-power energy recovery pathways and additional credits earned via CO_2 emission offset

Payback period (years)	Additional cash inflow required from the power-to-hydrogen pathway ($)	Additional cash inflow required from the power-to-power pathway ($)	Additional cash inflow required from CO_2 emissions offset ($)	Total additional cash inflow required ($)
8	637,709	6,135	16,194	660,038
9	564,665	5,432	14,339	584,436
10	506,742	4,875	12,868	524,485

The revenue earned under *price mechanism A* is 2.35 times the total revenue earned under *price mechanism B*. The price set for the emission offset credit and the power-to-power energy recovery pathway remains the same. Therefore, this difference is attributable to the higher selling price of \$16/kmol H_2 in *price mechanism A*, in comparison to \$6.64/kmol of H_2 in *price mechanism B*. In *price mechanism B*, the annual *Net Income* is \$284,045. The summation of 20 amortized capital cost installments throughout the project life is used as the cost to be recovered (C_{Total}). An internal rate of return (i) of 8% is used in (11.23). The value of the required payback period (n) is set as 8, 9 and 10, respectively, for the three cases. The annual additional cash flow ($Cash\ Flow_{Additional}$) required for the three cases are calculated to be \$660,038, \$584,436 and \$524,485, respectively. As the payback period decreases, the additional annual cash inflow increases. In order to better assess how this additional monetary benefit changes *price mechanism B*, the required cash inflow is split among the three revenue streams according to their percentage share in total annual revenue, as shown in Figure 11.2. In Table 11.5, the additional cash inflow required by each of the three revenue streams for the three different payback periods is given. As the majority of the share of revenue is earned via power-to-hydrogen, most of the additional cash inflow needs to be earned via this pathway.

The additional cash inflow required from each of the three revenue streams are added to the earnings calculated for *price mechanism B* previously. Now, by taking the ratio of the new revenue values and the total service provided by the energy hub, three new *premium price mechanisms* for each of three different payback periods are developed.

Using the hydrogen selling price of \$3.32/kg from *price mechanism B* as the baseline price, Figure 11.3 shows the required selling price of hydrogen for each of the three payback cases. The percentage increase in the price for payback periods of 8, 9 and 10 years with respect to the baseline price are 84.81%, 75.1% and 67.4%, respectively. The offer price of electricity produced by the 33-kW fuel cell module is the highest for a payback period of 8 years, at 17.97 cent/kW h. The previously set offer price of 9.3 cent/kW h in *price mechanism B* only accounts for the annual

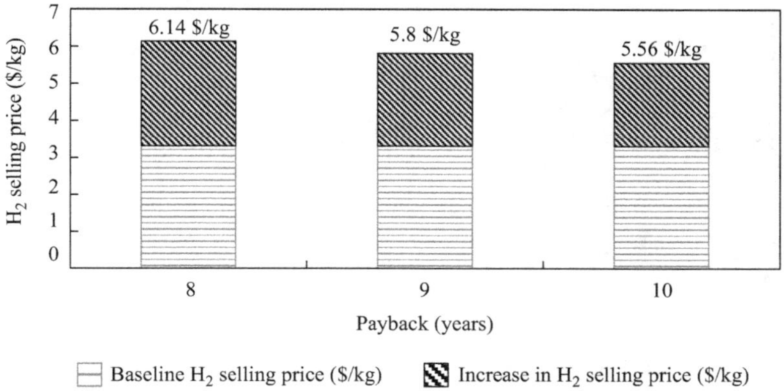

Figure 11.3 Comparison of increase in hydrogen selling price for payback periods of 8, 9 and 10 years

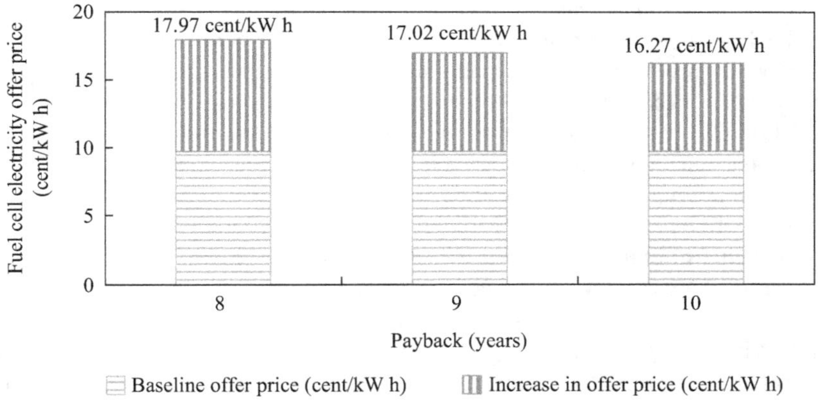

Figure 11.4 Comparison of increase in fuel cell electricity offer price for payback periods of 8, 9 and 10 years

amortized capital cost of the fuel cell module. The operating cost for producing and storing the hydrogen for the fuel cell is $4,201. For a payback period of 10 years, the additional cash inflow from the power-to-power energy recovery pathway is $4,875. Since the additional cash inflow for the 10 year payback case is the lowest, in each of the three set payback period cases, the operating cost incurred for the fuel cell power module is accounted for in all of the three new *premium price mechanisms* for the three payback periods. In Figure 11.4, the required offer price of electricity produced by the fuel cell module for payback periods of 8, 9 and 10 years is shown. This price is competitive with what is offered through the Feed In Tariff (FIT) program for landfill gas and biogas, which is between 16.8 and 17.1 ¢/kW h [20]. If the selling price of electricity was more competitive with the FIT

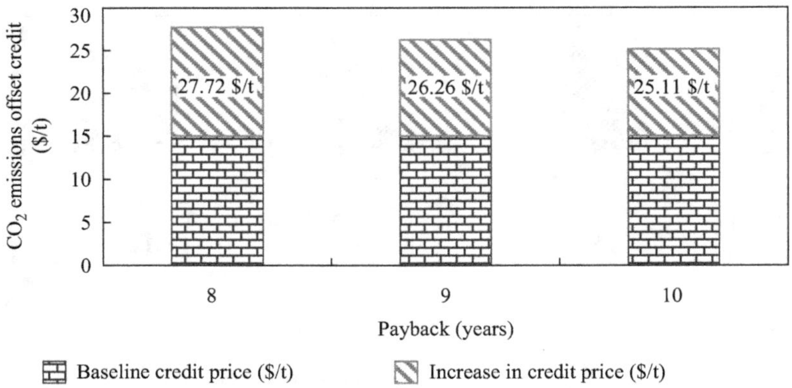

Figure 11.5 Comparison of emissions offset revenues for payback periods of 8, 9 and 10 years

program for solar and wind energy, which is as high as 38 ¢/kW h, the payback period would be substantially shorter.

The final revenue stream is the monetary credits earned by offsetting CO_2 emissions. Figure 11.5 shows the additional carbon price required for the energy hub to recover its cost within the three set payback periods. A baseline carbon price of \$15/t of CO_2 has been used in *price mechanism B*. This price has been based on the carbon price set in Alberta, Canada. Figure 11.5 shows that for faster payback periods, for example the carbon price required to achieve a payback period of 8 years needs to be at least ~\$28/t of CO_2.

The value of emissions credits effects the ability of industry to make a financial case for the use of green technology. This is important to note as the province of Ontario seeks feedback from industry to determine cap levels and permit credit values. With Ontario looking to reduce greenhouse gas emissions 15% by 2020 and California looking for a 10% reduction over the same period, the value of emissions credits will only rise. This bodes well for the use of power-to-gas to meet transportation and industrial needs.

11.4 Conclusion

A MILP optimization model of 2 MW energy hub producing electrolytic hydrogen for providing power-to-hydrogen and power-to-power energy services is developed in the GAMS software. The hydrogen is stored in an 89-kg tank at 172 bar. The power-to-hydrogen energy recovery pathway provides hydrogen demand to a fleet of 254 FCVs. The power-to-power energy recovery pathway explored in this study involves the sizing of the fuel cell module. It is seen that only a single 33 kW module fuel cell module can be installed at a refueling station with a maximum daily hydrogen demand of 670 kg. In addition to this, a comparative analysis on the economic feasibility of the energy hub has been carried out based on the three revenue streams, namely (1) revenues from power-to-hydrogen energy recovery pathway; (2) revenues from power-to-power energy recovery pathway and (3) revenues from offsetting CO_2 emissions when the water electrolysis method is used in place of the SMR process. When the energy hub is run for *price mechanism A*, a payback period of 1 year is achieved. Since 96.62% of the total revenues earned by the energy hub comes from selling hydrogen to the FCV end users, the high selling price of hydrogen ($8/kg) in *price mechanism A* leads to a short payback period. In order to assess, the operation of the energy hub under a more conservative pricing structure, the results from running the energy hub under *price mechanism A* has been used to determine the production price of hydrogen ($3.32/kg) for *price mechanism B*. The offer price of electricity (9.72 cent/kW h) and the CO_2 emission offset credit remains the same ($15/t of CO_2) in both *price mechanisms A and B*, only the hydrogen selling price is changed. For the same operating regime of the energy hub as that of in *price mechanism A*, a NPV analysis of the energy hub when run under *price mechanism B* shows that the energy hub has a negative NPV at the end of the 20-year project lifetime. Therefore, in this study, three *premium price mechanisms* are developed for payback periods of 8, 9 and 10 years. By using the net income earned by the energy hub for *price mechanism B*, the additional cash inflow required to achieve the set payback periods are calculated to be $660,038, $584,436 and $524,485, for 8, 9 and 10 year payback periods, respectively. The three additional cash inflows are added to the hydrogen selling price, electricity offer price and the CO_2 emission offset credit price according to the share of their contribution to the total revenue earned when the energy hub is run under *price mechanism B*. The hydrogen selling price in the three *premium price mechanisms* ranges from $5.56 to $6.14/kg. The electricity offer price ranges from 16.27 to 17.97 cent/kW h, and the CO_2 emission offset credit price ranges from $25.11 to $27.72/t of CO_2. The results show that the value of the hydrogen selling prices in the *premium price mechanisms* are well below typical market values of hydrogen, $16.72–$19.68/kg.

References

[1] M. Fowler, and U. Mukherjee, Adaptive energy ecosystems, *IEEE Can. Rev.* Spring (2014) 20–25.

[2] U. Mukherjee, M. Elsholkami, S. Walker, M. Fowler, A. Elkamel, and A. Hajimiragha, Optimal sizing of an electrolytic hydrogen production system using an existing natural gas infrastructure, *Int. J. Hydrogen Energy* 40.31 (2015) 976–991.

[3] M. Specht, M. Sterner, F. Baumgart, *et al.*, New Routes for the Production of SNG from Renewable Energy, *ForschungsVerbund Erneuerbare Energien*, 2010.

[4] M. Sterner, *Bioenergy and Renewable Power Methane in Integrated 100% Renewable Energy Systems*, Thesis, Kassel University Press, Kassel, Germany, 2009.

[5] Pacific Northwest National Laboratory, *Pathways to Commercial Success: Technologies and Products Supported by the Fuel Cell Technologies Office*, U.S. Department of Energy, Richland, 2015.

[6] R. Harvey, Interviewee, *Electrolyzer Specifications: Personal Communication with Hydrogenics*. [Interview]. 2014.

[7] G. Parks, R. Boyd, J. Cornish, and R. Remick. *Hydrogen Station Compression, Storage, and Dispensing Technical Status and Costs*. (No. NREL/BK-6A10-58564). National Renewable Energy Laboratory, 2014.

[8] National Renewable Energy Laboratory, *H2A Hydrogen Delivery Infrastructure Analysis Models and Conventional Pathway Options Analysis Results*, Interim Report No. DE-FG36-05GO15032, Office of Energy Efficiency & Renewable Energy, Washington D.C., 2008.

[9] Argonne National Laboratory, *Hydrogen Refueling Station Analysis Model*, U.S. Department of Energy, Lemont, IL, 2016.

[10] R. Abouatallah, Interviewee, *Fuel Cell Module Specifications: Personal Communication with Hydrogenics*. [Interview]. 2016.

[11] City of Markham, *Water Rate*, 2015.

[12] Independent Electricity System Operator. *Hourly Ontario Electricity Price*, 2015. Retrieved from http://www.ieso.ca/Pages/Power-Data/price.aspx.

[13] Ontario Energy Board, *Rate order – 2013 uniform electricity transmission rates*. Report No. EB-2012-0031, 2012.

[14] G. Bromaghim, K. Gibeault, J. Serfass, P. Serfass, and E. Wagner, *Hydrogen and Fuel Cells: The US Market Report*. Connecticut, Washington DC: National Hydrogen Association, 2010.

[15] A. Simons, and C. Bauer, *Lifecycle Assessment of Hydrogen Production*, White Paper, Laboratory for Energy Systems Analysis, Paul Scherrer Institute, 5232 Villigen-PSI, Switzerland, 2010.

[16] Greenhouse gas emissions associated with various methods of power generation in Ontario, Technical Report No. 20-22285, Ontario: Intrinsik Corp., 2016.

[17] P. Hill, and M. Penev, *Hydrogen Fueling Station in Honolulu, Hawaii Feasibility Analysis*, Office of Energy Efficiency & Renewable Energy (No. INL/EXT-14-31624), Honolulu, 2014.

[18] National Renewable Energy Laboratory, *Histogram of Fueling Amounts Vehicle and Infrastructure: National Renewable Energy Laboratory*, 2011. Retrieved from http://www.nrel.gov/hydrogen/docs/cdp/cdp_39.jpg.

[19] Huffington Post, *Alberta Extends Climate Change Rules, Including $15 Tonne Carbon Levy*, 2014. Available: http://www.huffingtonpost.ca/2014/12/19/alberta-climate-change_n_6357480.html [Accessed 2 March 2016].

[20] Independent Electricity System Operator, *FIT Price Schedule*, 2016. Available: http://fit.powerauthority.on.ca/fit-program/fit-program-pricing/fit-price-schedule [Accessed 15 March 2016].

Chapter 12

Smart multi-energy microgrids

Tomislav Capuder[1] and Tomislav Dragičević[2]

Abstract

It is trite to say that more needs to be done to save our environment. And the shift from fossil fuel to renewable energy sources is an essential move. This shift implies changes in the electricity sector, and also in the two main greenhouse gas producers – transport and heating. These changes alone will not work, unless the electricity powering them is produced from clean energy sources. Fine-tuning the interplay, between the variable and uncertain production and the flexibility via advanced control of demand, is the key in making a transition to a low-carbon society.

12.1 Introduction

We are becoming more and more aware of the problems created by today's energy production and consumption habits. Air pollution and CO_2 emissions are reaching historical peaks, while reports on polar caps reduction are warning us about the rising global temperature. Challenges of battling increasing CO_2 emissions are frequently present in environmental policies and even in the mainstream media due to their global impact; however, they are only a part of total emissions produced by any fossil fuel burning process. Each burning process also produces so-called local emissions or particles such as CO, NO_x, different sulphur oxides, etc. These are called local emissions since their range is up to 200 km, which mean after that they fall to the ground, they directly affect only the local population. Changes are needed and recent energy strategies and policies are usually promoting a shift from fossil fuel energy (such as coal, oil or gas) to renewable energy sources (RES) as a way of saving our planet's future. This change does not only imply changes in the electricity sector but also in the other two main producers of greenhouse gas

[1]Faculty of Electrical Engineering and Computing, Zagreb University, Croatia
[2]Faculty of Engineering and Science, Aalborg University, Denmark

emissions – transport and heating. Increasing the share of electric vehicles on the roads is going rather slow and the same is the case with highly efficient electricity-based heating, especially in European countries. It is important to keep in mind that these changes alone will not reduce pollution; if the electricity consumed by these new devices is not produced from clean energy sources, the 'shift' could actually have a negative impact on the environment. Fine-tuning the interplay, between variable and uncertain production on one side and flexibility which could be provided by advance control of demand on the other, is the key in making a transition to low-carbon society.

Energy produced from solar and wind is difficult to predict, both when and how much (referred to as uncertainty and variability). For this reason, power system operators face numerous challenges when integrating RES in the conventional power system structure (see e.g. [1,2]). For the electricity system to operate safely, and to enable the uninterrupted and secure supply to consumers, the system operators need to make sure there is a constant, second-to-second, balance between produced and consumed electricity. In reality, it means that future energy systems need to be designed as more flexible to be able to cope with these issues.

Addressing the above-stated problems is not an easy task. There are many ideas as to how the future energy systems should look like [3]. Although it is difficult to define only one common denominator of all the pathways proposed, most of them can be seen as concepts of planning and building a flexible system capable of responding to the variable and uncertain nature of renewable sources. Demand response is seen as one of the most promising solutions since it has a lot of hidden power balancing capacity which in many countries even exceeds the overall installed renewables. However, flexibility provision from load resources highly depends on human behaviour, making this capacity highly uncertain and variable in time. Therefore, integrating flexible load resources with local generation and storage within the entities called microgrids (environments like smart homes, commercial buildings and industrial facilities can all be categorized under the term microgrids) can largely compensate these effects and lead to better utilization of demand response potentials.

Microgrids are conventionally looked as electricity only systems, where installation of expensive energy storage systems is the only flexibility-enhancement option. On the other hand, multi-energy concepts allow exchange of energy between different energy vectors that operate at different time constants, which provides a higher degree of control freedom, implying unlocking of further flexibility potentials.

The concept presented in this chapter explains the idea of flexible and multi-energy microgrids (MEMs) and focuses on how one can create a sustainable low-carbon energy future by efficiently utilizing the existing sources, infrastructures and so-called energy vectors (energy vectors in this context relate to electricity, heat, gas, hydrogen, cooling and every other useful form of energy consumed today). This vision could change the perception of how most of us look at energy demand and supply. For instance, imagine that gas that warms up our houses is produced from photovoltaic power. Or that hydrogen fuelled, biofuelled and electric vehicles circulate together on our roads. Did you ever think that we can use CO_2 emitted from coal power stations and blend it with wind energy to produce synthetic gas which can then be transported to gas power plants, our gas boilers at home or even gas-driven vehicles? Benefits of this process are less in direct environmental savings, rather in creating additional value by increasing power system flexibility. Imagine that you do not need 'peakers' (such as gas power plants) or any other fossil-fuel-driven plants kept to provide reserve. Instead, wind power plants are operated so that excess production (that cannot be absorbed by power system demand) is converted to hydrogen or synthetic gas, further transported by existing gas infrastructure (such as the one in United Kingdom or Germany) and burned providing heat demand. By doing this, the firming capacity of wind (especially offshore wind) is increased, reducing the spilling of free green energy. There are many examples like this: chilled water can be generated from hot flues of micro power stations located in basements; wastewater can be a resource to simultaneously cool and heat the buildings in our districts and smart meters in our homes can indicate if it is cheaper to use gas or electricity to heat water for showering.

The presented vision of smart grid goes beyond electricity, challenges and rethinks the concept of energy, and particularly energy storage. It thus proposes a more realistic future for everyone's home, districts, cities and the whole energy system (see more in [4]). The proposed concepts in this chapter go beyond an idea and clearly show how and why multi-energy communities and microgrids can be very interesting to potential investors as they offer a reasonable return of investment and, in long term, present a business case resistant to future price volatility in different markets.

The second section of this chapter introduces and explains the basic concept of multi-energy systems (MESs), from system level to microgrid level; it also looks into drivers of distributed MESs, specifically at regulatory and market concepts important for the reader to understand operational principles of MEM. The third section discusses why MESs are flexible, where does the operational flexibility come from and shows how to model operation of MESs giving two examples as potential options for MES integration. The final section offers discussions and conclusions.

12.2 Understanding the idea behind flexible multi-energy communities

The two main drivers for providing operational system flexibility are coping with variability and uncertainty in both supply and demand. In this respect, flexibility has always been required, and flexible resources have always been present in power systems, as reflected in the classical unit commitment problem, whereby relevant reserves are indeed scheduled in advance to follow the load variation while also dealing with uncertainty in demand forecast and generation outages. In addition, flexible resources have always been used to provide real-time frequency control ancillary services to deal with either 'normal' supply-and-demand imbalances variations (as in the case of regulation and load following ancillary services) or contingency events (such as primary and secondary frequency response and tertiary reserve). It is therefore clear how RES, and low-carbon technologies (LCT), may add to flexibility requirements due to their contribution to both variability and uncertainty. It is not surprising that significant efforts are being paid to research the capability of the existing power systems to deal with increasing share of RES, given that variable and relatively unpredictable generation is challenging the usual operation practices for system control and balancing, from decrease in system inertia to increase in ramping requirements. The coordination of all resources that can provide operational flexibility needs to occur over different time frames, for instance from a day-ahead basis to real-time operation with the main goal of guaranteeing the supply–demand balance on a second-by-second basis. In a market environment, the market design and commercial arrangements must be such that the most economical resources are eventually deployed in any case by the system operator, which usually does so through a balancing spot market and various ancillary services markets or contracts. This highlights how the requirements and deployment of short-term flexibility are not a mere issue of system operation but have significant implications over different time scales and actions. Again, this is actually reflected in the way power systems work altogether [5].

Energy communities refer to a single system (or market) entity composed of aggregated commercially feasible technologies, such as distributed cogeneration system, electric heat pumps (EHPs), storage heaters, thermal storages, chillers and similar [6]. From the system perspective such communities act as demand response, however, with a significant difference from the classical flexible consumer response [7]. Usually, the smart grid ideas suggest that every consumer should participate in the future energy system by adjusting its consumption to the system needs. This is usually done by either shifting specific consumption time-of-use or by adjusting comfort level of the occupants. The majority of people, at least initially, will be very reluctant to exchange their comfort level for any kind of remunerations. On the contrary, in the multi-energy vision, there is no need for change in the behaviour pattern of the consumer and in the consumption demand. This means that the consumers do not have to switch off their appliances nor move their consumption to a more favourable period, as the capability of shifting between different energy vectors results in consumer always having the desired energy service. At the same time, these units provide additional flexibility in the form of different energy and ancillary services to the upstream power system [8].

The paradigm of flexible MESs relies on the interactions between different energy vectors (such as gas, heat, electricity, cooling and hydrogen). This means that at every moment, there is a choice from multiple input vectors and a possibility to select and to optimize the quantity of the input energy vector based on its price, availability or any other parameter one considers important. By converting this input vector (which is usually gas, electricity and hydrogen), different energy forms can be produced to fulfil the desired demand (electricity, heat, hot water and cooling). The first advantage of such an approach is obvious: if one has more choices, each with different price, the flexibility of such systems lays in selecting the least expensive one and making significant savings when compared to the traditional separate energy vector systems we have today. This will be further elaborated and supported with results in the third section of this chapter.

12.2.1 Drivers of distributed MES flexible operation

For all technologies, responsiveness and controllability are preconditioned with the existence of signals stimulating change of existing/conventional behaviour. Most of today's energy systems are rather passive in the sense that final consumers and distributed generation do not receive any type of signals (in this context the signals would be dynamic prices) to encourage them to act any differently than just serve the final demand or, as in case of for example local RES, produce all the electricity and receive incentives. Looking within our homes, most of us will find out that our heat demand comes from either gas boilers, district of city wide CHP or local electric heating. These devices are running following our heat demand; even when programmed to turn on only when we are home and our comfort requirements are higher, they do not coordinate with the energy system needs. In addition, most of us pay fixed energy tariffs. By charging us passive, fixed tariffs, retailers hedge their risks of procuring energy at the wholesale market (or bilaterally) and ensure constant income. To put it simple: if the income from sales is rather constant and predictable, it is easier to make profit when buying any type of merchandise. While such approach was acceptable in conventional energy systems, this paradigm has to change in two ways:

- In traditional systems, generation followed demand and was scheduled in a way to provide energy and ancillary services based on demand forecasts (including uncertainty and variability). On the contrary, in low-carbon energy system, a shift happens, requiring demand to adjust and be more able to follow generation (especially low-cost RES). Although RES, especially large-scale wind farms, can be controlled even for provision of frequency response, such behaviour underutilizes the usage of low-cost energy resources. It makes much more sense to optimize the power system operation so that both demand and generation are coordinated, maximizing utilization of both sides of energy equilibrium balance. Putting it simply: the concept of generation following demand shifts to demand following clean RES generation.
- The power system, as we know it for the past 150 years, relies on bulk energy resources, mostly conventional-fossil fuel-based types (coal, gas), and nuclear and hydro. While these units are large in both power and energy produced, they require large investments and a long time to be put into operation. Alternatively, each household (not to mention industry) could be both the

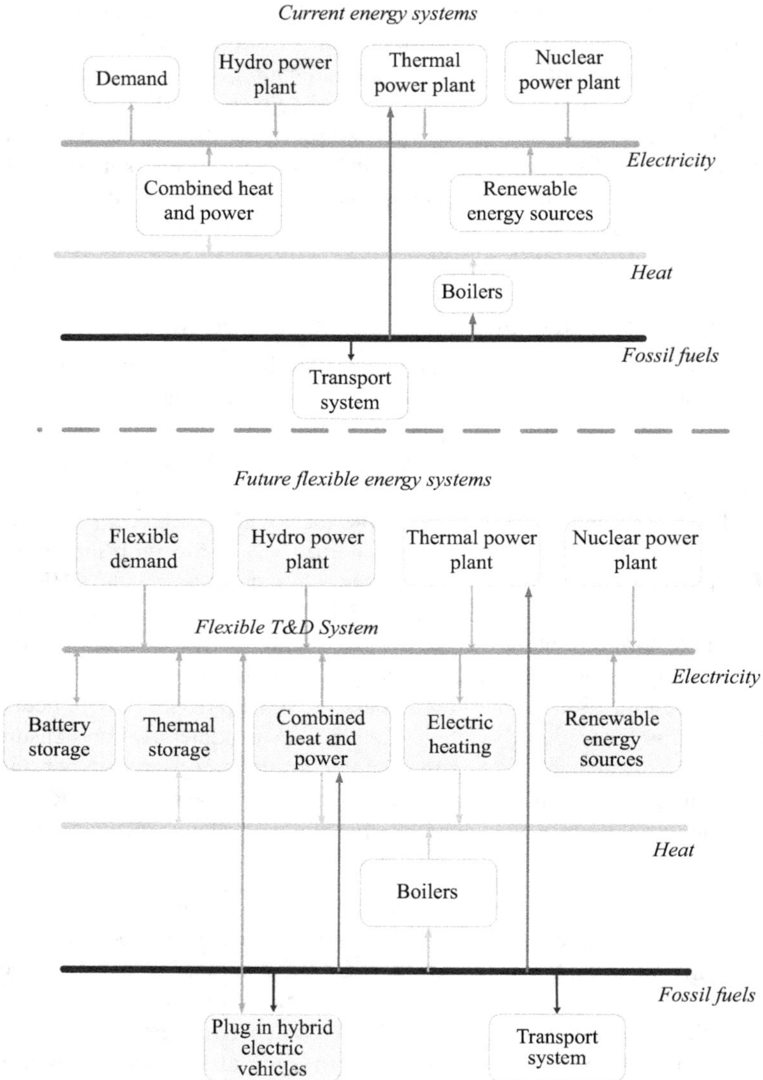

Figure 12.1 Concept of flexible future energy integrated system

producer and consumer of energy. Distributed energy is not new in the system and neither are business models suggested to coordinate their operation (such as virtual power plants and microgrids) or investment (such as energy communities). However, these models, in most cases, exist only in research articles

> or as pilot projects as sort of guerrilla attempts to make the transition to a cleaner energy system. At the end of 2016, European Commission brought the largest legislative package so far, called 'Clean Energy for all Europeans', in which it clearly promotes the shift to local energy production and distribution level prosumers (for the purpose of this chapter, term prosumer is used for a network user that both locally produces and consumes energy). The package suggests that by 2050, over 50% of total energy needs will come from end-users and sets pathways for creating frameworks for prosumer participation in system and market activities. We recommend that everyone read the entire 'Winter package' (or at least a number of summaries of most critical points for energy transition), as this is one of key energy documents for the upcoming period. However, the key aspect we would like to emphasize here is acknowledgement of importance of distributed energy and requests that they become active market participants. The reader should notice that distributed MES concepts fit perfectly into the new guidelines.

Integrated energy systems or MESs in fact present a transitional strategy towards all electric energy system. In the environment where both strategies of heat (currently around 40%–45% of total demand) and transport are suggesting that the future is going electric, MES seems to be a promising concept recognized in many studies as an efficient transition to a 100% renewable electric future. A general concept of this is shown in Figure 12.1.

12.3 Flexible operation of distributed multi-energy systems

MESs have the capability to change the usage of input energy vector (e.g. gas or electricity) to produce the desired energy vector demands (it can be multiple demands). The available literature explains this by the phrase spark spread [9]: it is defined as difference between energy vector costs taking into account efficiencies of conversion and storage devices used to produce the final energy output. Before providing models and explanatory results, simplified logic of operational advantages of MES are given using the example from Figure 12.2. Very often, MES operation is presented through the concept of Energy Hub (interested reader is directed to read more in the literature listed at the end of the chapter, such as [10,11]); however, for understanding the benefits of the concept, the proposed model is more intuitive. The MES example in Figure 12.2 should be understood as an idea and concept, a simplified way of understanding MES; not as a single operational unit (although similar do exist and are already installed and operational). In the example, the MES unit is composed of cogeneration unit (CHP), EHP, boiler and thermal energy storage (TES). The main idea of the figure is to have units which can produce multiple usable outputs and can provide flexibility services inside the lines, while the uncontrollable part, such as RES and consumers,

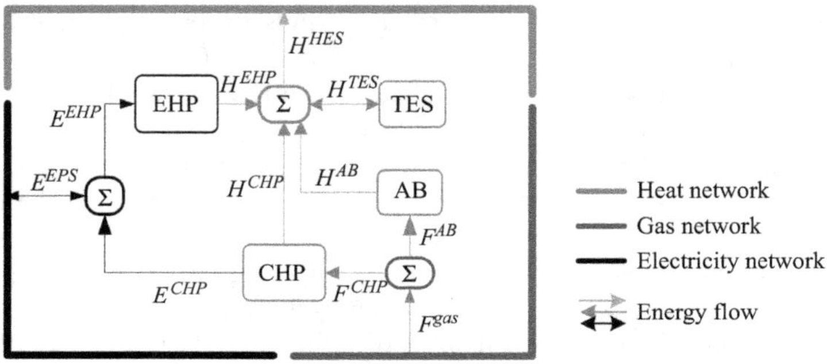

Figure 12.2 General concept of distributed multi-energy system (MES)

is presented by outside frame. In the context of this chapter, RES and demand are considered as uncontrollable only due to the fact that we are looking into other ways of providing flexibility and services; however, they should not in general be considered as passive system participants. The sum boxes present the controllers which would, interacting with different units, be able to provide flexibility from multiple energy vectors.

MES unit, presented by Figure 12.2, has a choice between two input energy vectors, gas and electricity, which can either be directly used by final consumption units or converted to desired energy demand. The desired input is selected based on prices. It should be mentioned here that gas and electricity markets are not cleared simultaneously since these two markets operate differently (although currently some recent studies look into simultaneous gas and electricity market clearing); therefore, signal prices are sent on different time frames. For the sake of simplicity, we show an example where gas market prices are constant throughout the day and electricity prices change. This reflects the current situation where even large consumers have contracted constant gas prices for longer periods, while these prices change daily at the wholesale market. On the other hand, electricity prices change on an hourly basis and are a result of day-ahead market clearing (known 12 h before the day of energy delivery).

Going back to the example when electricity price (defined as marginal clearing price or MCP for every hour at the day-ahead market, as shown in Figure 12.3) is high and spark spread is in favour of gas, MES unit will maximize usage of gas to produce the desired heat output. In fact, if the CHP unit is oversized compared to the current heat demand, it will produce excess heat energy which can then be stored for later usage. In addition, the surplus electricity produced will be sold at the power market, creating profit for the MES unit. If CHP unit is undersized and unable to produce the required heat demand, then electricity coproduced in CHP can be used for running electric heater (or in this case, drive the EHP) to produce the remaining heat demand.

When electricity MCP is low (as in the right graph in Figure 12.3), MES unit prefers to buy electricity from the market to supply its own electricity demand. It

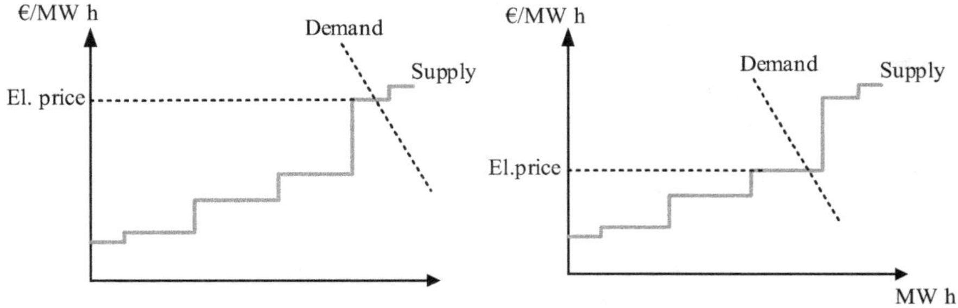

Figure 12.3 High and low market clearing price signal

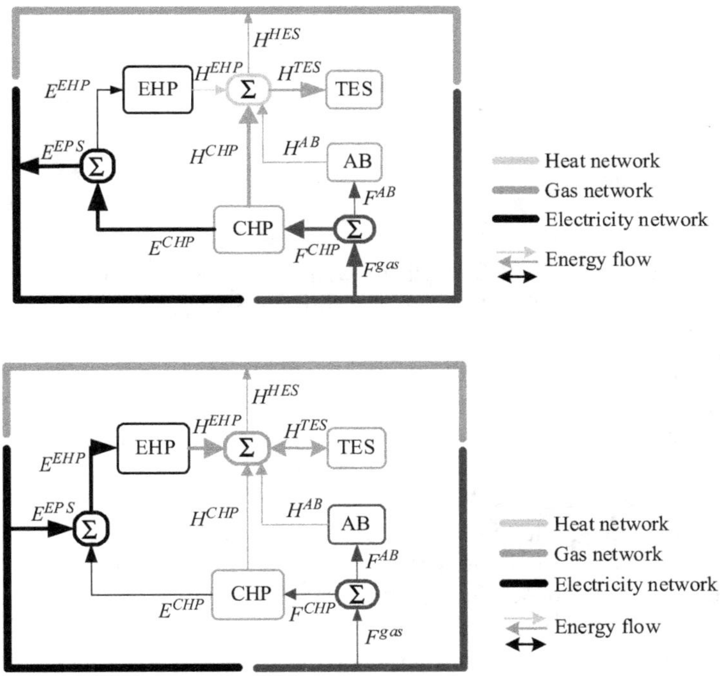

Figure 12.4 MES operation in scenarios of high (upper part) and low electricity prices (lower part of figure)

avoids running expensive gas units (such as CHP and boiler) and uses bought electricity to run the EHP to produce the desired heat demand.

Visualization of the MES operation when electricity prices are high is shown in upper part of Figure 12.4 and when electricity prices are low in lower segment of Figure 12.4.

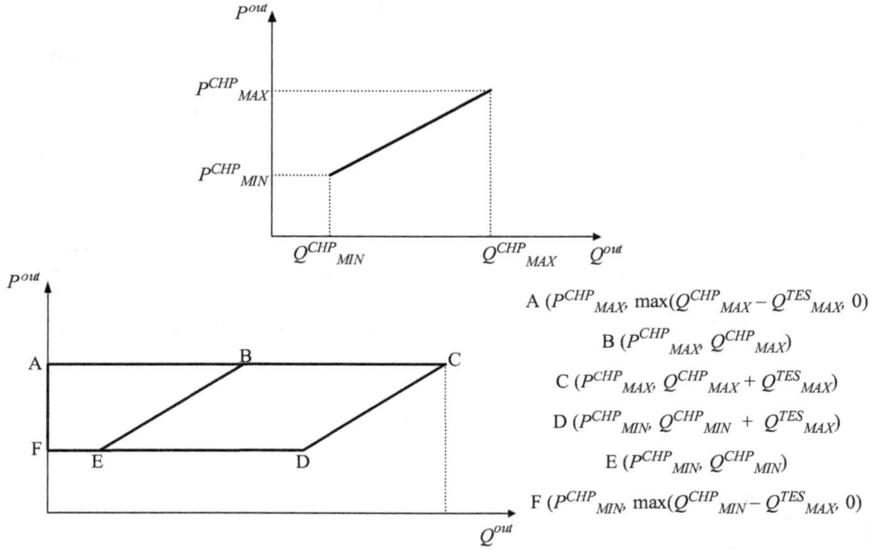

Figure 12.5 Operational flexibility areas of MES units

This basic concept will be further elaborated in next sections in two ways: (i) through graphical elaboration on why MES systems are capable of flexible operation; (ii) mathematical models and explanations of MES operational principles.

12.3.1 Where does the flexibility come from?

The simplest multi-energy device is a CHP producing simultaneously heat and electricity. CHP units are characterized by the type of turbine they use and in general there are three types: back-pressure steam turbine, extraction condensing steam turbine and gas turbines. The main difference, from the perspective of flexible operation aspect, is the heat to electricity ratio of those CHP turbines. Although extraction condensing steam turbine can operate more flexibly with regards to required demand, these units are almost never the choice for the investors at the distribution level due to higher costs. This means that distribution network level CHP units are almost always characterized with fixed power-to-heat ratio (see Figure 12.5 upper graph). In most cases, such CHP units operate so that they produce the desired heat demand, while electricity is treated as a by-product and is injected into the grid regardless of the needs or prices. One can easily imagine an early, windy, winter morning with high heat demand, lower electricity demand and high production from wind power plants – this is far from ideal case for the system operators and additional flexibility will be sought to mitigate potential misbalances between production and consumption. Coupling CHP unit with thermal storage enables decoupling of heat and electricity demand and this is visually shown in Figure 12.5 (lower graph). This means that the CHP units now do not have to strictly follow heat demand; it can already adjust its operation to market signals. Explanations of the points for the second diagram in Figure 12.5 (A, B, C, D, E and F) will be given as the modelling is explained in the next section.

The same logic can be easily applied to adding more units and coupling them into a MES system as the one in Figure 12.2. Feasible area of operation is wider and the operation of an MES unit becomes more flexible, meaning MES units are capable of fully decoupling different energy vectors (electricity, heat and gas) and responding to different price signals.

The authors would like to add that the presented approach aims at explaining basic principles; to properly model, plan and operate MES systems, extensive knowledge is required on local consumption, local energy networks, different energy markets and future development.

12.3.2 Multi-energy community modelling

Different understandings of energy community concept can be found in the literature [12,13]. For the purpose of this chapter, we will refer to energy community as a group of distribution system level consumers with different energy demands being supplied by MES units. A group of consumers, located close to each other and coordinating their operation by scheduling production and consumption, is often called a microgrid. For this reason, terms MEM and energy community will be used as equivalent terms.

In this section, we will show two key aspects of multi-energy units providing flexibility. The first aspect will be demonstrated through a centrally operated MES community composed of different units with a goal of demonstrating how much more efficient it is to provide different energy demand through coupling units then it is when each energy demand is produced independently. The second aspect will be shown through a corrective action scheduling model enabling self-balancing based on new information closer to real time of delivery. This flexibility, seen as the capability of self-balancing (or reducing energy community deviations towards the power system from originally scheduled dispatch), contributes to reliable system operation.

12.3.2.1 Example 1

For the first case analysis, we go back to the concept shown in Figure 12.2 (and Figure 12.4), where MES unit can be composed of CHP, boiler, EHP and TES units. Further, we recognize that the final consumer demand can be supplied in multiple ways or options, namely we look into the following possibilities:

- Option 1 (O1): Gas is being burned in a gas boiler, while electricity is supplied from the upstream power system. This is the most common way how energy is supplied today to the final consumers and represents fully decoupling of energy vectors.
- Option 2 (O2): Heat demand is being supplied by electric heating, either by running an electric heater or EHP. The electricity demand is supplied from the upstream system. O2 is in many national strategies seen as the final step

towards fully decarbonized energy systems, where both heat and transportation are being electrified.

- Option 3 (O3): A more flexible option is adding thermal storage couple with electric heating, and we analyse this in O3.
- Option 4 (O4): Central CHP unit is installed, producing heat and electricity. In this chapter, we consider CHP units being fuelled by natural gas (same for boilers). Recently, CHP units fuelled with biogas and biomass have gained a lot of attention, especially in Denmark where majority of heat demand is being supplied this way. Regardless of the fuel MES concepts, modelling aspects and understanding of benefits are similar; adjustments are needed only in terms of CHP and boiler units technical and economic parameters (such as cost of fuel of efficiency of electric/heat conversion). In O4, gas boiler and electric grid serve as auxiliary sources for supplying the consumers.
- Option 5 (O5): Similar to O4, however, in case CHP is additionally coupled with TES, we call this O5.
- Option 6 (O6): Electricity and heat demand are provided by coordinating CHP and EHP operation.
- Option 7 (O7): Same units are analysed as in O6, with the addition of TES to gain additional flexibility option.

Details of these analyses, such as optimal sizing of each unit, explanations of models, etc., can be found in [14,15], and we recommend readers interested in this area to look them up for better understanding. In this chapter, we will focus only on modelling aspects and reductions in annual operational cost as indicators of flexibility.

The main idea driving the optimization problem is minimization of operational costs; the expenses are gas bought to supply either CHP or gas boiler electricity bought for final consumption and operation of EHP. Additionally, profit can be made by selling excess electricity back on the market. This can be expressed as

$$\text{minimize} \sum_{t=1}^{T} \left[C_t^{gas} + C_t^{electricity} \right] \tag{12.1}$$

Each unit of MES should be modelled with corresponding maximum and minimum value limits and belonging efficiencies. For CHP, this can be expressed as in the following equations:

$$E_t^{CHP} = F_t^{CHP} * \eta_t^{elec} \tag{12.2}$$

$$H_t^{CHP} = F_t^{CHP} * \eta_t^{therm} \tag{12.3}$$

$$I_t^{CHP} * E^{CHP\_MIN} \leq E_t^{CHP} \leq I_t^{CHP} * E_t^{CHP\_MAX} \tag{12.4}$$

$$-ramp \leq E_t^{CHP} - E_{t-1}^{CHP} \leq ramp \tag{12.5}$$

Equations (12.2) and (12.3) model the conversion of gas to electricity and heat, (12.4) defines the minimum and maximum power that can be produced by CHP through binary variable I^{CHP} (indicating if CHP is even being operated or not), while (12.5) models the capability of CHP unit to change its output between two consecutive time steps.

A similar approach to modelling can be applied to modelling of EHP; (12.6) describes the relationship between heat output and electricity input through the relevant Coefficient of Performance (*COP*) value. Heat production from EHP is limited by a constraint upper limit, H^{EHP_MAX}.

$$H_t^{EHP} = E_t^{EHP} * COP_t^{EHP} \tag{12.6}$$

$$0 \leq H_t^{EHP} \leq H^{EHP\_MAX} \tag{12.7}$$

An interesting aspect of EHP operation is that it can be powered from electricity flow coming from the electrical grid (E^{EHP_G}) but also from CHP in cases where CHP produces low-cost electricity as by-product to heat demand (E^{EHP_CHP}). These aspects are shown by the following equations:

$$E_t^{EHP} = E_t^{EHP\_CHP} + E_t^{EHP\_G} \tag{12.8}$$

$$E_t^{EHP\_CHP} = E_t^{CHP} * \alpha_t \tag{12.9}$$

The gas boiler is modelled as in the following equations:

$$H_t^{BOILER} = \eta_t^{BOILER} * F_t^{BOILER} \tag{12.10}$$

$$0 \leq H_t^{BOILER} \leq H^{BOILER\_MAX} \tag{12.11}$$

Finally, electricity and heat balance of production and consumption need to be obeyed, and they are modelled as in [13,14]; they can be understood as the Kirchhoff nodal laws: the sum of all energy produced locally, consumed locally and exchanged with the grid is equal to zero:

$$E^{DEMAND} = (1 - \alpha_t) * E_t^{CHP} + E_t^{EXCHANGE} - E_t^{EHP\_G} \tag{12.12}$$

$$H_t^{TES} = H_{t-1}^{TES} + H_t^{CHP} + H_t^{BOILER} + H_t^{EHP} - H_t^{DEMAND} \tag{12.13}$$

The presented equations are a general model of all seven listed options of MES units. In case a specific unit is not included in an option, its variables take on the value of 0 for all time steps *t*. For example, in O5 only CHP, TES and boiler compose MES unit – this means that all variables modelling EHP will take on the value 0. By doing this, the above-presented model becomes unified and valid for all options O1–O7.

The reader should be aware that analysing operational costs for MES, or in this case operational cost savings, is subject to specific market conditions (volatile market prices), consumption of a specific energy community as well as size of each unit (such as CHP, EHP, TES and boiler). These factors have been researched in details in papers [14,15], and we encourage the reader to look for them for deeper understanding of these issues.

Operational costs of seven MES options, O1–O7, described in Figure 12.2 are shown in Figure 12.6. The values come from optimizing operational costs of each option, over 1-year time horizon for energy community of 1,000 houses. All unit

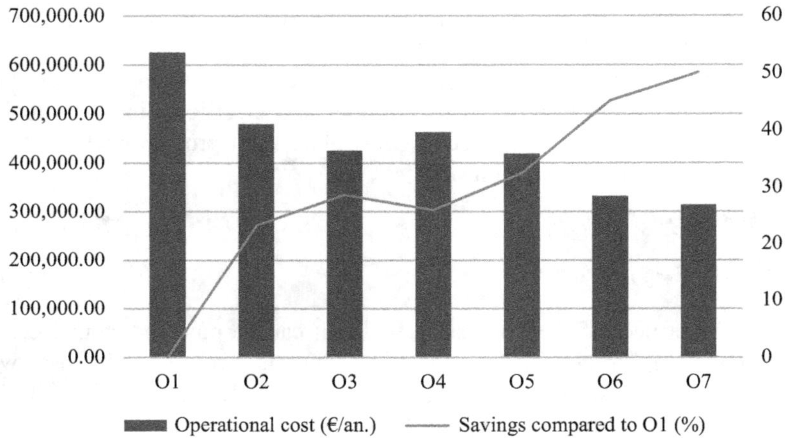

Figure 12.6 Operational costs for seven different MES options

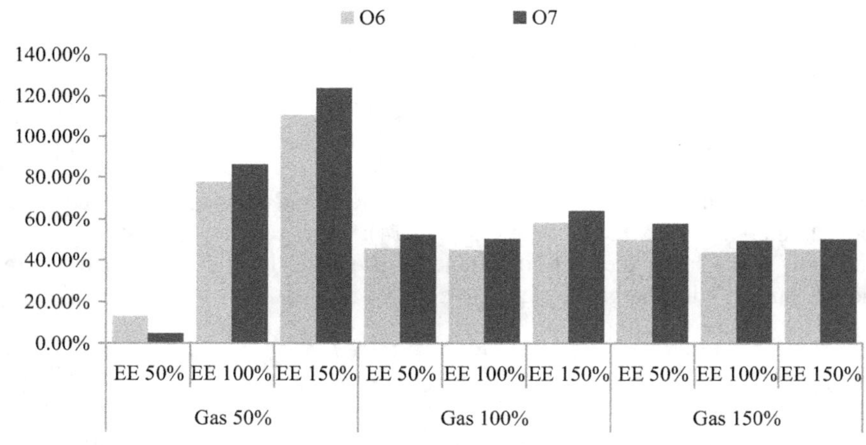

Figure 12.7 MES operational costs for changing market conditions

sizes are dimension to have maximum net present value (NPV) over the lifespan of 15 years (we consider this to be the optimal size of each MES option) [16].

When looking at the results, the value of 50% lower operational costs for coordinated MES operation as oppose to separate energy production really sticks out making us wonder how inefficiently we produce our energy demand, and are we overpaying our energy bills. Since these results might be valid only for specific conditions, Figure 12.7 demonstrates that regardless of the gas and electricity market prices (for the presented analyses we vary today's prices by 50% up and down, creating six additional scenarios), coupling energy vectors is always a more preferable option for providing desired demand.

12.3.2.2 Example 2

In the second example, the focus is on real-time operation and short-term flexibility of a MEM. In the example, a distributed residential community is modelled where each house has a flexible unit installed. This residential community can consist of any number of households, and each household can be equipped with different energy sources and have various demand curves (heating, cooling and electricity); in the presented example, 300 households are modelled. Depending on the community configuration, each household can be supplied by either household CHP unit, household EHP unit (cooling and heating), household auxiliary boilers, household heat storages and household installed RES units (PV panels).

This energy community will act as a single-power market/system entity called a balancing group (each energy community can be a separate balancing group, or multiple communities can have contracts and agreements defining them as a one balancing group with a single balancing group leader). This means that such a group is responsible for all deviations caused by behaviour of its members. More precisely, in case the community is not able to accurately forecast its operation on a day-ahead basis, it will be responsible for all the deviations from the announced schedule. This means that it will have to either self-balance these deviations or pay third parties to do this for them. A general idea is to have a central coordinator of community activities which receives new information as soon as available and makes corrective decisions based on this new information. There are different sources of uncertainty within energy community. For example, it is rather difficult to forecast electricity production of each house's PV system. Since this forecasting needs to be done at least 18 h before first kW h from PV is produced and injected into the grid (this is how the day-ahead market currently operates), the errors between forecasted and actually delivered can be quite high. Even more, they will probably be even higher as the day goes by (more time will elapse since the initial forecast was made). The same goes for all energy demands: cooling, heating and electricity. However, during the day, new forecasts can be made, new information can be available and central operator of such community can consider forecasting errors and be able to reschedule the operating points of flexible units. A concept of such central coordinator is shown in Figure 12.8.

For test-case analyses, we selected a community of 300 households. The entire model with comprehensive results can be found in [17,18]. The logic of modelling MEM is very similar to that as in Example 1 (with a small difference on values of efficiencies of different units). The difference is the corrective aspect of the algorithm, repeating the above-explained model in every time-step. By doing this, it uses new information and reduced the deviation from the original schedule.

Since goals of making a transition to LCT focus both on reducing energy costs as well as the environmental footprint, we will show the impact of implementing such corrective algorithms in those two aspects. Figure 12.9 shows the results for such analyses.

On a day-ahead level MEM, or energy community, will optimize its schedule and set operational points for its dispatchable units. This is shown as *no MPC*

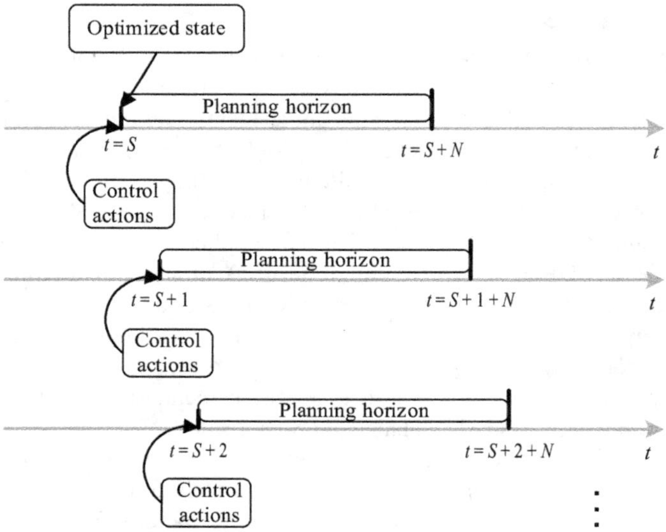

Figure 12.8 Corrective concept for optimal operation of energy community

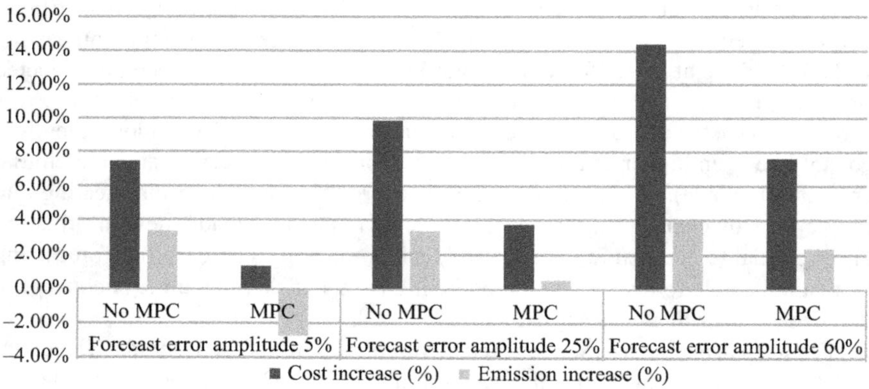

Figure 12.9 Impact of corrective actions on operational cost and environment

results in Figure 12.9. However, since this is done before noon on a day-1 of delivery, there is a high probability that both demand and RES actual real-time production will be different than originally forecasted. This means that if the microgrid sticks with the original schedule, it will create imbalances in the system. These imbalances will be penalized since other power system entities need to provide balancing services to maintain electricity balance equilibrium.

The corrective algorithm enables corrections of the original schedule based on the new information and forecasts. On each-time step (this can be different values, from 15 min, half an hour, hour – based on market setup), the optimization procedure is repeated, updating the operational points of flexible units and minimizing the error between real-time operation and the schedule announced at the day-ahead market. By doing this, imbalances can be significantly reduced. In the results, this is referred to as MPC.

Figure 12.9 should be understood in that way: if there is no corrective actions scenario call, no MPC will occur, while in case energy community has the capability to adjust, MPC scenario will occur. It is easy to notice that MPC scenario shows improvements in both aspects of reducing the operational costs and CO_2 emission impact.

12.4 Concluding remarks

Although one might argue that strategies for making a transition to zero emission energy systems exist, there is still an important contribution to be made in terms of relevant transitions paths, concepts and, most importantly, business cases for making different technologies interesting to both investors and energy system operators. The idea of all energy sectors going electric makes sense from a long-term perspective; however, the maturity of technologies required to do that, such as batteries and wide adaptation of electric heating and electric vehicles, does not seem to yet be at the stage where a simple switch is possible. It is somehow often neglected that we produce, transport and use our energy inefficiently, and that there is significant potential for reducing both costs and environmental impact by reducing these losses.

In this chapter, we present concepts of multi-energy communities/microgrids and focus on techno-economic and environmental aspects. While the idea of energy being produced at the proximity where it will be consumed is supported by regulatory guidelines of European Commission, there are multiple aspects missing to integrate local prosumers making them flexible system participants. From a technical point of view, integrating prosumers and local flexible consumption/production makes sense only if they can react to signals and adopt their behaviour to system needs. In multi-energy environment, this can be done by shifting the production of the desired energy output from one fuel to the other. By doing this, MES systems do not affect final consumers comfort (as would for example shifting electricity consumption to another period or preheating the room during favourable electricity prices). In fact, by responding to changes and signals, MES units benefit both in terms of financial savings and environmental impact reductions. We further stress the importance of developing and implementing adjustable control algorithms and introducing closer to real-time rescheduling, which is one of critical aspects when empowering new system users. In the context of developing energy communities, it should not be enough that they invest into LCT; they need to be stimulated to act the drivers of flexibility instead of creators of additional flexibility needs.

References

[1] Pavić I, Capuder T, and Kuzle I. Low carbon technologies as providers of operational flexibility in future power systems. Appl Energy [Internet]. 2016;168:724–38. Available from: http://linkinghub.elsevier.com/retrieve/pii/S0306261916301118.

[2] Pavic I, Capuder T, and Kuzle I. A comprehensive approach for maximizing flexibility benefits of electric vehicles. IEEE Syst J. 2017. doi:10.1109/JSYST.2017.2730234.

[3] Mathiesen BV, Lund H, and Karlsson K. 100% Renewable energy systems, climate mitigation and economic growth. Appl Energy [Internet]. 2011 Feb [cited 2014 Jul 25];88(2):488–501. Available from: http://linkinghub.elsevier.com/retrieve/pii/S0306261910000644.

[4] Mancarella P. MES (multi-energy systems): an overview of concepts and evaluation models. Energy. 2014;65:1–17.

[5] Pavić I, Capuder T, and Kuzle I. Value of flexible electric vehicles in providing spinning reserve services. Appl Energy [Internet]. 2015;157:60–74. Available from: http://linkinghub.elsevier.com/retrieve/pii/S0306261915009101.

[6] Mehleri ED, Sarimveis H, Markatos NC, and Papageorgiou LG. A mathematical programming approach for optimal design of distributed energy systems at the neighbourhood level. Energy [Internet]. 2012 Aug [cited 2013 Nov 7];44(1):96–104. Available from: http://linkinghub.elsevier.com/retrieve/pii/S036054421200103X.

[7] Mancarella P, and Chicco G. Real-time demand response from energy shifting in distributed multi-generation. IEEE Trans Smart Grid. 2013;4(4):1928–38.

[8] Mancarella P, and Chicco G. Integrated energy and ancillary services provision in multi-energy systems. 2013 IREP Symposium-Bulk Power System Dynamics and Control. 2013. p. 1–19.

[9] Dall'Anese E, Mancarella P, and Monti A. Unlocking flexibility: Integrated optimization and control of multienergy systems. IEEE Power Energy Mag. 2017;15:43–52.

[10] Geidl M, and Andresson G. Optimal power flow of multiple energy carriers. IEEE Trans Power Syst. 2007;22(1):145–55.

[11] Chicco G, and Mancarella P. A unified model for energy and environmental performance assessment of natural gas-fueled poly-generation systems. Energy Convers Manag [Internet]. 2008 Aug [cited 2013 Apr 9];49(8):2069–77. Available from: http://linkinghub.elsevier.com/retrieve/pii/S0196890408000824.

[12] Walker G. What are the barriers and incentives for community-owned means of energy production and use? Energy Policy. 2008;36(12):4401–5.

[13] European Comimission. Energy for the Future: Renewable Sources of Energy; White Paper for a Community Strategy and Action Plan COM(97) 599 final. 1997.

[14] Capuder T, and Mancarella P. Techno-economic and environmental modelling and optimization of flexible distributed multi-generation options. Energy. 2014;71:516–33.

[15] Capuder T, and Mancarella P. Assessing the Benefits of Coordinated Operation of Aggregated Distributed Multi-Energy Generation. Power System Computation Conference (PSCC) 2016. 2016. p. 1–7.

[16] Martinez Cesena EA, Capuder T, and Mancarella P. Flexible distributed multienergy generation system expansion planning under uncertainty. IEEE Trans Smart Grid. 2016;7(1):1–10.

[17] Holjevac N, Capuder T, and Kuzle I. Adaptive control for evaluation of flexibility benefits in microgrid systems. Energy [Internet]. 2014; Available from: http://dx.doi.org/10.1016/j.energy.2015.04.031.

[18] Holjevac N, Capuder T, Zhang N, Kuzle I, and Kang C. Corrective receding horizon scheduling of flexible distributed multi-energy microgrids. Appl. Energy. 2017;207:176–194.

Suggested literature on other multi-energy aspects

[19] Chicco G, and Mancarella P. From cogeneration to trigeneration: profitable alternatives in a competitive market. IEEE Trans Energy Convers. 2006;21(1):265–72.

[20] Wang H, and Mancarella P. Towards sustainable urban energy systems: high resolution modelling of electricity and heat demand profiles. Powercon. 2016. p. 1–6.

[21] Clegg S, and Mancarella P. Integrated electrical and gas network flexibility assessment in low-carbon multi-energy systems. IEEE Trans Sustain Energy. 2016;7(2):718–31.

Chapter 13

Conservation and demand management in community energy systems*

Jessie Ma[1] *and Bala Venkatesh*[2]

Abstract

Community Energy Systems (CESs) are localized systems that can generate, deliver, and/or store energy, which can come in different forms, including electricity, natural gas, and district heating. These can be operated in islanded mode or tied into the main grid, either continually or for backup purposes. Since CESs are by definition small-scale, even small deviations from forecasts can be much more costly to users as those costs of overbuilding or underbuilding are shared among a much smaller group of consumers (rather than the much larger pool across the larger system). Accurate peak load forecasts are very difficult, and they are especially difficult for CESs because inaccuracies cannot be smoothed across a larger base. Conservation and demand management can be efficient tools to smooth over inevitable deviations from forecasts.

The conservation model proposed in this chapter would **target conservation at the most elastic (price sensitive) consumers only during narrowly defined peak periods in order to increase utilization of fixed assets and drive down unit costs**. This would reduce the overall capacity requirements of the system, and these savings would be saved among all users.

The three main elements of this model are to (1) lower the peak in order to defer capacity expansions; (2) increase utilization in order to reduce unit costs and rates; and (3) target conservation efforts at the most elastic (price sensitive) consumers so that conservation is procured at the lowest possible cost. Conservation can be achieved through a combination of disincentives for consumption during very narrow peak periods and incentives for consumption during off-peak periods. Together, these have the effect of flattening the demand curve.

*Disclaimer: Any opinions, findings, conclusions or recommendations are those of the authors and do not reflect the views of the Independent Electricity System Operator, its employees or its administration.

[1]Centre for Urban Energy, Ryerson University, Canada

[2]Department of Electrical & Computer Engineering, Faculty of Engineering and Architectural Science, Ryerson University, Canada

13.1 Introduction

As localized energy solutions gain more attention, it becomes increasingly more important that Community Energy Systems (CESs) be better understood so that they can be designed and operated in a manner most beneficial to society. Integrated CES combine electricity microgrids with other energy needs and fuels [1]. CES can take different forms depending on the local needs, resources, and constraints. For example, it could involve district heating systems comprising "pipes, heating equipment, cooling equipment, heat exchangers, as well as the social infrastructure for management and settlement" [2]. It could be operated in isolated islanded mode, as would be more common in remote communities, or be grid-connected—either under normal operating mode or during contingency situations, using the grid as backup.

There are multiple potential purposes and benefits of CES. CESs are ideal for serving remote communities in which infrastructure to deliver energy is not available or sufficient, and energy sources are available and accepted locally. These could also include communities in the developing world, when transmission and distribution infrastructure is sparse and prohibitively expensive to build. CES could allow these communities to leapfrog over the building of transmission and distribution and instead go straight to local CES and microgrids, thereby enabling and hastening economic and social development. This would mirror their progress in communications technologies, where telephone landlines are bypassed, and consumers go directly to using mobile devices. In the developed world, CES could be useful for improving resiliency, reliability, and security. When the regional grid is unavailable, local energy solutions could ensure that consumers' needs are still met. Furthermore, if the community is involved in the planning of CES, there is greater social acceptance of the energy infrastructure and systems, thereby avoiding the NIMBYism that has stymied large energy infrastructure projects.

Due to the localized nature of CES, it is critical for CES to be able to absorb momentary fluctuations in supply and demand. This balance is more difficult for small-scale systems like CES because there is a much smaller pool available to absorb fluctuations. If the CES incorporates nondispatchable generation sources such as wind or solar, this task becomes increasingly more challenging.

As CESs become less dependent on the traditional grid, communities will increasingly need the ability to balance local generation with local loads. Conservation and demand management can play an important role in maintaining this balance in certain situations. This chapter will explore the role of conservation in CESs and suggest models to optimize desired outcomes.

Conservation can be a flexible and inexpensive resource to ensure the smooth operation of CES. It does not require any significant capital infrastructure investments, as a new power plant would, and could rely on existing smart grid technologies.

13.2 Role of conservation

Conservation can be used to maximize societal benefits by minimizing the capital costs in CES.

13.2.1 Definitions and terminology

Conservation and demand management can take various forms. Some key concepts in the electricity context are summarized in Table 13.1.

In the interests of simplicity for this paper, the term "conservation" is used to include all forms of demand side management outlined in Table 13.1. This will also allow for a more comprehensive and holistic view of different demand side means of achieving objectives, rather than having language and terminology restrict the range of tools available.

13.2.2 Conservation goals and system philosophy

It is of paramount importance to first define the goals of the CES. Only after these goals are clearly defined can conservation efforts be shaped to service those goals. And then, conservation would be pursued only if it is the most favorable option to achieve those goals. Therefore, in revisiting the underlying CES philosophy, the authors propose the model, shown in Figure 13.1, in which conservation is a tool to maximize societal benefit.

Figure 13.1 maps the underlying philosophy of CES with the overarching goal as maximizing society benefit. This can be achieved through two channels: minimizing service costs and maximizing service benefits. Service costs comprise fixed costs and variable costs. Fixed costs can be lowered by more fully utilizing the fixed asset base. Conservation can be a tool to flatten the load profile and dispense with the need for the infrastructure serving the only peak levels—these are the assets that cost the most and are used the least.

Traditional conservation programs tend to focus on the variable costs (through behavioral change) and efficiency. This model focuses on the reducing fixed costs.

Table 13.1 Demand response, conservation, and energy efficiency for electricity

Reduce peak	**Reduce volume (total energy consumed)**	
Demand response	**Conservation**	**Energy efficiency**
Changes in electric usage in response to financial signals	• Using less energy • Usually behavioral	• Using energy more effectively • Usually technological
Can defer building new peaking generation, transmission and distribution infrastructure	• Can save on generation fuel costs • Fixed costs still apply	
Lower carbon emissions, since peaking plants are typically fossil fuels	Reduced emissions if displaced fuel is carbon-based	
Demand-billed customers: industrial, large commercial	Energy-billed customers: residential, small commercial	Both demand and energy-billed customers
Requires measurement and communication equipment	Usually education, awareness programs	Usually incentives to upgrade

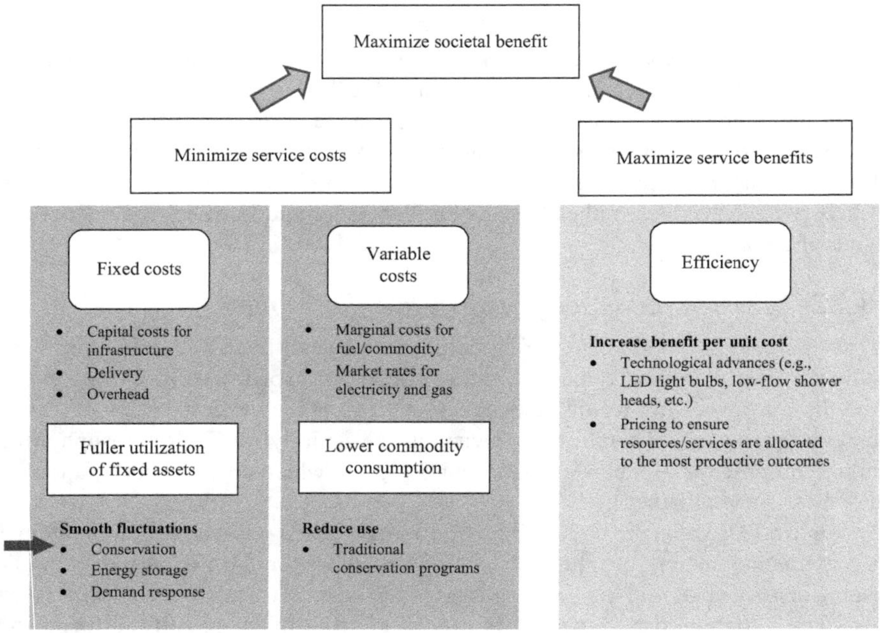

Figure 13.1 Model of underlying system philosophy for electricity and gas services

Conservation is one way to reduce peak. This can be achieved by shifting some of the peak demand to nonpeak hours while keeping the total consumption constant. Offsetting peak demand flattens the demand curve, maximizes the utilization rate of current infrastructure, and lowers both costs and rates for consumers.

13.2.3 Proposed model

The basic premise of this model is that **conservation would be targeted at the most elastic (price sensitive) consumers only during narrowly defined peak periods in order to increase utilization of fixed assets and drive down unit costs**. This would reduce the overall capacity requirements of the system, and these savings would be saved among all users.

There are three fundamental elements of this conservation model:

1. **Peak**—Conservation at different times yield different results. Many of the stated objectives of conservation, including infrastructure deferral and reduction of greenhouse gas emissions, require conservation to occur during peak time periods in order to meet their objectives. The *time* at which conservation happens is critical.
2. **Utilization**—The utilization of fixed assets—or the proportion of time over which an asset is being used—can have a profound impact on consumer costs.

Conservation, if not deployed carefully, could have the unintended effect of lowering utilization, stranding assets, and pushing up rates for fixed costs.

3. **Elasticity**—Only the most elastic (price sensitive) consumers would be targeted. This would ensure that conservation would be procured at the lowest possible cost. Typically, the most elastic consumers on a system are industrial rather than residential.

Conservation efforts during the peak times would be pursued only when

- Savings due to reduced or deferred infrastructure exceed the administrative program costs (i.e., it passes a net benefits test)
- Conservation yields higher returns per unit cost than alternative means that also shift demand (e.g., storage, demand response, etc.)

Through reducing peaks and maximizing asset utilization, multiple benefits can be achieved. Unit prices would be lowered for all consumers in a sustainable manner by reimagining our cost structure. System expansions could be deferred. Lowering peak consumption adds a measure of control and certainty for including conservation and demand management in CES planning so that costly and unnecessary overbuilding is less likely to happen. It also builds some flexibility into the system as future supply and demand situations in CES are highly uncertain.

13.2.4 Utilization rates

This model aims to improve the utilization rates of fixed assets, thereby lowering the unit costs borne by all consumers. The utilization rate is the percentage of time that an asset is being used. Figures 13.2 and 13.3 depict the annual consumption patterns for electricity and gas, respectively, at a system-wide level. These can be scaled down to the CES level, but fluctuations would be expected to be more pronounced and utilization rates lower. The dark grey areas approximately represent the times when the assets are being used, while the light grey areas are when the assets are paid for but unused. Utilization rates are simply the dark grey area divided by the dark and light grey areas combined.

The utilization rates for electricity in Ontario and gas in Ohio were calculated to be approximately 55% and 34%, respectively. The ideal utilization rate is 100%, however, that is not practical for regions like Ontario, where there are large seasonal fluctuations in temperature. This model aims to move utilities closer to 100%, and existing consumption patterns show that there is certainly potential for increasing utilization rates.

The goal of the proposed conservation model is to both

- Increase the dark grey region and decrease the light grey region relative to each other—that would lead to more efficient use of existing assets
- Lower the height of the shaded region—that would reduce the amount of fixed assets needed to provide the service

For ease of calculations, the areas were computed using fitted curves of representative frequency demand curves (i.e., annual demand data was reordered from

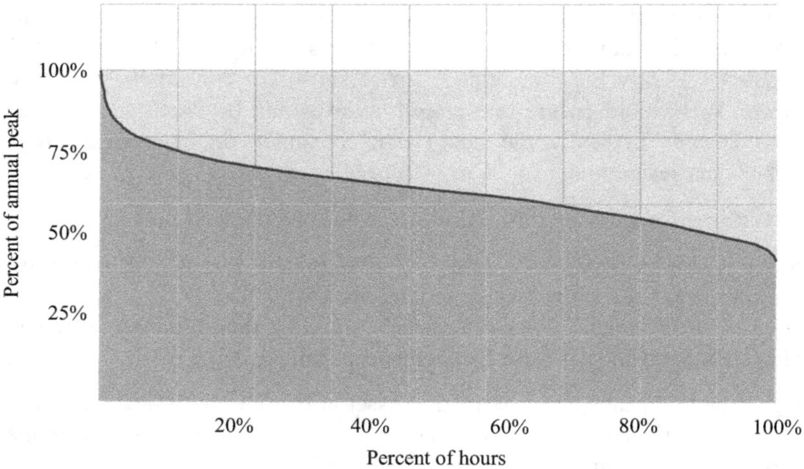

Figure 13.2 Annual electricity frequency demand pattern in Ontario

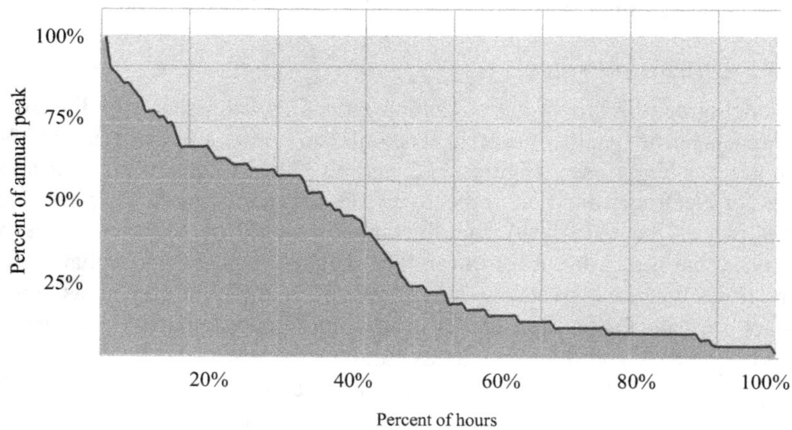

Figure 13.3 Annual natural gas frequency demand pattern in Ohio

largest to smallest). Also, utilization rates are "best case" figures as utility systems are built with an additional buffer above the peaks to account for emergency circumstances. Further, no publicly available data for natural gas consumption in Ontario was found, so Ohio was used instead.

13.2.5 Coincident peaks

Prime opportunities for integration of conservation efforts in CES are during coincident peaks—namely, in **winter for gas and electricity**—in regions similar to

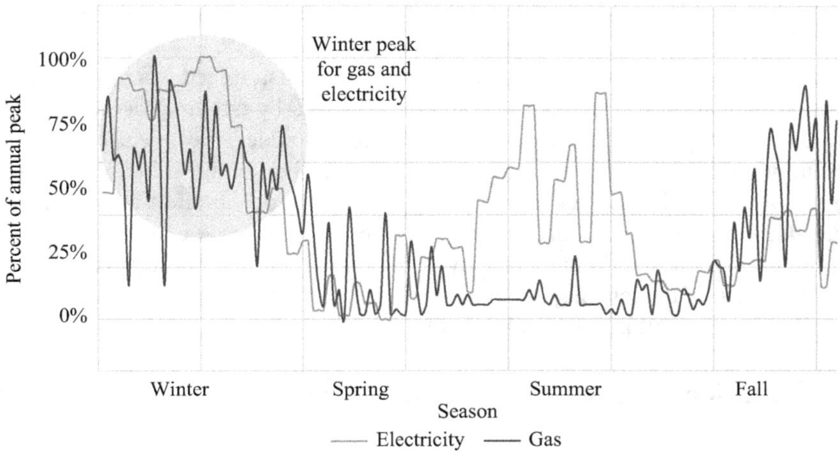

Figure 13.4 Annual demand patterns of electricity and gas

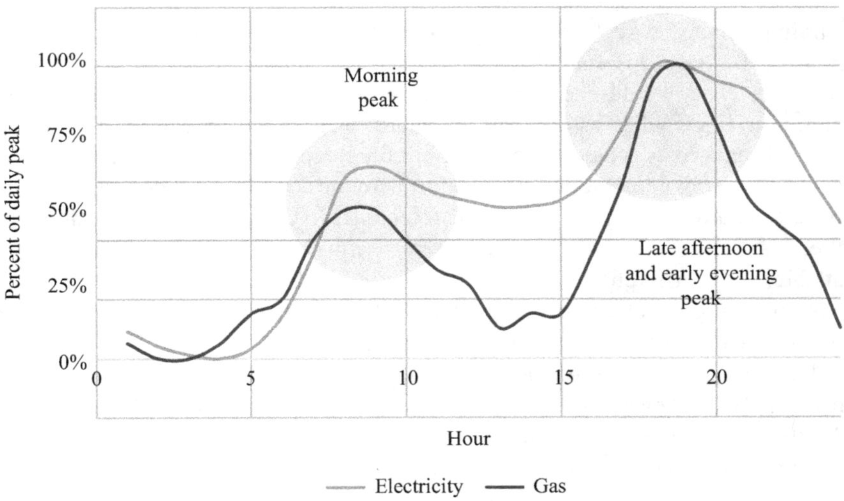

Figure 13.5 Daily demand patterns for electricity and gas

Ontario. Figure 13.4 shows the superimposed demand patterns of electricity and gas consumption.

Peak days for electricity and gas depend on climate and weather. Generally, the electricity peak demand day in Ontario occurs around the hottest or coldest day of the year, depending on which season is more extreme. For natural gas, the peak demand usually occurs on a cold winter day.

Further, peaks over the course of a day also coincide between electricity and gas, Figure 13.5 shows the superimposed daily demand patterns for electricity and gas. Both services experience bimodal peaks: once in the morning, and another in the late afternoon and early evening.

Presently, it is common for utilities to offer customized energy audits for industrial consumers. These audits could be enhanced by targeting coincident peak periods. Existing audits are typically designed to reduce the overall volume of consumption, whether it is through energy efficient light bulbs or better insulation. These audits could also include measures to reduce consumption at system peak times. Newer smart technologies are becoming more affordable and could present new opportunities to automatically lower usage at peak periods. Special attention should be paid to coincident peak times, namely, winter for gas and electricity, and mornings and late afternoons/early evenings.

13.3 Implementation of conservation for CES

CES operators could use a combination of "carrots and sticks"—incentives to consume outside of peak periods, and disincentives to consume at peak times.

13.3.1 Disincentives to consume at peak times

A straightforward way to encourage conservation during the peak times is by increasing the rate differential between the small number of peak hours and all other times. This would encourage conservation during peak times and shift consumption to times when assets have available capacity. Overall, utilization will be improved, unit costs would decrease, additional capacity would be freed for new consumption, and construction of new infrastructure would be deferred.

Rates could be designed so that the total costs to consumers are the same; what matters is the *relative* prices between peak times and times when capacity is available. This difference needs to be wide enough to encourage targeted conservation during beneficial times. As the peak decreases and utilization increases, unit prices will lower for the entire buyer pool.

Traditional conservation programs target variable costs through volumetric conservation. However, there have been limited efforts to directly encourage the full utilization of fixed assets, which would lower rates for all consumers.

Community energy can be varied and take many forms in regards to the customers it is serving [3]. Although conservation for higher utilization still applies, implementation may need to be adaptable based on the consumers that are served by a community energy supplier. If there are fewer high volume elastic consumers or they are not present at all, different disincentives will need to be developed.

Capital costs can be borne by customers through a fixed charge, a variable rate, a rate based on peak demand, or some combination of the three. For simplicity, we assume that customers are charged a fixed charge in this section.

$$
\begin{aligned}
\textit{Minimize fixed charges} &= \frac{P_N \cdot R}{C} \\
&= \frac{(P_0 - \Delta_P) \cdot R}{C}
\end{aligned}
$$

where P_0 is the peak system demand without using adjusted rates (MW for electricity; m^3 for gas), P_N is the new peak system demand using adjusted rates (MW for electricity; m^3 for gas), Δ_P is the change in system demand using adjusted rates $(P_0 - P_N)$ (MW for electricity; m^3 for gas), R is the rate of avoided capacity (\$/MW for electricity; \$/m^3 for gas), and C is the total number of consumers.

To minimize the new rates to the remaining customers, Δ_P should be maximized. (This assumes that customers all pay a fixed charge linked to capital infrastructure.) In other words, the largest consumers of the service should be targeted to curtail consumption during peak periods. This will result in minimizing prices for society as a whole.

This can be achieved through a number of interventions, including rates (as described in this paper), energy storage, or demand response. A traditional method is a public appeal to curtail consumption. However, adequate reductions in consumption cannot be guaranteed, and there is no economic incentive for curtailment or disincentive for continued consumption.

The ideal level of Δ_P should be derived from the value of no longer requiring Δ_P of assets to be built, owned, and/or maintained. One way of determining that rate for peak users is assigning the value to keep those assets serving the peak to those users requiring them to be in place. That way, the "user pays principle" will be upheld.

A sensitivity analysis can be done to find the optimal value of Δ_P/P_0 (i.e., the percentage decrease in peak consumption). There will be declining returns as Δ_P/P_0 increases because targeted loads become less and less elastic (price sensitive).

This model could be designed to be revenue neutral. The purpose is not to increase overall revenues to the utility, but to shift demand in order to avoid building and maintaining infrastructure that is unused for much of the rest of the year. Those savings can then be shared among all consumers through lower rates. When applied to a community energy situation, the smaller consumer base requires a smaller distribution network. In this case, the need to overbuild is lessened reducing the need to shift demand as much.

13.3.2 Incentives to consume outside of peak times

Further to the "stick" approach of higher differential rates described in Section 13.3.1, a "carrot" approach of encouraging demand during times of excess capacity could be implemented. This could be in the form of lower rates, points, or even periods when electricity or gas is free.

These are examples of similar existing programs that could be adapted to the electricity and gas context in CES.

- TXU Energy, a utility in Texas, offers free electricity overnight [3]. Although this program is motivated by managing the variable costs (rather than the fixed costs), it still has the impact of lowering the peak.
- Enel in Italy offers incentives if customers limit their consumption below a prescribed threshold during peak hours [3].
- Baltimore gas and electric gives credits to its customers for reduced consumption during peak periods [3].

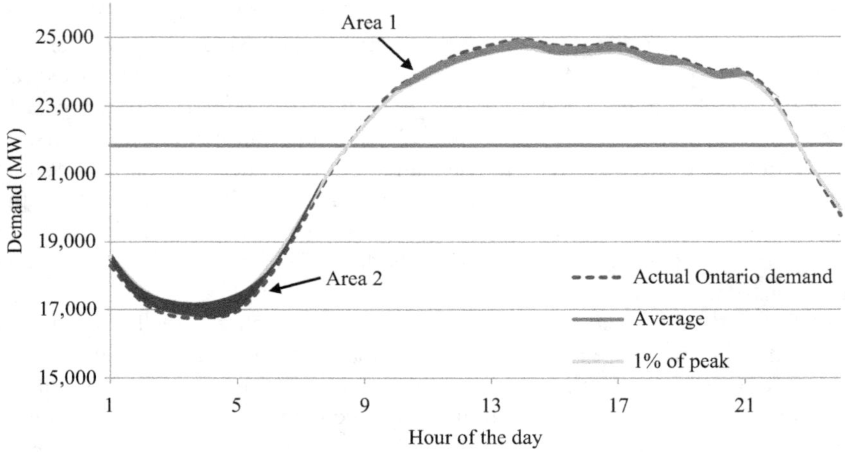

Figure 13.6 Illustrative example of potential for peak reduction of Ontario electricity demand

Further, Bay Area Rapid Transit, the public transportation authority in San Francisco, is testing a new program that would award riders points for using public transit during off-peak periods [4]. These points could then be redeemed for small cash rewards. This concept is transferrable to electricity and gas.

13.3.3 New managed system demand patterns

The combination of disincentives to consume at peak periods and incentives to consume outside of peak periods would modify system demand patterns to lower the peaks and raise the troughs.

To show how this might look in practice, we assume that seasonal variations would persist since services will still be required for heating and cooling, however, there could be opportunity to shift consumption over the course of a day. Figure 13.6 depicts an illustrative example if 1% of the peak load were shifted on July 13, 2013 for electricity in Ontario. The 25,000 MW peak would then be lowered to 22,500 MW, freeing up 2,500 MW in capacity.

Ontario's Independent Electricity System Operation (IESO) pegs avoided capacity costs at system peak at \$3.83/kW-year for transmission and \$4.73/kW-year for distribution [5]. Therefore, a 2,500 MW reduction in capacity would translate to \$21.4 million a year.

To create the graph in Figure 13.6, the hourly consumption of the peak demand day was plotted along with the average for that day. The portion of the curve above the average was lowered to achieve a desired peak reduction—in this case, by 1%. This reduction was offset by raising the segment of the curve below the average by the same amount. In this way, the total consumption and average are kept constant,

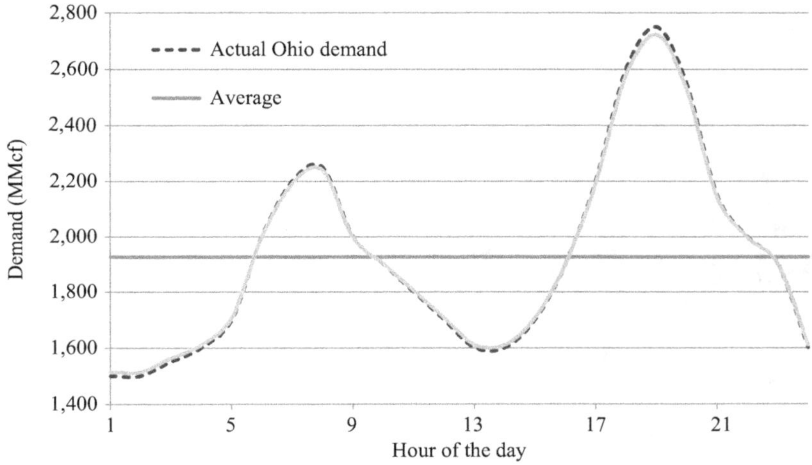

Figure 13.7 Illustrative example of potential for peak reduction of Ohio natural gas demand

but the time of consumption changes. In Figure 13.6, Area 1, the area between the actual and reduced curves *above* the average is equal to Area 2, the areas between the actual and reduced curves *below* the average.

Figure 13.7 shows an illustrative example of the same conservation model applied to natural gas.

Data to create the gas graph was extrapolated from related sources [6]. Furthermore, these are idealized graphs; in practice, the energy curtailed (Area 1) and the energy increase (Area 2) could differ due to losses from storage technologies.

It should be noted that peak usage occurs for a very small fraction of time, but infrastructure is built to accommodate this peak. This small peak period is when conservation should be targeted. The frequency distribution of Ontario electricity demand was also plotted using IESO power data [7] in Figure 13.8. The graph shows that approximately 12% of capacity is used for 1% of the time.

A variation on this model is to target only the large volume users (e.g., top 5%, 10%, or 25% of the market) with peak pricing as they are more likely to be elastic (i.e., responsive to price). This can also ensure that rates for low volume users, which usually include vulnerable customers such as people with low-income and seniors, will be lower.

While the examples are shown for the Ontario-wide system, similarly shaped graphs can be created for CES at a much smaller, localized scale.

13.3.4 Implementation

Implementation of this conservation and demand management model can take different forms. Declining costs for emerging technologies such as smart grid and

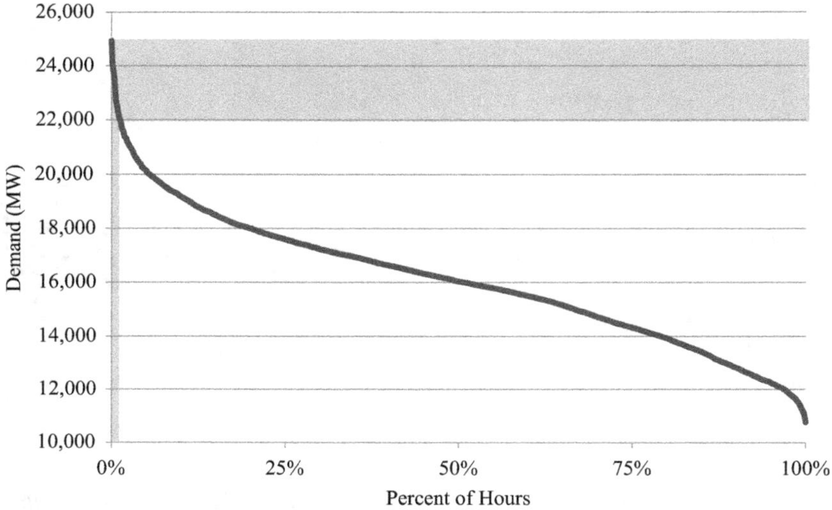

Figure 13.8 Frequency distribution of Ontario electricity demand

energy storage allows for a greater range of feasible solutions. Possibilities include

- **Energy storage**—Storage units strategically located downstream of assets nearing capacity are ideal for the proposed conservation model. During peak times, the storage unit can discharge, thereby alleviating the peak demand. During off-peak times, the storage unit can recharge, thereby increasing utilization. Examples of storage units of electricity include batteries, flywheels, and gravitational devices. For gas, a simple tank or reservoir could work.
- **Smart grids**—Intelligent devices that can measure, respond to, and communicate data can be used to manage conservation and demand. These would facilitate demand reductions during peak times by communicating higher prices or incentives to curtail. During off-peak times, additional consumption could be encouraged through lower prices or other incentives. Also, it is not necessary for the same consumers who respond during these two periods; different consumers can respond, so long as they are all downstream of the system bottleneck they are trying to alleviate.
- **Audits for industrial consumers**—Utilities often offer engineering audits for their large industrial customers to assist in conservation. However, these audits are usually performed independently for electricity and gas, and they typically look at volumetric reductions. There is an opportunity for integrating these audits by conducting the audits for both electricity and gas together, thereby ensuring that conservation in one service will not inadvertently increase consumption in the other. Furthermore, these audits could be improved by introducing a time element and targeting conservation at peak periods.

Table 13.2 Pressures on future demand for electricity and gas

	Upward pressures	Downward pressures
Electricity	• Increased use of electric vehicles • Switching from gas to electricity for heating • More extreme weather from climate change requires more heating and cooling • Growing population	• Technological advances and improved energy efficiency • Economy transitioning from manufacturing to knowledge-based activity • Higher electricity prices • Increasing distributed generation • Spread of energy storage and microgrids
Gas	• More extreme weather from climate change requires more heating • More district heating systems • Growing population • Lower gas prices	• Switching from gas to electricity for heating • Technological advances and improved energy efficiency • Growth in geothermal heating

13.3.5 Future scenarios

This model is designed to withstand different future scenarios. Planning for future demands of electricity and gas is extremely difficult, and this complexity is compounded for CES since smaller pools have less ability to smooth over deviations from forecasts. Most long-term forecasts by central authorities have been inaccurate. The costs of these inaccurate forecasts are borne by all users through rates due to overbuilding or service interruptions due to underbuilding. Therefore, it is important to design models that are nimble and flexible. Table 13.2 shows various upward and downward pressures on future demands for electricity and gas.

Should demand increase, this model is a way to delay new construction by using existing infrastructure more efficiently. Should demand decrease, then this model would enable the decommissioning of unused, expensive assets. Should demand remain relatively constant, the asset base would stay the same.

13.4 Conclusions

Conservation can be used as an effective tool to maximize societal benefit by enabling fuller utilization of fixed assets in CES. This managed demand would ease operation of small-scale CES and lower long-term costs by deferring infrastructure investments.

This model targets **conservation at the most elastic (price sensitive) consumers only during narrowly defined peak periods in order to increase utilization of fixed assets and drive down unit costs**. This would reduce the overall capacity requirements of the system, and these savings would be saved among all users.

The three key elements of this conservation model are

1. Reducing peak consumption to lower total fixed costs
2. Maximize utilization in order to lower rates for fixed costs
3. Target conservation at the most elastic (price sensitive) consumers so that conservation is procured at the lowest possible cost.

Conservation programs typically focus on the variable costs—the market rates for electricity, or volumes for gas and water—and not the fixed costs. But these fixed capital investments can be quite significant. Conservation can be used to ensure that existing fixed assets are used optimally and that utilities do not build more than is needed. Savings can then be passed onto all consumers in the form of lower rates.

Demand can be managed through "carrots and sticks": incentives in the form of lower rates or rewards during periods outside of peak times; and disincentives in the form of temporarily higher rate differential during very narrow peak periods.

Acknowledgments

The authors thank Nima Alibabaei, Afarin Amirirad, Benjamin Fischer, and Shiva Motamedi for their research assistance.

The authors are also grateful to the stakeholders and experts who have generously offered their feedback.

References

[1] Natural Resources Canada, "Community Energy Systems." [Online]. Available: https://www.nrcan.gc.ca/sites/www.nrcan.gc.ca/files/canmetenergy/files/pubs/Community%2BEnergy%2BSystems.pdf. [Accessed 24 May 2017].

[2] Wise, "Benefits of Community Energy Projects," Embark. [Online]. Available: http://www.embark.com.au/display/public/content/Benefits+of+community+energy+projects. [Accessed 28 November 2016].

[3] C. Krauss and D. Cardwell, "A Texas Utility Offers a Nighttime Special: Free Electricity," 9 November 2015. [Online]. Available: http://www.nytimes.com/2015/11/09/business/energy-environment/a-texas-utility-offers-a-nighttime-special-free-electricity.html. [Accessed 13 October 2016].

[4] V. Ferreira, "The future of public transit? Bribing subway riders with cash payouts so they'll avoid rush hour," 6 October 2016. [Online]. Available: http://news.nationalpost.com/toronto/the-future-of-public-transportation-bribing-your-subway-riders-with-up-to-100-a-month-to-avoid-rush-hour. [Accessed 13 October 2016].

[5] Independent Electricity System Operator, "Conservation & Demand Management Energy Efficiency Cost Effectiveness Guide," 2015. [Online]. Available: http://www.ieso.ca/Documents/conservation/LDC-Toolkit/Guidelines-and-Tools/CDM-EE-Cost-Effectiveness-Test-Guide-v2-20150326.pdf. [Accessed 2 October 2016].

[6] "Ohio Gas Demand Profiles from PLEXOS Integrated Energy Model," 9 June 2015. [Online]. Available: https://www.misoenergy.org/Library/Repository/Meeting%20Material/Stakeholder/Workshops. [Accessed 13 October 2016].

[7] Independent Electricity System Operator, "Data Directory." [Online]. Available: http://www.ieso.ca/Pages/Power-Data/Data-Directory.aspx. [Accessed 18 July 2016].

Index

www.ingramcontent.com/pod-product-compliance
Lightning Source LLC
LaVergne TN
LVHW011914200926
848484LV00002B/618